SEVENTH EDITION

Reinforced Concrete Design

George F. Limbrunner, PE
Hudson Valley Community College (Emeritus)

Abi O. Aghayere, PE
Rochester Institute of Technology

Prentice Hall
Upper Saddle River, New Jersey
Columbus, Ohio

Library of Congress Cataloging-in-Publication Data

Limbrunner, George F.
    Reinforced concrete design/George F. Limbrunner, Abi O. Aghayere.—7th ed.
        p. cm.
    ISBN-13: 978-0-13-504435-3
    ISBN-10: 0-13-504435-9
    1. Reinforced concrete.   I. Aghayere, Abi O.   II. Title.
    TA444.L44 2010
    624.1′8341—dc22

                                                                    2008039708

Vice President and Executive Publisher: Vernon R. Anthony
Acquisitions Editor: Eric Krassow
Editorial Assistant: Sonya Kottcamp
Production Manager: Wanda Rockwell
Creative Director: Jayne Conte
Cover Designer: Bruce Kenselaar
Cover photo: Corbis Corporation
Director of Marketing: David Gesell
Manager, Rights and Permissions: Zina Arabia
Manager, Visual Research: Beth Brenzel
Manager, Cover Visual Research & Permissions: Karen Sanatar
Image Permission Coordinator: Jan Marc Quisumbing
Executive Marketing Manager: Derril Trakalo
Senior Marketing Coordinator: Alicia Dysert

This book was set in TimesTen Roman by Aptara®, Inc., and was printed and bound by Hamilton Printing Co. The
cover was printed by DPC.

Pearson Education Ltd., London   Pearson Education Australia Pty. Limited
Pearson Education Singapore Pte. Ltd.   Pearson Education North Asia Ltd. , Hong Kong
Pearson Education Canada, Inc.   Pearson Educación de Mexico, S.A. de C.V.
Pearson Education—Japan   Pearson Education Malaysia Pte. Ltd.

Prentice Hall
is an imprint of

www.pearsonhighered.com

10 9 8 7 6 5 4
ISBN-13: 978-0-13-504435-3
ISBN-10:   0-13-504435-9

NOTICE TO THE READER

Preface

The primary objective of *Reinforced Concrete Design*, Seventh Edition, remains the same as that of the previous editions: to provide a basic understanding of the strength and behavior of reinforced concrete members and simple reinforced concrete structural systems.

With relevant reinforced concrete research and literature continuing to become available at a rapid rate, it is the intent of this book to translate this vast amount of information and data into an integrated source that reflects the latest information available. It is not intended to be a comprehensive theoretical treatise of the subject, because it is believed that such a document could easily obscure the fundamentals emphasized in engineering technology programs. In addition, it is believed that adequate comprehensive books on reinforced concrete design do exist for those who seek the theoretical background, the research studies, and more rigorous applications.

This seventh edition has been prepared with the primary objective of updating its contents to conform to the latest *Building Code Requirements for Structural Concrete* (ACI 318-08) of the American Concrete Institute. Because the ACI Code serves as the design standard in the United States, it is strongly recommended that the code be used as a companion publication to this book.

In addition to the necessary changes to conform to the new code, some sections have been edited and some new homework problems have been added. A section on the design for torsion and a section on the design of shear walls have been added. Answers to selected problems are furnished at the back of the text.

Throughout the seven editions, the text content has remained primarily an elementary, noncalculus, practical approach to the design and analysis of reinforced concrete structural members using numerous examples and a step-by-step solution format. In addition, there are chapters that provide a conceptual approach on such topics as prestressed concrete and detailing of reinforced concrete structures. The metric system (SI), the use of which is gradually gaining momentum in the reinforced concrete design and construction field in the United States, is introduced in Appendix C with several example problems.

Form design is an important consideration in most structural design problems involving concrete members, and Chapter 12 illustrates procedures for the design of job-built forms for slabs, beams, and columns. Appropriate tables are included that will expedite the design process.

This book has been thoroughly tested over the years in engineering technology programs and should serve as a valuable design guide and resource for technologists, technicians, engineering and architectural students, and design engineers. In addition, it will aid engineers and architects preparing for state licensing examinations for professional registration.

As in the past, appreciation is extended to students, past and present, and colleagues who, with their constructive comments, criticisms, and enthusiasm, have provided input and encouragement for this edition.

Additional Resources

To access supplementary materials online, instructors need to request an instructor access code. Go to **www.pearsonhighered.com/irc**, where you can register for an instructor access code. Within 48 hours after registering, you will receive a confirming e-mail, including an instructor access code. Once you have received your code, go to the site and log on for full instructions on downloading the materials you wish to use.

George F. Limbrunner
Abi O. Aghayere

Contents

CHAPTER 3   REINFORCED CONCRETE BEAMS: T-BEAMS AND DOUBLY REINFORCED BEAMS   72

CHAPTER 4   SHEAR IN BEAMS   120

CHAPTER 5   DEVELOPMENT, SPLICES, AND SIMPLE-SPAN BAR CUTOFFS   162

CHAPTER 6   CONTINUOUS CONSTRUCTION DESIGN CONSIDERATIONS   199

CHAPTER 7   SERVICEABILITY   227

CHAPTER 8   WALLS   245

CHAPTER 12   CONCRETE FORMWORK   414

CHAPTER 13   DETAILING REINFORCED CONCRETE STRUCTURES   464

APPENDIX A   TABLES AND DIAGRAMS   482

APPENDIX B   SUPPLEMENTARY AIDS AND GUIDELINES   502

APPENDIX C   METRICATION   508

APPENDIX D   ANSWERS TO SELECTED PROBLEMS   525

INDEX   528

Chapter 1

Materials and Mechanics of Bending

1-1 CONCRETE

Concrete consists primarily of a mixture of cement and fine and coarse aggregates (sand, gravel, crushed rock, and/or other materials) to which water has been added as a necessary ingredient for the chemical reaction of curing. The bulk of the mixture consists of the fine and coarse aggregates. The resulting concrete strength and durability are a function of the proportions of the mix as well as other factors, such as the concrete placing, finishing, and curing history.

The compressive strength of concrete is relatively high. Yet it is a relatively brittle material, the tensile strength of which is small compared with its compressive strength. Hence steel reinforcing rods (which have high tensile and compressive strength) are used in combination with the concrete; the steel will resist the tension and the concrete the compression. *Reinforced concrete* is the result of this combination of steel and concrete. In many instances, steel and concrete are positioned in members so that they both resist compression.

1-2   THE ACI BUILDING CODE

The design and construction of reinforced concrete buildings is controlled by the *Building Code Requirements for Structural Concrete* (ACI 318-08) of the American Concrete Institute (ACI) [1]. The use of the term *code* in this text refers to the ACI Code unless otherwise stipulated. The code is revised, updated, and reissued on a 3-year cycle. The code itself has no legal status. It has been incorporated into the building codes of almost all states and municipalities throughout the United States, however. When so incorporated, it has official sanction, becomes a legal document, and is part of the law controlling reinforced concrete design and construction in a particular area.

1-3   CEMENT AND WATER

Structural concrete uses, almost exclusively, hydraulic cement. With this cement, water is necessary for the chemical reaction of *hydration*. In the process of hydration, the cement sets and bonds the fresh concrete into one mass. *Portland cement*, which originated in England, is undoubtedly the most common form of cement. Portland cement consists chiefly of calcium and aluminum silicates. The raw materials are limestones, which provide calcium oxide (CaO), and clays or shales, which furnish silicon dioxide (SiO_2) and aluminum oxide (Al_2O_3). Following processing, cement is marketed in bulk or in 94-lb (1-ft^3) bags.

In fresh concrete, the ratio of the amount of water to the amount of cement, by weight, is termed the *water/cement ratio*. This ratio can also be expressed in terms of gallons of water per bag of cement. For complete hydration of the cement in a mix, a water/cement ratio of 0.35 to 0.40 (4 to 4½ gal/bag) is required. To increase the *workability* of the concrete (the ease with which it can be mixed, handled, and placed), higher water/cement ratios are normally used.

1-4   AGGREGATES

In ordinary structural concretes, the aggregates occupy approximately 70% to 75% of the volume of the hardened mass. Gradation of aggregate size to produce close packing is desirable because, in general, the more densely the aggregate can be packed, the better are the strength and durability.

Aggregates are classified as fine or coarse. *Fine aggregate* is generally sand and may be categorized as consisting of particles that will pass a No. 4 sieve (four openings per linear inch). *Coarse aggregate* consists of particles that would be retained on a No. 4 sieve. The maximum size of coarse aggregate in reinforced concrete is governed by various ACI Code requirements. These requirements are established primarily to ensure that the concrete can be placed with ease into the forms without any danger of jam-up between adjacent bars or between bars and the sides of the forms.

1-5   CONCRETE IN COMPRESSION

The theory and techniques relative to the design and proportioning of concrete mixes, as well as the placing, finishing, and curing of concrete, are outside the scope of this book and are adequately discussed in many other publications [2–5]. Field testing, quality control, and inspection are also adequately covered elsewhere. This is not to imply that these are of less importance in overall concrete construction technology but only to reiterate that the objective of this book is to deal with the design and analysis of reinforced concrete members.

We are concerned primarily with how a reinforced concrete member behaves when subjected to load. It is generally accepted that the behavior of a reinforced concrete member under load depends on the stress–strain relationship of the materials, as well as the type of stress to which it is subjected. With concrete used principally in compression, the compressive stress–strain curve is of primary interest.

The compressive strength of concrete is denoted f_c' and is assigned the units *pounds per square inch* (psi). For calculations, f_c' is frequently used with the units *kips per square inch* (ksi).

A test that has been standardized by the American Society for Testing and Materials (ASTM C39) [6] is used to determine the compressive strength (f_c') of concrete. The test involves compression loading to failure of a specimen cylinder of concrete. The compressive strength so determined is the highest compressive stress to which the specimen is subjected. Note in Figure 1-1 that f_c' is not the stress that exists in the specimen at failure but that which occurs at a strain of about 0.002. Currently, 28-day concrete strengths (f_c') range from 2500 to 9000 psi, with 3000 to 4000 psi being common for reinforced concrete structures and 5000 to 6000 psi being common for prestressed concrete members. Concretes of much higher strengths have been achieved under laboratory conditions. The curves shown in Figure 1-1 represent the result of compression tests on 28-day standard cylinders for varying design mixes.

A review of the stress–strain curves for different-strength concretes reveals that the maximum compressive strength is generally achieved at a unit strain of approximately 0.002 in./in. Stress then decreases, accompanied by additional strain. Higher-strength concretes are more brittle and will fracture at a lower maximum strain than will the lower-strength concretes. The initial slope of the curve varies, unlike that of steel, and only approximates a straight line. For steel, where stresses are below the

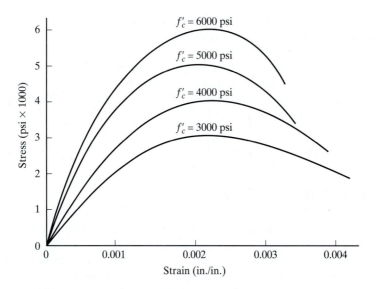

FIGURE 1-1   Typical stress–strain curves for concrete.

yield point and the material behaves elastically, the stress–strain plot will be a straight line. The slope of the straight line is the modulus of elasticity. For concrete, however, we observe that the straight-line portion of the plot is very short, if it exists at all. Therefore, there exists no constant value of modulus of elasticity for a given concrete because the stress–strain ratio is not constant. It may also be observed that the slope of the initial portion of the curve (if it approximates a straight line) varies with concretes of different strengths. Even if we assume a straight-line portion, the modulus of elasticity is different for concretes of different strengths. At low and moderate stresses (up to about $0.5f'_c$), concrete is commonly assumed to behave elastically.

The ACI Code, Section 8.5.1, provides the accepted empirical expression for *modulus of elasticity*:

$$E_c = w_c^{1.5} 33 \sqrt{f'_c}$$

where

E_c = modulus of elasticity of concrete in compression (psi)

w_c = unit weight of concrete (lb/ft^3)

f'_c = compressive strength of concrete (psi)

This expression is valid for concretes having w_c between 90 and 160 lb/ft^3. For normal-weight concrete, the unit weight w_c will vary with the mix proportions and with the character and size of the aggregates. If the unit weight is taken as 144 lb/ft^3, the resulting expression for modulus of elasticity is

$$E_c = 57,000 \sqrt{f'_c} \qquad \text{(see Table A-6 for values of } E_c\text{)}$$

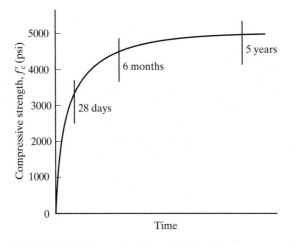

FIGURE 1-2   Strength–time relationship for concrete.

It should also be noted that the stress–strain curve for the same-strength concrete may be of different shapes if the condition of loading varies appreciably. With different *rates of strain* (loading), we will have different-shape curves. Generally, the maximum strength of a given concrete is smaller at slower rates of strain.

Concrete strength varies with time, and the specified concrete strength is usually that strength that occurs 28 days after the placing of concrete. A typical strength–time curve for normal stone concrete is shown in Figure 1-2. Generally, concrete attains approximately 70% of its 28-day strength in 7 days and approximately 85% to 90% in 14 days.

Concrete, under load, exhibits a phenomenon termed *creep*. This is the property by which concrete continues to deform (or strain) over long periods of time while under constant load. Creep occurs at a decreasing rate over a period of time and may cease after several years. Generally, high-strength concretes exhibit less creep than do lower-strength concretes. The magnitude of the creep deformations is proportional to the magnitude of the applied load as well as to the length of time of load application.

1-6   CONCRETE IN TENSION

The tensile and compressive strengths of concrete are not proportional, and an increase in compressive strength is accompanied by an appreciably smaller percentage increase in tensile strength. According to the ACI Code Commentary, the tensile strength of normal-weight concrete in flexure is about 10% to 15% of the compressive strength.

The true tensile strength of concrete is difficult to determine. The *split-cylinder test* (ASTM C496) [6] has been used to determine the tensile strength of lightweight aggregate concrete and is generally accepted as a good measure of the true tensile strength. The split-cylinder test uses a standard 6-in.-diameter, 12-in.-long cylinder placed on its side in a testing machine. A compressive line load is applied uniformly along the length of the cylinder, with support furnished along the full length of the bottom of the cylinder. The compressive load produces a transverse tensile stress, and the cylinder will split in half along a diameter when its tensile strength is reached.

The tensile stress at which splitting occurs is referred to as the *splitting tensile strength*, f_{ct}, and may be calculated by the following expression derived from the theory of elasticity:

$$f_{ct} = \frac{2P}{\pi LD}$$

where

f_{ct} = splitting tensile strength of lightweight aggregate concrete (psi)

P = applied load at splitting (lb)

L = length of cylinder (in.)

D = diameter of cylinder (in.)

Another common approach has been to use the *modulus of rupture*, f_r (which is the maximum tensile bending stress in a plain concrete test beam at failure), as a measure of tensile strength (ASTM C78) [6]. The moment that produces a tensile stress just equal to the modulus of rupture is termed the *cracking moment*, M_{cr}, and may be calculated using methods discussed in Section 1-8. The ACI Code recommends that the modulus of rupture f_r be taken as $7.5\lambda \sqrt{f_c'}$, where f_c' is in psi. Greek lowercase lambda (λ) is a modification factor reflecting the lower tensile strength of lightweight concrete relative to normal-weight concrete. The values for λ are as follows:

Normal-weight concrete—1.0
Sand-lightweight concrete—0.85
All-lightweight concrete—0.75

Interpolation between these values is permitted. See ACI Code Section 8.6.1. for details. If the average splitting tensile strength f_{ct} is specified, then $\lambda = f_{ct}/(6.7\sqrt{f_c'}) \leq 1.0$.

1-7   REINFORCING STEEL

Concrete cannot withstand very much tensile stress without cracking; therefore, tensile reinforcement must be embedded in the concrete to overcome this deficiency. In the United States, this reinforcement is in the form of steel reinforcing bars or welded wire reinforcing composed of steel wire. In addition, reinforcing in the form of structural steel shapes, steel pipe, steel tubing, and high-strength steel tendons is permitted by the ACI Code. Many other approaches have been taken in the search for an economical reinforcement for concrete. Principal among these are the fiber-reinforced concretes, where the reinforcement is obtained through the use of short fibers of steel or other materials, such as fiberglass. For the purpose of this book, our discussion will primarily include steel reinforcing bars and welded wire reinforcing. High-strength steel tendons are used mainly in prestressed concrete construction (see Chapter 11).

The specifications for steel reinforcement published by the ASTM are generally accepted for the steel used in reinforced concrete construction in the United States and are identified in the ACI Code, Section 3.5.

The steel bars used for reinforcing are, almost exclusively, round deformed bars with some form of patterned ribbed projections rolled onto their surfaces. The patterns vary depending on the producer, but all patterns should conform to ASTM specifications. Steel reinforcing bars are readily available in straight lengths of 60 ft. Smaller sizes are also available in coil stock for use in automatic bending machines. The bars vary in designation from No. 3 through No. 11, with two additional bars, No. 14 and No. 18.

For bars No. 3 through No. 8, the designation represents the bar diameter in eighths of an inch. The No. 9, No. 10, and No. 11 bars have diameters that provide areas equal to 1-in.-square bars, $1\frac{1}{8}$-in.-square bars, and $1\frac{1}{4}$-in.-square bars, respectively. The No. 14 and No. 18 bars correspond to $1\frac{1}{2}$-in.-square bars and 2-in.-square bars, respectively, and are commonly available only by special order. Round, plain reinforcing bars are permitted for spirals (lateral reinforcing) in concrete compression members.

ASTM specifications require that identification marks be rolled onto the bar to provide the following information: a letter or symbol indicating the producer's mill, a number indicating the size of the bar, a symbol or letter indicating the type of steel from which the bar was rolled, and for grade 60 bars, either the number 60 or a single continuous longitudinal line (called a *grade line*) through at least five deformation spaces. The *grade* indicates the minimum specified yield stress in ksi. For instance, a grade 60 steel bar has a minimum specified yield stress of 60 ksi. No symbol indicating grade is rolled onto grade 40 or 50 steel bars. Grade 75 bars can have either two grade lines through at least five deformation spaces or the grade mark 75. Reference [7] is an excellent resource covering the various aspects of bar identification.

Reinforcing bars are usually made from newly manufactured steel (billet steel). Steel types and ASTM specification numbers for bars are tabulated in Table A-1. Note that ASTM A615, which is billet steel, is available in grades 40, 60, and 75. (The full range of bar sizes is not available in grades 40 and 75, however.) ASTM A706, low-alloy steel, which was developed to satisfy the requirement for reinforcing bars with controlled tensile properties and controlled chemical composition for weldability, is available in only one grade. Tables A-2 and A-3 contain useful information on cross-sectional areas of bars.

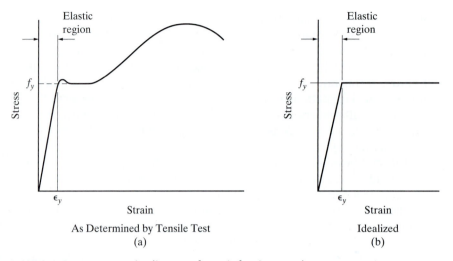

FIGURE 1-3   Stress–strain diagram for reinforcing steel.

The most useful physical properties of reinforcing steel for reinforced concrete design calculations are yield stress (f_y) and modulus of elasticity. A typical stress–strain diagram for reinforcing steel is shown in Figure 1-3a. The idealized stress–strain diagram of Figure 1-3b is discussed in Chapter 2.

The yield stress (or yield point) of steel is determined through procedures governed by ASTM standards. For practical purposes, the yield stress may be thought of as that stress at which the steel exhibits increasing strain with no increase in stress. The yield stress of the steel will usually be one of the known (or given) quantities in a reinforced concrete design or analysis problem. See Table A-1 for the range of f_y.

The modulus of elasticity of carbon reinforcing steel (the slope of the stress–strain curve in the elastic region) varies over a very small range and has been adopted as 29,000,000 psi (ACI Code, Section 8.5.2).

Unhindered corrosion of reinforcing steel will lead to cracking and spalling of the concrete in which it is embedded. Quality concrete, under normal conditions, provides good protection against corrosion for steel embedded in the concrete with adequate cover (minimum requirements are discussed in Chapter 2). This protection is attributed to, among other factors, the high alkalinity of the concrete. Where reinforced concrete structures (or parts of structures) are subjected to corrosive conditions, however, some type of corrosion protection system should be used to prevent deterioration. Examples of such structures are bridge decks, parking garage decks, wastewater treatment plants, and industrial and chemical processing facilities.

One method used to minimize the corrosion of the reinforcing steel is to coat the bars with a suitable protective coating. The protective coating can be a nonmetallic material such as epoxy or a metallic material such as zinc (galvanizing). The ACI Code requires epoxy-coated reinforcing bars to comply with ASTM A775 or ASTM

A934 and galvanized bars to comply with ASTM A767. The bars to be epoxy coated or zinc coated (galvanized) must meet the code requirements for uncoated bars as tabulated in Table A-1.

Welded wire reinforcing (WWR) (commonly called *mesh*) is another type of reinforcement. It consists of cold-drawn wire in orthogonal patterns, square or rectangular, resistance welded at all intersections. It may be supplied in either rolls or sheets, depending on wire size. WWR with wire diameters larger than about $\frac{1}{4}$ in. is usually available only in sheets.

Both plain and deformed WWR products are available. Plain WWR must conform to ASTM A185 and be made of wire conforming to ASTM A82. Deformed WWR must conform to ASTM A497 and be made of wire conforming to ASTM A496. Both materials have a yield strength of 70,000 psi. For both materials, the code has assigned a yield strength value of 60,000 psi but makes provision for the use of higher-yield strengths provided the stress corresponds to a strain of 0.35%. The deformed wire is usually more expensive, but it can be expected to have an improved bond with the concrete.

A rational method of designating wire sizes to replace the formerly used gauge system has been adopted by the wire industry. Plain wires are described by the letter

PHOTO 1-1 Concrete construction in progress. Note formwork, reinforcing bars, and pumping of concrete.

W followed by a number equal to 100 times the cross-sectional area of the wire in square inches. Deformed wire sizes are similarly described, but the letter D is used. Thus a W9 wire has an area of 0.090 in.^{2} and a D8 wire has an area of 0.080 in.^{2} A W8 wire has the same cross-sectional area as the D8 but is plain rather than deformed. Sizes between full numbers are given by decimals, such as W9.5.

Generally, the material is indicated by the symbol WWR, followed by spacings first of longitudinal wires, then of transverse wires, and last by the sizes of longitudinal and transverse wires. Thus WWR6 × 12-W16 × W8 indicates a plain WWR with 6-in. longitudinal spacing, 12-in. transverse spacing, and a cross-sectional area equal to 0.16 in.^{2} for the longitudinal wires and 0.08 in.^{2} for the transverse wires.

Additional information about WWR, as well as tables relating size number with wire diameter, area, and weight, may be obtained through the Wire Reinforcement Institute [8] or the Concrete Reinforcing Steel Institute [8 and 9]. ACI 318-08 contains a useful chart that gives area (in.^{2}/ft) for various WWR spacings (see Appendix E).

Most concrete is reinforced in some way to resist tensile forces. Some structural elements, particularly footings, are sometimes made of *plain concrete*, however. Plain concrete is defined as structural concrete with no reinforcement or with less reinforcement than the minimum amount specified for reinforced concrete. Plain concrete is discussed further in Chapter 10.

1-8   BEAMS: MECHANICS OF BENDING REVIEW

The concept of bending stresses in homogeneous elastic beams is generally discussed at great length in all strength of materials textbooks and courses. Beams composed of material such as steel or timber are categorized as homogeneous, with each exhibiting elastic behavior up to some limiting point. Within the limits of elastic behavior, the internal bending stress distribution developed at any cross section is linear (straight line), varying from zero at the neutral axis to a maximum at the outer fibers.

The accepted expression for the maximum bending stress in a beam is termed the *flexure formula*, where

$$f_b = \frac{Mc}{I}$$

where

f_b = calculated bending stress at the outer fiber of the cross section

M = the applied moment

  c = distance from the neutral axis to the outside tension or compression fiber of the beam

  I = moment of inertia of the cross section about the neutral axis.

The flexure formula represents the relationship between bending stress, bending moment, and the geometric properties of the beam cross section. By rearranging the flexure formula, the maximum moment that may be applied to the beam cross section, called the *resisting moment, M_R,* may be found:

$$M_R = \frac{F_b I}{c}$$

where F_b = the allowable bending stress.

This procedure is straightforward for a beam of known cross section for which the moment of inertia can easily be found. For a reinforced concrete beam, however, the use of the flexure formula presents some complications, because the beam is not homogeneous and concrete does not behave elastically over its full range of strength. As a result, a somewhat different approach that uses the beam's internal bending stress distribution is recommended. This approach is termed the *internal couple method.*

Recall from strength of materials that a couple is a pure moment composed of two equal, opposite, and parallel forces separated by a distance called the *moment arm,* which is commonly denoted Z. In the internal couple method, the couple represents an internal resisting moment and is composed of a compressive force C above the neutral axis (assuming a single-span, simply supported beam that develops compressive stress above the neutral axis) and a parallel internal tensile force T below the neutral axis.

As with all couples, and because the forces acting on any cross section of the beam must be in equilibrium, C must equal T. The internal couple must be equal and opposite to the bending moment at the same location, which is computed from the external loads. It represents a couple developed by the bending action of the beam.

The internal couple method of determining beam strength is more general and may be applied to homogeneous or nonhomogeneous beams having linear (straight-line) or nonlinear stress distributions. For reinforced concrete beams, it has the advantage of using the basic resistance pattern found in the beam.

The following three analysis examples dealing with plain (unreinforced) concrete beams provide an introduction to the internal couple method. Note that the unreinforced beams are considered homogeneous and elastic. This is valid if the moment is small and tensile bending stresses in the concrete are low (less than the tensile bending strength of the concrete) with no cracking of the concrete developing. For this condition, the entire beam cross section carries bending stresses. Therefore, the analysis for bending stresses in the uncracked beam can be based on the properties of the gross cross-sectional area using the elastic-based flexure formula. The use of the flexure formula is valid as long as the maximum tensile stress in the concrete does not exceed the modulus of rupture f_r. If a moment is applied that causes the maximum tensile stress just to reach the modulus of rupture, the cross section will be on the verge of cracking. This moment is called the *cracking moment, M_{cr}.*

These examples use both the internal couple approach and the flexure formula approach so that the results may be compared.

Example 1-1

A normal-weight plain concrete beam is 6 in. × 12 in. in cross section, as shown in Figure 1-4. The beam is simply supported on a span of 4 ft and is subjected to a midspan concentrated load of 4500 lb. Assume $f'_c = 3000$ psi.

a.   Calculate the maximum concrete tensile stress using the internal couple method.

b.   Repeat part (a) using the flexure formula approach.

c.   Compare the maximum concrete tensile stress with the value for modulus of rupture f_r using the ACI-recommended value based on f'_c.

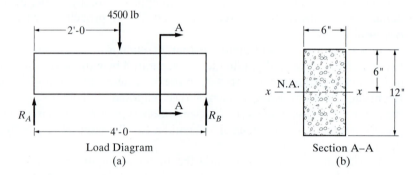

Load Diagram
(a)

Section A–A
(b)

FIGURE 1-4   Sketch for Example 1-1.

Solution:

Calculate the weight of the beam (weight per unit length):

$$\text{weight of beam} = \text{volume per unit length} \times \text{unit weight}$$

$$= \frac{6\text{in.}(12\text{ in.})}{144\text{ in.}^2/\text{ft}^2}(150\text{ lb/ft}^3)$$

$$= 75\text{ lb/ft}$$

Calculate the maximum applied moment:

$$M_{\max} = \frac{PL}{4} + \frac{wL^2}{8}$$

$$= \frac{4500\text{ lb}(4\text{ ft})}{4} + \frac{75\text{ lb/ft}(4\text{ ft})^2}{8}$$

$$= 4650\text{ ft-lb}$$

a.   Internal couple method

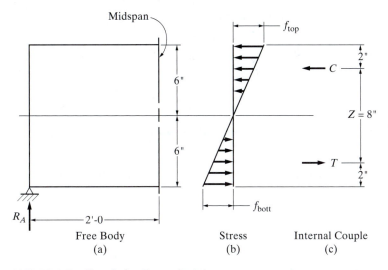

Free Body (a)   Stress (b)   Internal Couple (c)

FIGURE 1-5   Sketch for Example 1-1.

1. Because the beam is homogeneous, elastic, and symmetrical with respect to both the X–X and Y–Y axes, the neutral axis (N.A.) is at midheight. Stresses and strains vary linearly from zero at the neutral axis (which is also the centroidal axis) to a maximum at the outer fiber. As the member is subjected to positive moment, the area above the N.A. is stressed in compression and the area below the N.A. is stressed in tension. These stresses result from the bending behavior of the member and are shown in Figure 1-5.

2. C represents the resultant compressive force above the N.A. T represents the resultant tensile force below the N.A. C and T each act at the centroid of their respective triangles of stress distribution. Therefore $Z = 8$ in. C and T must be equal (since $\Sigma H_F = 0$). The two forces act together to form the internal couple (or internal resisting moment) of magnitude CZ or TZ.

3. The internal resisting moment must equal the bending moment due to external loads at any section. Therefore

$$M = CZ = TZ$$

$$4650 \text{ ft-lb } (12 \text{ in./ft}) = C \ (8 \text{ in.})$$

from which

$$C = 6975 \text{ lb} = T$$

4. C = average stress × area of beam on which stress acts

$$C = \tfrac{1}{2} f_{top} \ (6 \text{ in.})(6 \text{ in.}) = 6975 \text{ lb}$$

Solving for f_{top} yields

$$f_{top} = 388 \text{ psi} = f_{bott}$$

b.   Flexure formula approach

$$I = \frac{bh^3}{12} = \frac{6(12^3)}{12} = 864 \text{ in.}^4$$

$$f_{top} = f_{bott} = \frac{Mc}{I} = \frac{4650(12)(6)}{864} = 388 \text{ psi}$$

c.   The ACI-recommended value for the modulus of rupture (based on f_c') is

$$f_r = 7.5 \lambda \sqrt{f_c'} = 7.5 (1.0)\sqrt{3000}$$

$$f_r = 411 \text{ psi}$$

The calculated tensile stress (f_{bott}) of 388 psi is about 6% below the modulus of rupture, the stress at which flexural cracking would be expected.

Example 1-1 is based on elastic theory and assumes the following: (1) a plane section before bending remains a plane section after bending (the variation in strain throughout the depth of the member is linear from zero at the neutral axis), and (2) the modulus of elasticity is constant; therefore, stress is proportional to strain and the stress distribution throughout the depth of the beam is also linear from zero at the neutral axis to a maximum at the outer fibers.

The internal couple approach may also be used to find the moment strength (resisting moment) of a beam.

Example 1-2

Calculate the cracking moment M_{cr} for the plain concrete beam shown in Figure 1-6. Assume normal-weight concrete and $f_c' = 4000$ psi.

a.   Use the internal couple method.
b.   Check using the flexure formula.

Solution:

The moment that produces a tensile stress just equal to the modulus of rupture f_r is called the cracking moment, M_{cr}. The modulus of rupture for normal-weight concrete is calculated from ACI Equation (9-10):

$$f_r = 7.5\sqrt{f_c'} = 7.5\sqrt{4000} = 474 \text{ psi}$$

For convenience, we will use force units of kips (1 kip = 1000 lb). Therefore, $f_r = 0.474$ ksi.

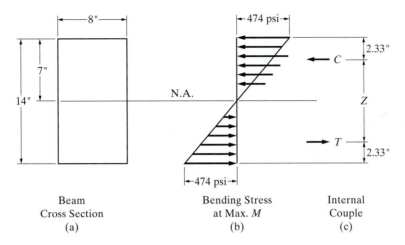

Beam
Cross Section
(a)
        Bending Stress
at Max. M
(b)
        Internal
Couple
(c)

FIGURE 1-6   Sketch for Example 1-2.

a.   Using the internal couple method

$$Z = 14 - 2(2.33) = 9.34 \text{ in.}$$

$$C = T = \tfrac{1}{2}(0.474)(8)(7) = 13.27 \text{ kips}$$

$$M_{cr} = CZ = TZ = \frac{13.27(9.34)}{12} = 10.33 \text{ ft-kips}$$

b.   Check using the flexure formula

$$f = \frac{Mc}{I}$$

$$M_R = M_{cr} = \frac{f_r I}{c}$$

$$I = \frac{bh^3}{12} = \frac{8(14)^3}{12} = 1829 \text{ in.}^4$$

$$M_{cr} = \frac{f_r I}{c} = \frac{0.474(1829)}{7(12)} = 10.32 \text{ ft-kips}$$

The internal couple method may also be used to analyze irregularly shaped cross sections, although for homogeneous beams it is more cumbersome than the use of the flexure formula.

Example 1-3 _____

Calculate the cracking moment (resisting moment) for the T-shaped unreinforced concrete beam shown in Figure 1-7. Use $f'_c = 4000$ psi. Assume positive

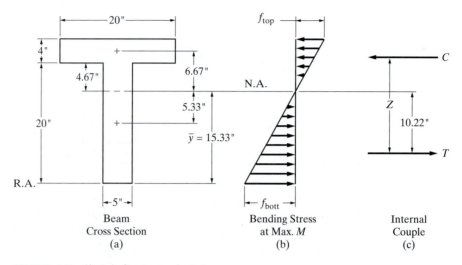

FIGURE 1-7   Sketch for Example 1-3.

moment (compression in the top). Use the internal couple method and check using the flexure formula.

Solution:

The neutral axis must be located so that the strain and stress diagrams may be defined. The location of the neutral axis with respect to the noted reference axis is calculated from

$$\bar{y} = \frac{\Sigma(Ay)}{\Sigma A}$$

$$= \frac{4(20)(22) + 5(20)(10)}{4(20) + 5(20)}$$

$$= 15.33 \text{ in.}$$

The bottom of the cross section is stressed in tension. Note that the stress at the bottom will be numerically larger than at the top because of the relative distances from the N.A. The stress at the bottom of the cross section will be set equal to the modulus of rupture ($\lambda = 1.0$ for normal-weight concrete):

$$f_{bott} = f_r = 7.5\lambda \sqrt{f_c'} = 7.5(1.0)\sqrt{4000} = 474 \text{ psi} = 0.474 \text{ ksi}$$

Using similar triangles in Figure 1-7b, the stress at the top of the flange is

$$f_{top} = \frac{8.67}{15.33}(0.474) = 0.268 \text{ ksi}$$

Similarly, the stress at the bottom of the flange is

$$f_{\text{bott of flange}} = \frac{4.67}{15.33}(0.474) = 0.1444 \text{ ksi}$$

The total tensile force can be evaluated as follows:

$$T = \text{average stress} \times \text{area}$$

$$= \tfrac{1}{2}(0.474)(15.33)(5) = 18.17 \text{ kips}$$

and its location below the N.A. is calculated from

$$\tfrac{2}{3}(15.33) = 10.22 \text{ in. (below the N.A.)}$$

The compressive force will be broken up into components because of the irregular area, as shown in Figure 1-8. Referring to both Figure 1-7 and Figure 1-8, the component internal compressive forces, component internal couples, and M_R may now be evaluated. The component forces are first calculated:

$$C_1 = 0.1444(20)(4) = 11.55 \text{ kips}$$

$$C_2 = \tfrac{1}{2}(0.1236)(20)(4) = 4.94 \text{ kips}$$

$$C_3 = \tfrac{1}{2}(0.1444)(5)(4.67) = 1.686 \text{ kips}$$

$$\text{total } C = C_1 + C_2 + C_3 = 18.18 \text{ kips}$$

$$C \approx T \quad \text{(O.K.)}$$

Next we calculate the moment arm distance from each component compressive force to the tensile force T:

$$Z_1 = 10.22 + 4.67 + \tfrac{1}{2}(4.00) = 16.89 \text{ in.}$$

$$Z_2 = 10.22 + 4.67 + \tfrac{2}{3}(4.00) = 17.56 \text{ in.}$$

$$Z_3 = 10.22 + \tfrac{2}{3}(4.67) = 13.33 \text{ in.}$$

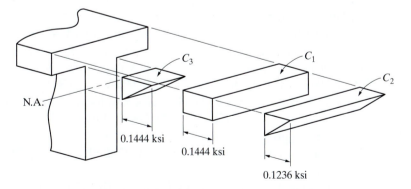

FIGURE 1-8   Component compression forces for Example 1-3.

The magnitudes of the component internal couples are then calculated from force × moment arm as follows:

$$M_{R_1} = 11.55(16.89) = 195.1 \text{ in.-kips}$$

$$M_{R_2} = 4.94(17.56) = 86.7 \text{ in.-kips}$$

$$M_{R_3} = 1.686(13.33) = 22.5 \text{ in.-kips}$$

$$M_{cr} = M_R = M_{R_1} = M_{R_2} = M_{R_3} = 304 \text{ in.-kips}$$

Check using the flexure formula. The moment of inertia is calculated using the transfer formula from statics:

$$I = \Sigma I_o + \Sigma Ad^2$$

$$I = \tfrac{1}{12}(20)(4^3) + \tfrac{1}{12}(5)(20^3) + 4(20)(6.67^2) + 5(20)(5.33^2)$$

$$= 9840 \text{ in.}^4$$

$$M_{cr} = M_R = \frac{f_r I}{c} = \frac{0.474(9840)}{15.33} = 304 \text{ in.-kips} \qquad \text{(Checks O.K.)}$$

As mentioned previously, the three examples are for plain, unreinforced, and uncracked concrete beams that are considered homogeneous and elastic within the bending stress limit of the modulus of rupture. The internal couple method is also applicable to nonhomogeneous beams with nonlinear stress distributions of any shape, however. Because reinforced concrete beams are nonhomogeneous, the flexure formula is not directly applicable. Therefore the basic approach used for reinforced concrete beams is the internal couple method.

REFERENCES

[1] *Building Code Requirements for Structural Concrete* (ACI 318-08). American Concrete Institute, P.O. Box 9094, Farmington Hills, MI 48333-9094, 2008.

[2] ACI Committee 211. *Standard Practice for Selecting Proportions for Normal, Heavyweight, and Mass Concrete* (ACI 211.1-91). American Concrete Institute, P.O. Box 9094, Farmington Hills, MI 48333-9094, 1991. (Reapproved 2002.)

[3] George E. Troxell, Harmer E. Davis, and Joe W. Kelly. *Composition and Properties of Concrete*, 2nd ed. New York: McGraw-Hill Book Company, 1968.

[4] *Design and Control of Concrete Mixtures*, 14th ed. Engineering Bulletin of the Portland Cement Association, 5420 Old Orchard Road, Skokie, IL 60077, 2002.

[5] Joseph J. Waddell, ed. *Concrete Construction Handbook*, 3rd ed. New York: McGraw-Hill Book Company, 1993.

[6] *ASTM Standards.* American Society for Testing and Materials, 100 Barr Harbor Drive, West Conshohocken, PA 19428-2959.

[7] *Manual of Standard Practice,* 27th ed. Concrete Reinforcing Steel Institute, 933 North Plum Grove Road, Schaumburg, IL 60173.

[8] *Manual of Standard Practice,* 7th ed. Wire Reinforcement Institute, 942 Main Street, Suite 300, Hartford, CT 06103, 2006.

[9] Concrete Reinforcing Steel Institute, 933 North Plum Grove Road, Schaumburg, IL 60173.

PROBLEMS

Note: In the following problems, assume plain concrete to have a weight of 145 pcf (conservative) unless otherwise noted.

1-1. The unit weight of normal-weight reinforced concrete is commonly assumed to be 150 lb/ft^3. Find the weight per lineal foot (lb/ft) for a normal weight reinforced concrete beam that:

(a) Has a rectangular cross section 16 in. wide and 28 in. deep.

(b) Has a cross section as shown in the accompanying diagram.

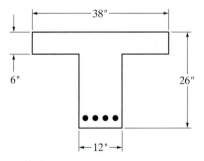

PROBLEM 1-1

1-2. Develop a spreadsheet application that will display in a table the values of modulus of elasticity E_c for concrete having unit weight ranging from 95 pcf to 155 pcf (in steps of 5 pcf) and compressive strength ranging from 3500 psi to 7000 psi (in steps of 500 psi). Display the modulus of elasticity rounded to the nearest 1000 psi.

1-3. A normal-weight concrete test beam 6 in. by 6 in. in cross section and supported on a simple span of 24 in. was loaded with a point load at midspan. The beam failed at a load of 2100 lb. Using this information, determine the modulus of rupture f_r of the concrete and compare with the ACI-recommended value based on an assumed concrete strength f'_c of 3000 psi.

1-4. A plain concrete beam has cross-sectional dimensions of 10 in. by 10 in. The concrete is known to have a modulus of rupture f_r of 350 psi. The beam spans between simple supports. Determine the span length at which this beam will fail due to its own weight. Assume a unit weight of 145 pcf.

1-5. The normal-weight plain concrete beam shown is on a simple span of 10 ft. It carries a dead load (which includes the weight of the beam) of 0.5 kip/ft. There is a concentrated load of 2 kips located at midspan. Use f'_c = 4000 psi. Compute the maximum bending stress. Use the internal couple method and check with the flexure formula.

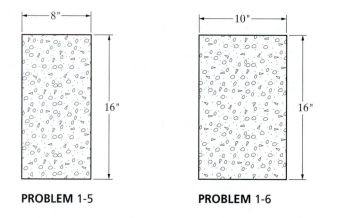

PROBLEM 1-5   PROBLEM 1-6

1-6. Calculate the cracking moment (resisting moment) for the unreinforced concrete beam shown. Assume normal-weight concrete with f'_c = 3000 psi. Use the internal couple method and check with the flexure formula.

1-7. Develop a spreadsheet application to solve Problem 1-6. Set up the spreadsheet so a table will be generated in which the width of the beam varies from 8 in. to 16 in. (1-in. increments) and the depth varies from 12 in. to 24 in. (1-in. increments.) The spreadsheet should allow the user to input any value for f'_c between 3000 psi and 8000 psi.

1-8. Rework Example 1-3 but invert the beam so that the flange is on the bottom and the web extends vertically upward. Calculate the cracking moment using the internal couple method and check using the flexure formula. Assume positive moment.

1-9. Calculate the cracking moment (resisting moment) for the U-shaped unreinforced concrete beam shown. Assume normal-weight concrete with f'_c = 3500 psi. Use the internal couple method and check with the flexure formula. Assume positive moment.

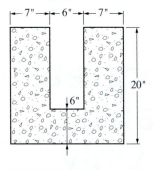

PROBLEM 1-9

1-10. The plain concrete beam shown is used on a 12-ft simple span. The concrete is normal weight with $f'_c = 3000$ psi Assume positive moment.

(a) Calculate the cracking moment.

(b) Calculate the value of the concentrated load P at midspan that would cause the concrete beam to crack. (Be sure to include the weight of the beam.)

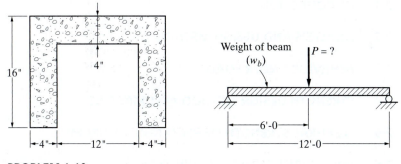

PROBLEM 1-10

Chapter 2

Rectangular Reinforced Concrete Beams and Slabs: Tension Steel Only

2-1 INTRODUCTION

When a beam is subjected to bending moments (also termed *flexure*), bending strains are produced. Under positive moment (as normally defined), compressive strains are produced in the top of the beam and tensile strains are produced in the bottom. These *strains* produce *stresses* in the beam, compression in the top, and tension in the bottom. Bending members must therefore be able to resist both tensile and compressive stresses.

For a concrete flexural member (beam, wall, slab, and so on) to have any significant load-carrying capacity, its basic inability to resist tensile stresses must be overcome. By embedding reinforcement (usually deformed steel bars) in the tension zones, a *reinforced concrete* member is created. When properly designed and constructed, members composed of these materials perform very adequately when subjected to flexure.

Initially, we will consider simply supported single-span beams that, as they carry only positive moment (tension in the bottom), will be reinforced with steel bars placed near the bottom of the beam.

2-2 ANALYSIS AND DESIGN METHOD

In the beam examples in Chapter 1, we assumed both a straight-line strain distribution and straight-line stress distribution from the neutral axis to the outer fibers. This, in effect, stated that stress was proportional to strain. This analysis is sometimes called *elastic design*.

As stated in Chapter 1, elastic design is considered valid for the homogeneous plain concrete beam as long as the tensile stress does not exceed the modulus of rupture, that stress at which tensile cracking commences. With homogeneous materials used in construction, such as structural steel and timber, the limit of stress–strain proportionality is generally termed the *proportional limit*. Note that the modulus of rupture for the plain concrete beam may be considered analogous to the proportional limit for structural steel and timber with respect to the limit of stress–strain proportionality.

With structural steel, the proportional limit and yield stress have nearly the same value, and when using the allowable stress design (ASD) method, an allowable bending stress is determined by applying a factor of safety to the yield stress.

With timber, the determination of an allowable bending stress is less straightforward, but it may be thought of as some fraction of the breaking bending stress. Using the allowable bending stress and the assumed linear stress–strain relationship, both the analysis and design of timber members and structural steel members (using the ASD method) are performed by a method that is similar to that used in the Chapter 1 examples.

Even though a reinforced concrete beam was known to be a nonhomogeneous member, for many years the elastic behavior approach was considered valid for concrete design, and it was known as the *working stress design* (WSD) method. The basic assumptions for the WSD method were as follows: (1) A plane section before bending remains a plane section after bending; (2) Hooke's law (stress is proportional to strain) applies to both the steel and the concrete; (3) the tensile strength of concrete is zero and the reinforcing steel carries all the tension; and (4) the bond between the concrete and the steel is perfect, so no slip occurs.

Based on these assumptions, the flexure formula was still used even though the beam was nonhomogeneous. This was accomplished by theoretically transforming one material into another based on the ratio of the concrete and steel moduli of elasticity.

Although the WSD method was convenient and was used for many years, it has been replaced with a more modern and realistic approach for the analysis and design of reinforced concrete. One basis for this approach is that at some point in the loading, the proportional stress–strain relationship for the compressive concrete ceases to exist. When first developed, this method was called the *ultimate strength design* (USD) method. Since then, the name has been changed to the *strength design method*.

The assumptions for the strength design method are similar to those itemized for the WSD method, with one notable exception. Research has indicated that the compressive concrete stress is approximately proportional to strain up to only moderate loads. With an increase in load, the approximate proportionality ceases to exist, and the compressive stress diagram takes a shape similar to the concrete compressive stress–strain curve of Figure 1-1. Additional assumptions for strength design are discussed in Section 2-4.

A major difference between ASD and strength design lies in the way the applied loads (i.e., *service loads*—the loads that are specified in the general building code) are handled and in the determination of the capacity (strength) of the reinforced concrete members. In the strength design method, service loads are amplified using load factors. Members are then designed so that their practical strength at failure, which is somewhat less than the true strength at failure, is sufficient to resist the amplified loads. The strength at failure is still commonly called the *ultimate strength*, and the load at or near failure is commonly called the *ultimate load*. The stress pattern assumed for strength design is such that predicted strengths are in substantial agreement with test results.

2-3   BEHAVIOR UNDER LOAD

Before discussing the strength design method, let us review the behavior of a long-span, rectangular reinforced concrete beam as the load on the beam increases from zero to the magnitude that would cause failure. The reinforced concrete simple beam of Figure 2-1 is assumed subjected to downward loading, which will cause positive moment in the beam. Steel reinforcing, three bars in this example, is located near the bottom of the beam, which is the tension side. Note that the overall depth of the beam is designated h, whereas the location of the steel, referenced to the compression face, is defined by the *effective depth, d*. The effective depth is measured to the centroid of the reinforcing steel. In this example, the centroid is at the center of the single layer of bars. If there are multiple layers of bars, then the effective depth is measured from the compression face to the centroid of the bar group.

   At very small loads, assuming that the concrete has not cracked, both concrete and steel will resist the tension, and concrete alone will resist the compression. The stress distribution will be as shown in Figure 2-1. The strain variation will be linear

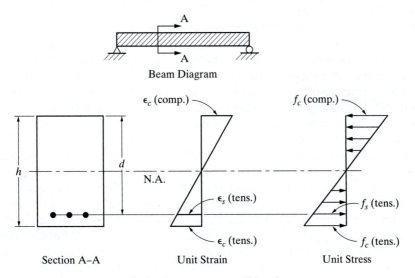

Beam Diagram

FIGURE 2-1   Flexural behavior at very small loads.

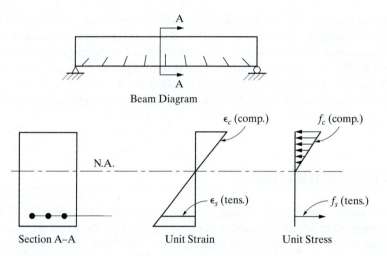

FIGURE 2-2   Flexural behavior at moderate loads.

from the neutral axis to the outer fiber. Note that stresses also vary linearly from zero at the neutral axis and are, for all practical purposes, proportional to strains. This will be the case when stresses are low (below the modulus of rupture).

At moderate loads, the tensile strength of the concrete will be exceeded, and the concrete will crack (hairline cracks) in the manner shown in Figure 2-2. Because the concrete cannot transmit any tension across a crack, the steel bars will then resist the entire tension. The stress distribution at or near a cracked section then becomes as shown in Figure 2-2. This stress pattern exists up to approximately a concrete stress f_c of about $f_c'/2$. The concrete compressive stress is still assumed to be proportional to the concrete strain.

With further load increase, the compressive strains and stresses will increase; they will cease to be proportional, however, and some nonlinear stress curve will result on the compression side of the beam. This stress curve above the neutral axis will be essentially the same shape as the concrete stress–strain curve (see Figure 1-1). The stress and strain distribution that exists at or near the ultimate load is shown in Figure 2-3. Eventually, the ultimate capacity of the beam will be reached and the beam will fail. The actual mechanism of the failure is discussed later in this chapter.

At this point the reader may well recognize that the process of attaining the ultimate capacity of a member is irreversible. The member has cracked and deflected significantly; the steel has yielded and will not return to its original length. If other members in the structure have similarly reached their ultimate capacities, the structure itself is probably crumbling and in a state of distress or partial ruin, even though it may not have completely collapsed. Naturally, although we cannot ensure that this state will never be reached, factors are introduced to create the commonly accepted margins of safety. Nevertheless the ultimate capacities of members are, at present, the basis for reinforced concrete analysis and design. In this text, it is in such a context that we will speak of failures of members.

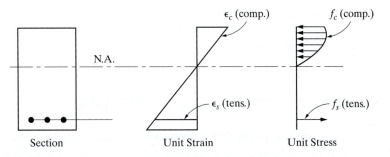

FIGURE 2-3   Flexural behavior near ultimate load.

2-4   STRENGTH DESIGN METHOD ASSUMPTIONS

The development of the strength design approach depends on the following basic assumptions:

1.   A plane section before bending remains a plane section after bending. That is, the strain throughout the depth of the member varies linearly from zero at the neutral axis. Tests have shown this assumption to be essentially correct.

2.   Stresses and strains are approximately proportional only up to moderate loads (assuming that the concrete stress does not exceed approximately $f'_c/2$). When the load is increased and approaches an ultimate load, stresses and strains are no longer proportional. Hence the variation in concrete stress is no longer linear.

3.   In calculating the ultimate moment capacity of a beam, the tensile strength of the concrete is neglected.

4.   The maximum usable concrete compressive strain at the extreme fiber is assumed equal to 0.003. This value is based on extensive testing, which indicated that the flexural concrete strain at failure for rectangular beams generally ranges from 0.003 to 0.004 in./in. Hence the assumption that the concrete is about to crush when the maximum strain reaches 0.003 is slightly conservative.

5.   The steel is assumed to be uniformly strained to the strain that exists at the level of the centroid of the steel. Also, if the strain in the steel (ϵ_s) is less than the yield strain of the steel (ϵ_y), the stress in the steel is $E_s\epsilon_s$. This assumes that for stresses less than f_y, the steel stress is proportional to strain. For strains equal to or greater than ϵ_y, the stress in the reinforcement will be considered independent of strain and equal to f_y. See the idealized stress–strain diagram for steel shown in Figure 1-3b.

6.   The bond between the steel and concrete is perfect and no slip occurs.

Assumptions 4 and 5 constitute what may be termed *code criteria* with respect to failure. The true ultimate strength of a member will be somewhat greater than that computed using these assumptions. The strength method of design and analysis of the ACI Code is based on these criteria, however, and consequently so is our basis for bending member design and analysis.

2-5   FLEXURAL STRENGTH OF RECTANGULAR BEAMS

Based on the assumptions previously stated, we can now examine the strains, stresses, and forces that exist in a reinforced concrete beam subjected to its *ultimate moment*, that is, the moment that exists just prior to the failure of the beam. In Figure 2-4, the assumed beam has a width b and an effective depth d; it is reinforced with a steel area of A_s. (A_s is the total *cross-sectional area* of tension steel present.)

Based on the preceding assumptions, it is possible that a beam may be loaded to the point where the maximum tensile steel unit stress equals its yield stress (as a limit) and the concrete compressive strain is less than 0.003 in./in. It is also possible that in another beam, the maximum concrete compressive strain will equal 0.003 in./in. and the tensile steel unit stress will be less than its yield stress f_y. When either condition occurs, it implies a specific mode of failure, which will be discussed later.

As stated previously, the compressive stress distribution above the neutral axis for a flexural member is similar to the concrete compressive stress–strain curve as

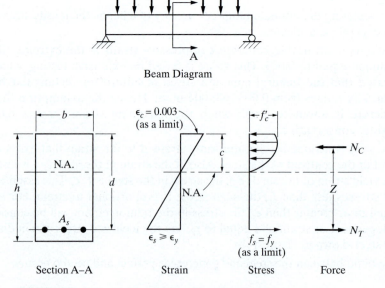

FIGURE 2-4   Beam subjected to ultimate moment.

depicted in Figure 1-1. As may be observed in Figure 2-4, the ultimate compressive stress f'_c does not occur at the outer fiber, neither is the shape of the curve the same for different-strength concretes. Actually, the magnitudes of the compressive concrete stresses are defined by some irregular curve, which could vary not only from concrete to concrete but also from beam to beam. Present theories accept that, at ultimate moment, compressive stresses and strains in concrete are not proportional. Although strains are assumed linear, with maximum strain of 0.003 in./in. at the extreme outer compressive fiber, the maximum concrete compressive stress f'_c develops at some intermediate level near, but not at, the extreme outer fiber.

The flexural strength or resisting moment of a rectangular beam is created by the development of these internal stresses that, in turn, may be represented as internal forces. As observed in Figure 2-4, N_C represents a theoretical internal resultant compressive force that in effect constitutes the total internal compression above the neutral axis. N_T represents a theoretical internal resultant tensile force that in effect constitutes the total internal tension below the neutral axis.

These two forces, which are parallel, equal, and opposite and separated by a distance Z, constitute an internal resisting couple whose maximum value may be termed the *nominal moment strength* of the bending member. As a limit, this nominal moment strength must be capable of resisting the design bending moment induced by the applied loads. Consequently, if we wish to design a beam for a prescribed loading condition, we must arrange its concrete dimensions and the steel reinforcements so that it is capable of developing a moment strength at least equal to the maximum bending moment induced by the loads.

The determination of the moment strength is complex because of the shape of the compressive stress diagram above the neutral axis. Not only is N_C difficult to evaluate but its location relative to the tensile steel is difficult to establish. Because the moment strength is actually a function of the magnitude of N_C and Z, however, it is not really necessary to know the exact shape of the compressive stress distribution above the neutral axis. To determine the moment strength, it is necessary to know only (1) the total resultant compressive force N_C in the concrete and (2) its location from the outer compressive fiber (from which the distance Z may be established). These two values may easily be established by replacing the unknown complex compressive stress distribution by a fictitious one of simple geometrical shape, provided the fictitious distribution results in the same total compressive force N_C applied at the same location as in the actual distribution when it is at the point of failure.

2-6   EQUIVALENT STRESS DISTRIBUTION

For purposes of simplification and practical application, a fictitious but equivalent rectangular concrete stress distribution was proposed by Whitney [1] and subsequently adopted by the ACI Code (Sections 10.2.6 and 10.2.7). The ACI Code also

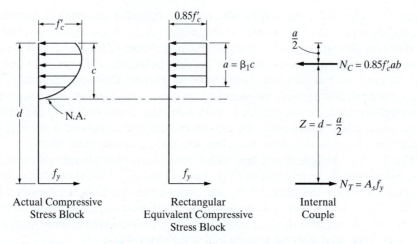

Actual Compressive Stress Block	Rectangular Equivalent Compressive Stress Block	Internal Couple

FIGURE 2-5   Equivalent stress block for strength design and analysis.

stipulates that other compressive stress distribution shapes may be used, provided results are in substantial agreement with comprehensive test results. Because of the simplicity of the rectangular shape, however, it has become the more widely used fictitious stress distribution for design purposes.

With respect to this equivalent stress distribution as shown in Figure 2-5, the average stress intensity is taken as $0.85 f_c'$ and is assumed to act over the upper area of the beam cross section defined by the width b and a depth of a. The magnitude of a may be determined by

$$a = \beta_1 c$$

where

c = distance from the outer compressive fiber to the neutral axis

β_1 = a factor that is a function of the strength of the concrete as follows and as shown in Figure 2-6:

For 2500 psi $\leq f_c' \leq$ 4000 psi: $\beta_1 = 0.85$.

For $f_c' >$ 4000 psi:

$$\beta_1 = 0.85 - \frac{0.05(f_c' - 4000)}{1000} \geq 0.65$$

It is in no way maintained that the compressive stresses are actually distributed in this most unlikely manner. It is maintained, however, that this equivalent rectangular distribution gives results close to those of the complex actual stress distribution. An isometric view of the accepted internal relationships is shown in Figure 2-7.

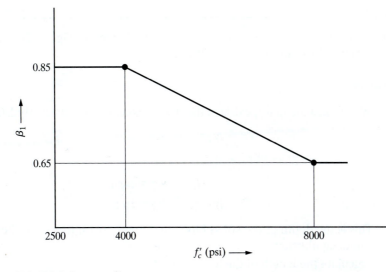

FIGURE 2-6 β_1 v. f_c'.

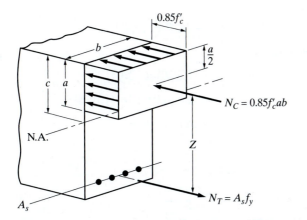

FIGURE 2-7   Equivalent stress block for strength design and analysis.

Using the equivalent stress distribution in combination with the strength design assumptions, we may now determine the nominal moment strength M_n of rectangular reinforced concrete beams that are reinforced for tension only.

The nominal moment strength determination is based on the assumption that the member will have the exact dimensions and material properties used in the design computations. As discussed in Section 2-9, the nominal moment strength will be further reduced when it is used in practical analysis and design work.

Example 2-1

Determine M_n for a beam of cross section shown in Figure 2-8, where $f'_c = 4000$ psi. Assume A615 grade 60 steel.

Solution:

1.   We will assume that f_y exists in the steel, subject to later check. By $\Sigma H = 0$,

$$N_C = N_T$$

$$(0.85 f'_c)ab = A_s f_y$$

$$a = \frac{A_s f_y}{0.85 f'_c b} = \frac{2.37(60)}{0.85(4)(10)} = 4.18 \text{ in.}$$

This is the depth of the stress block that must exist if there is to be horizontal equilibrium.

2.   Calculate the length of the lever arm, Z:

$$Z = d - \frac{a}{2} = 23 - \frac{4.18}{2} = 20.9 \text{ in.}$$

3.   Calculate M_n:

$$M_n = N_C Z \text{ or } N_T Z$$

Based on the concrete,

$$M_n = N_C Z = (0.85 f'_c)ab \,(20.9)$$

$$= 0.85(4.0)(4.18)(10)(20.9)$$

$$= 2970 \text{ in.-kips}$$

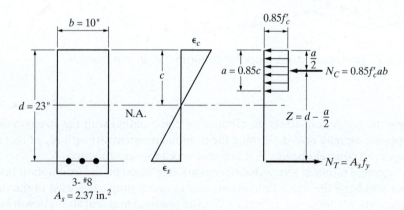

FIGURE 2-8   Sketch for Example 2-1.

$$\frac{2970 \text{ in.-kips}}{12 \text{ in./ft}} = 248 \text{ ft-kips}$$

or, based on the steel,

$$M_n = A_s f_y Z$$

$$= 2.37(60)(20.9)$$

$$= 2970 \text{ in.-kips}$$

4. In the foregoing computations, the assumption was made that the steel reached its yield strain (and therefore its yield stress) before the concrete reached its "ultimate" (by definition) strain of 0.003. This assumption will now be checked by calculating the strain ϵ_t in the steel when the concrete strain reaches 0.003. ϵ_t is defined as the net tensile strain at the centroid of the extreme tension steel at nominal strength.

    Referring to Figure 2-9, we may locate the neutral axis as follows:

$$a = \beta_1 c \qquad \text{(ACI Code, Section 10.2.7)}$$

$$\beta_1 = 0.85 \qquad \text{because } f_c' = 4000 \text{ psi}$$

Therefore

$$c = \frac{a}{0.85} = \frac{4.18}{0.85} = 4.92 \text{ in.}$$

By similar triangles in the strain diagram, we may find the strain in the steel when the concrete strain is 0.003:

$$\frac{0.003}{c} = \frac{\epsilon_t}{d - c}$$

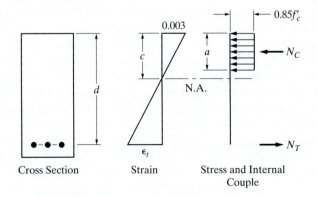

Cross Section   Strain   Stress and Internal
                                              Couple

FIGURE 2-9   Steel strain check.

Then

$$\epsilon_t = \frac{d - c}{c}(0.003) = \frac{23 - 4.92}{4.92}(0.003)$$

$$= 0.011 \text{ in./in.}$$

The strain at which the steel yields (ϵ_y) may be determined from the basic definition of the modulus of elasticity, E = stress/strain:

$$\epsilon_y = \frac{f_y}{E_s} = \frac{60,000}{29,000,000} = 0.00207 \text{ in./in. (see Table A-1)}$$

This represents the strain in the steel when the stress first reaches 60,000 psi.

Because the computed strain in the steel (0.011) is greater than the yield strain (0.00207), the steel reaches its yield stress before the concrete reaches its strain of 0.003, and the assumption that the stress in the steel is equal to the yield stress was correct. (See assumption 5 in Section 2-4 and the idealized stress–strain diagram for steel shown in Figure 1-3b.)

2-7   BALANCED, BRITTLE, AND DUCTILE FAILURE MODES

In Figure 2-9, the *ratio* between the steel strain and the maximum concrete strain is fixed once the neutral axis is established. The location of the neutral axis will vary based on the amount of tension steel in the cross section, as the stress block is just deep enough to ensure that the resultant compressive force is equal to the resultant tensile force ($\Sigma H = 0$). If more tension bars are added to the bottom of a reinforced concrete cross section, the depth of the compressive stress block will be greater, and therefore the neutral axis will be lower. Referring again to Figure 2-9, if there were just enough steel to put the neutral axis at a location where the yield strain in the steel and the maximum concrete strain of 0.003 existed at the same time, the cross section would be said to be *balanced* (see the ACI Code, Section 10.3.2). The amount of steel required to create this condition is relatively large. However, the balanced condition is the dividing line between two distinct types of reinforced concrete beams that are characterized by their failure modes. If a beam has more steel than is required to create the balanced condition, the beam will fail in a *brittle mode*. The additional steel will cause the neutral axis to be low (see Figure 2-10). This will in turn cause the concrete to reach a strain of 0.003 *before* the steel yields. Should more moment (and therefore strain) be applied to the beam cross section, failure will be initiated by a sudden crushing of the concrete. The brittle failure mode is undesirable.

If a beam has less steel than is required to create the balanced condition, the beam will fail in a *ductile mode*. The neutral axis will be higher than the balanced neutral axis, and the steel will reach its yield strain (and therefore its yield stress) before the concrete reaches a strain of 0.003. Figure 2-10 shows these variations in neutral axis

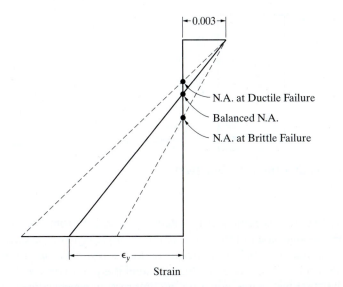

FIGURE 2-10   Strain distribution and failure modes in flexural members.

location for beams that are on the verge of failure. In each case, the concrete has been strained to 0.003. Following the ductile failure mode case through to failure, we see that a slight additional load will cause the steel to stretch a considerable amount. The strains in the concrete and the steel continue to increase. The tensile force is not increasing, however, as the steel stress has reached f_y and is not increasing. Because the compressive force cannot increase ($\Sigma H = 0$), and because the concrete strain and therefore its stress are increasing, the area under compression must be decreasing and the neutral axis must rise. This process continues until the reduced area fails in compression as a secondary effect. This failure due to yielding is a gradual one, with the beam showing greatly increased deflection after the steel reaches the yield point; hence, there is adequate warning of impending failure. The steel, being ductile, will not actually pull apart even at failure of the beam. The ductile failure mode is desirable and is required by the ACI Code, as discussed in the following section.

2-8   DUCTILITY REQUIREMENTS

Although failure due to yielding of steel is gradual, with adequate warning of collapse, failure due to crushing of concrete is sudden and without warning. We have seen that flexural members with less steel than is required to produce the balanced condition will fail by yielding of the steel due to the strain in the steel exceeding the yield strain. The ACI Code (Section 10.3.4) defines a section as *tension controlled* when the net tensile strain ϵ_t in the extreme tension steel is equal to or greater than 0.005 when the concrete has reached its assumed strain limit of 0.003. With reference

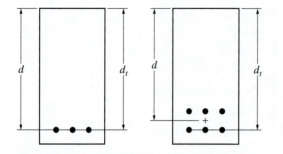

FIGURE 2-11   Definition of extreme tension steel.

to Figure 2-11, note that the extreme tension steel is located d_t from the extreme compression face. For a single layer of steel, $d_t = d$. For multiple layers of steel, $d_t > d$.

Further, ACI Code (Section 10.3.3) defines a section as *compression controlled* when the net tensile strain ϵ_t in the extreme tension steel is equal to or less than the yield strain ϵ_y of the steel just as the concrete in compression reaches its assumed strain limit of 0.003. See Table A-1 for values of ϵ_y. (The code permits ϵ_y to be taken as 0.002 for grade 60 steel.)

For non-prestressed flexural members subjected to little or no axial load, the net tensile strain ϵ_t at nominal strength shall not be less than 0.004. Because all values of ϵ_y for current reinforcing steel bars are less than 0.004, this ensures a tension-controlled flexural member, one that will exhibit ductility and fail by yielding of the steel.

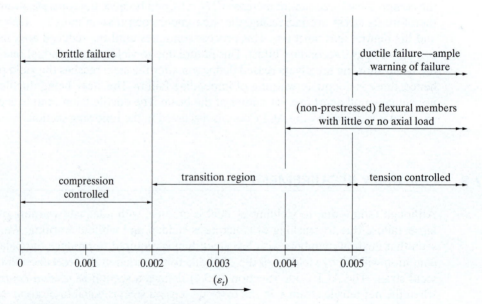

FIGURE 2-12   Strain–limit definitions.

Sections in which ϵ_t falls between the compression-controlled strain limit ϵ_y and the tension-controlled limit 0.005 constitute a transition region between compression-controlled and tension-controlled sections. Whether a section is tension controlled, compression controlled, or in the transition region has implications, which are discussed shortly. See Figure 2-12 for a graphical representation of the strain limit definitions.

Example 2-2

For the beam cross section of Example 2-1, determine the amount of steel A_s required to cause the strain in the tension steel ϵ_t to be 0.005 just as the maximum strain in the concrete reaches 0.003. $f'_c = 3000$ psi and $f_y = 60,000$ psi.

Solution:

The cross section, strain, stress, and internal couple are shown in Figure 2-13. Note that $d_t = d$.

1. Determine the location of the neutral axis. By similar triangles:

$$\frac{c}{0.003} = \frac{23 - c}{0.005}$$

$$0.005c = 0.003(23 - c)$$

$$0.005c + 0.003c = 0.003(23) = 0.0690$$

from which

$$c = \frac{0.0690}{0.008} = 8.63 \text{ in.}$$

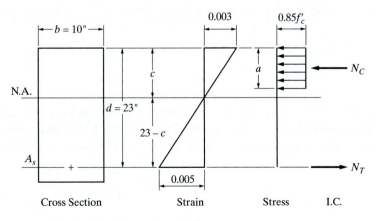

FIGURE 2-13 Sketch for Example 2-2.

2.   Determine N_C:

$$a = \beta_1 c = 0.85(8.63) = 7.34 \text{ in.}$$

$$N_C = 0.85 f'_c \, ab$$

$$= 0.85(3 \text{ kips/in.}^2)(7.34 \text{ in.})(10 \text{ in.}) = 187.2 \text{ kips}$$

3.   Determine A_s:

$$N_C = N_T = 187.2 \text{ kips}$$

$$N_T = A_s f_y$$

$$A_s = \frac{N_T}{f_y} = \frac{187.2 \text{ kips}}{60 \text{ kips/in.}^2} = 3.12 \text{ in.}^2$$

For ϵ_t to reach 0.005 just as the maximum concrete strain reaches 0.003, A_s must be 3.12 in.^{2}. This is the maximum amount of steel for the section to be a tension-controlled section. Any steel in excess of 3.12 in.^{2} will cause ϵ_t to be less than 0.005. Recall that the code does not allow $\epsilon_t < 0.004$ in a non-prestressed flexural member with little or no axial load.

At this point, note again that heavily reinforced beams are less efficient than their more lightly reinforced counterparts. One structural reason for this is that, for a given beam size, an increase in A_s is accompanied by a decrease in the lever arm of the internal couple ($Z = d - a/2$). This may be illustrated by doubling the tension steel for the beam of Example 2-1 and recalculating M_n:

$$A_s = 2(2.37) = 4.74 \text{ in.}^2 \qquad (100\% \text{ increase})$$

$$a = \frac{A_s f_y}{0.85 f'_c b} = \frac{4.74(60)}{0.85(4)(10)} = 8.36 \text{ in.}$$

$$M_n = 0.85 f'_c ab \left(d - \frac{a}{2} \right) = 0.85(4)(8.36)(10)\left(23 - \frac{8.36}{2} \right) = 446 \text{ ft-kips}$$

For the beam of Example 2-1, M_n was 248 ft-kips; therefore,

$$\frac{446 - 248}{248} \times 100 = 80\% \text{ increase}$$

Section 10.5.1 of the ACI Code also establishes a lower limit on the amount of tension reinforcement for flexural members. The code states that where tensile reinforcement is required by analysis, the steel area A_s shall not be less than that given by

$$A_{s,min} = \frac{3\sqrt{f'_c}}{f_y} b_w d \geq \frac{200}{f_y} b_w d$$

(Note that for rectangular beams, $b_w = b$.)

$A_{s,\min}$ is conveniently calculated using Table A-5, where the larger of $3\sqrt{f_c'}/f_y$ and $200/f_y$ is tabulated. Note that $A_{s,\min}$ is the product of the tabulated value and $b_w d$.

The lower limit guards against sudden failure essentially by ensuring that a beam with a very small amount of tensile reinforcement has a greater moment strength as a reinforced concrete section than that of the corresponding plain concrete section computed from its modulus of rupture. Alternatively, it is satisfactory to provide an area of tensile reinforcement that is one-third greater than that required by analysis (ACI Code, Section 10.5.3). This requirement applies especially to grade beams, wall beams, and other deep flexural members where the minimum reinforcement requirement specified in ACI 10.5.1 would result in an excessively large amount of steel.

Minimum required reinforcement in structural slabs (see Section 2-13) is governed by the required shrinkage and temperature steel as outlined in the ACI Code, Section 7.12.

2-9 STRENGTH REQUIREMENTS

The basic criterion for strength design may be expressed as

$$\text{strength furnished} \geq \text{strength required}$$

All members and all sections of members must be proportioned to meet this criterion.

The required strength may be expressed in terms of design loads or their related moments, shears, and forces. Design loads may be defined as service loads multiplied by the appropriate *load factors*. (When the word *design* is used as an adjective [e.g., design load] throughout the ACI Code, it indicates that load factors are included.) The subscript u is used to indicate design loads, moments, shears, and forces.

The ACI Code, Section 9.2, specifies load factors to be used and load combinations to be investigated. Loads to be considered are dead loads, live loads, fluid loads, loads due to weight and pressure of soil, and snow loads, among others. For applications in this text, we will consider only dead load, live load, and loads due to weight and pressure of soil. For the combination of dead load and live load, the general representation is as follows:

$$U = 1.2D + 1.6L \geq 1.4D$$

where U is defined as the required strength to resist factored loads or related internal moments and forces, D is the service dead load, and L is the service live load. (The term *service load* generally refers to the load specified in applicable building codes as representing minimum requirements.) The factors 1.2 and 1.6 (and 1.4) represent load factors. The load factors are part of the overall safety provision in reinforced concrete structures and are meant to reflect the variability in load effects. Dead loads can be more accurately estimated than live loads, and hence a lower load factor is used for dead load. Live loads have a greater variability than dead loads, and hence a higher load factor is used.

A second part of the overall safety provision, provided in the ACI Code, Section 9.3, is the reduction of the theoretical capacity of a structural element by a *strength-reduction factor* ϕ. This provides for the possibility that small adverse variations in material strengths, workmanship, and dimensions, although within acceptable tolerances and limits of good practice, may combine to result in undercapacity. In effect, the nominal strength of a member, when multiplied by the ϕ factor, will furnish us with a practical strength that is obviously less than the nominal strength.

The ACI Code, Section 9.3, provides for these variables by using the following ϕ factors:

Tension-controlled sections	0.90
Compression-controlled sections	
spirally reinforced	0.75
other reinforced members	0.65
Shear and torsion	0.75
Bearing on concrete	0.65

Additionally, for sections in which ϵ_t is between the limits for tension-controlled sections and compression-controlled sections, ϕ shall be permitted to be linearly increased from that for compression-controlled sections to 0.90 as ϵ_t increases from the compression-controlled strain limit to 0.005.

In each case, the practical strength of a reinforced concrete member will be the product of the nominal strength and the ϕ factor. Therefore, in terms of moment, we can say that

$$\text{practical moment strength} = \phi M_n$$

The individual values assigned to ϕ factors are determined using statistical studies of structural resistances and loading variabilities tempered by subjective factors reflecting the consequences of a structural failure.

The variation in ϕ for tension-controlled sections, transition sections, and compression-controlled (nonspirally reinforced) sections is shown in Figure 2-14. For the transition section, an expression for ϕ as a function of ϵ_t can be developed:

$$\phi = 0.65 + \left(\frac{0.90 - 0.65}{0.005 - \epsilon_y}\right)(\epsilon_t - \epsilon_y)$$

This is valid for the range of $\epsilon_y \leq \epsilon_t \leq 0.005$. Assuming $\epsilon_y = 0.002$, this expression becomes:

$$\phi = 0.65 + (\epsilon_t - 0.002)\left(\frac{250}{3}\right)$$

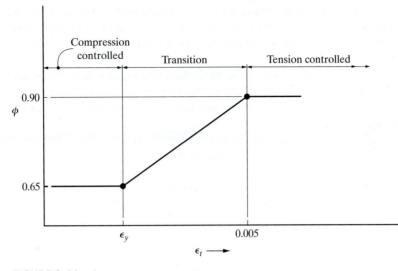

FIGURE 2-14 ϕ v. ϵ_t.

The calculation of ϵ_t can also be simplified using the depth of the stress block a, which is commonly calculated in analysis problems. Refer to Figure 2-8 and set $d_t = d$.

$$\epsilon_t = \frac{d_t - c}{c}(0.003)$$

Substitute $c = a/\beta_1$:

$$\epsilon_t = \frac{d_t - \dfrac{a}{\beta_1}}{\dfrac{a}{\beta_1}}(0.003) = \frac{0.003\beta_1 d_t}{a} - 0.003 \tag{2-1}$$

2-10   RECTANGULAR BEAM ANALYSIS FOR MOMENT (TENSION REINFORCEMENT ONLY)

The *flexural analysis problem* is characterized by knowing precisely what comprises the cross section of a beam. That is, the following data are *known*: tension bar size and number (or A_s), beam width (b), effective depth (d) or total depth (h), f'_c, and f_y. To be found, basically, is the beam strength, although this may be manifested in various ways: Find ϕM_n, check the adequacy of the given beam, or find an allowable load that the beam can carry. The *flexural design problem*, on the other hand, requires the determination of one or more of the dimensions of the cross section or the determination of the main tension steel to use. It will be important to recognize the differences between these two types of problems because the methods of solution are different.

To expedite reinforced concrete analysis and design calculations, use is frequently made of tables of pertinent quantities. Tables find their greatest use in the design process, but because they are also useful for analysis, they are developed here.

In Example 2-1, we calculated the nominal moment strength M_n for a rectangular reinforced concrete section (tension steel only). Based on concrete:

$$M_n = 0.85f_c'ab\left(d - \frac{a}{2}\right)$$

and

$$a = \frac{A_s f_y}{0.85f_c'b}$$

We will now develop a modified expression for M_n (and ϕM_n), which will then be adapted for table use. It is convenient to use the concept of *reinforcement ratio* ρ (lowercase Greek "rho"), the ratio of tension steel area to effective concrete area:

$$\rho = \frac{A_s}{bd}$$

from which

$$A_s = \rho bd$$

Substituting this into the expression for a:

$$a = \frac{A_s f_y}{(0.85f_c')b} = \frac{\rho bd f_y}{(0.85f_c')b} = \frac{\rho d f_y}{0.85f_c'}$$

Arbitrarily define ω (omega):

$$\omega = \rho\frac{f_y}{f_c'}$$

Then

$$a = \frac{\omega d}{0.85}$$

Substitute into the ϕM_n expression:

$$\phi M_n = \phi(0.85f_c')(b)\frac{\omega d}{0.85}\left[d - \frac{\omega d}{2(0.85)}\right]$$

Simplify and rearrange:

$$\phi M_n = \phi b d^2 f_c' \omega (1 - 0.59\omega)$$

Arbitrarily define $\bar{k}$:

$$\bar{k} = f_c' \omega (1 - 0.59\omega)$$

The term $\bar{k}$ is sometimes called the *coefficient of resistance*. It varies with ρ, f_c', and f_y. The general expression for ϕM_n can now be written as

$$\phi M_n = \phi b d^2 \bar{k}$$

Before introducing the tables, we discuss another important quantity that depends on these same three factors and that is included in the tables.

In Example 2-1, we calculated the strain ϵ_t in the extreme tension steel at nominal strength (when the maximum concrete strain reaches 0.003). Assuming one layer of steel ($d = d_t$), it can be shown that ϵ_t is fixed if ρ, f_c', and f_y are known. Referring to Figure 2-8 and Example 2-1:

$$\epsilon_t = \frac{d - c}{c}(0.003)$$

Also,

$$a = \beta_1 c$$

From which

$$c = \frac{a}{\beta_1}$$

We previously developed

$$a = \frac{\rho d f_y}{0.85 f_c'}$$

By substitution,

$$c = \frac{\rho d f_y}{0.85 f_c' \beta_1}$$

Again by substitution,

$$\epsilon_t = \frac{d - c}{c}(0.003) = \frac{\left(d - \dfrac{\rho d f_y}{0.85 f_c' \beta_1}\right)}{\left(\dfrac{\rho d f_y}{0.85 f_c' \beta_1}\right)}(0.003)$$

From which

$$\epsilon_t = \frac{0.00255 f_c' \beta_1}{\rho f_y} - 0.003 \qquad (2\text{-}2)$$

Note, again, that this is based on the assumption that $d = d_t$. For multiple layers of steel, use basic principles or Equation (2-1).

Tables A-7 through A-11 give the values of the coefficient of resistance $\bar{k}$ for values of ρ and various combinations of f_c' and f_y. The maximum tabulated value of ρ is that which corresponds to an ϵ_t value of 0.004, the minimum value allowed by code for non-prestressed flexural members with little or no axial load (ACI Code, Section 10.3.5). Values of ϵ_t are tabulated between 0.004 and 0.005 because in this transition region ϕ must be determined (ACI Code, Section 9.3.2.2). When $\epsilon_t > 0.005$, sections are tension-controlled and $\phi = 0.90$. Note that for the tables, it is assumed that $d = d_t$, as it would be for one layer of steel. If there are multiple layers of steel (see Figure 2-11), then $d_t > d$ and ϵ_t will be larger than the tabulated ϵ_t. Therefore, the tables are conservative with regard to the determination of ϕ based on ϵ_t.

Example 2-3

Determine if the beam shown in Figure 2-15 is adequate as governed by the ACI Code (318-08). The loads shown are service loads. The uniformly distributed load is: DL = 0.65 kip/ft, LL = 0.80 kip/ft. The dead load excludes the beam weight. The point load is a live load; $f_c' = 4000$ psi, $f_y = 60,000$ psi.

Solution:

A logical approach to this type of problem is to compare the practical moment strength (ϕM_n) with the applied design moment resulting from the *factored*

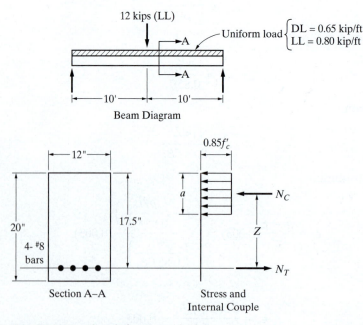

FIGURE 2-15   Sketch for Example 2-3.

loads. This latter moment will be noted M_u. If the beam is adequate for the moment, $\phi M_n \geq M_u$. The procedure outlined here for ϕM_n is summarized in Section 2-11.

Determination of ϕM_n

1. Given:

$$f_y = 60,000 \text{ psi}$$

$$b = 12 \text{ in.}$$

$$d = 17.5 \text{ in.}$$

$$A_s = 3.16 \text{ in.}^2 \quad \text{(Table A-2)}$$

$$f_c' = 4000 \text{ psi}$$

2. To be found: ϕM_n and M_u.

3.

$$\rho = \frac{A_s}{bd} = \frac{3.16}{12(17.5)} = 0.01505$$

4. From Table A-5:

$$A_{s,min} = 0.0033b_w d$$

$$= 0.0033(12)(17.5) = 0.69 \text{ in.}^2 \qquad \text{(O.K.)}$$

$$3.16 \text{ in.}^2 > 0.69 \text{ in.}^2$$

5. From Table A-10, ϵ_t is not tabulated; therefore $\epsilon_t > 0.005$, the section is tension-controlled, and $\phi = 0.90$. Also from Table A-10, $\bar{k} = 0.7809$. (It is conservative to use the lower tabulated value. An interpolation between 0.7809 and 0.7853 could be done, but it is not warranted.)

6. Calculate ϕM_n from $\phi N_C Z$ or $\phi N_T Z$:

$$a = \frac{A_s f_y}{0.85 f_c' b} = \frac{3.16(60)}{0.85(4)(12)} = 4.65 \text{ in.}$$

$$Z = d - \frac{a}{2} = 17.5 - \frac{4.65}{2} = 15.18 \text{ in.}$$

Based on steel:

$$M_n = A_s f_y Z$$

$$= 3.16(60)(15.18) = 2880 \text{ in.-kips}$$

$$\frac{2880}{12} = 240 \text{ ft-kips}$$

$$\phi M_n = 0.90(240) = 216 \text{ ft-kips}$$

or

Calculate ϕM_n using the coefficient of resistance $\overline{k}$:

$$\phi M_n = \phi b d^2 \overline{k}$$

$$= \frac{0.90(12 \text{ in.})(17.5 \text{ in.})^2(0.7809 \text{ kips/in.}^2)}{12 \text{ in./ft}} = 215 \text{ ft-kips}$$

Find M_u: Because the given service dead load excluded the beam weight, it will now be calculated. As the beam weight is a uniformly distributed load, it will be found in terms of weight per linear foot (kips/ft):

$$\text{beam weight} = \text{beam volume per foot of length} \times 0.150 \text{ kip/ft}^3$$

$$= \frac{20 \text{ in.}(12 \text{ in.})}{144 \dfrac{\text{in.}^2}{\text{ft}^2}} \times 1 \text{ ft} \times 0.150 \text{ kip/ft}^3$$

$$= 0.250 \text{ kip (per linear foot)}$$

$$= 0.250 \text{ kip/ft}$$

Then summarizing the loads,

superimposed service uniform dead load = 0.65 kip/ft

total service uniform dead load = 0.250 + 0.65 = 0.90 kip/ft = w_{DL}

total service uniform live load = 0.80 kip/ft = w_{LL}

Total factored uniform load:

$$w_u = 1.2 w_{\text{DL}} + 1.6 w_{\text{LL}}$$

$$= 1.2(0.90) + 1.6(0.80) = 2.36 \text{ kips/ft}$$

Superimposed concentrated live load = 12 kips = P_{LL}.
Total factored concentrated load:

$$P_u = 1.6 P_{\text{LL}} = 1.6(12) = 19.2 \text{ kips}$$

$$M_u = \frac{w_u \ell^2}{8} + \frac{P_u \ell}{4}$$

$$= \frac{2.36(20)^2}{8} + \frac{19.2(20)}{4}$$

$$= 214 \text{ ft-kips} < 215 \text{ ft-kips}$$

Therefore the beam is adequate.

2-11 SUMMARY OF PROCEDURE FOR RECTANGULAR BEAM ANALYSIS FOR ϕM_n (TENSION REINFORCEMENT ONLY)

1. List the known quantities. Use a sketch.
2. Determine what is to be found. (An "analysis" may require any of the following to be found: ϕM_n, allowable service live load or dead load, maximum allowable span.)
3. Calculate the reinforcement ratio:

$$\rho = \frac{A_s}{bd}$$

4. Calculate $A_{s,\min}$ and compare with A_s (use Table A-5.)
5. Determine ϵ_t by calculation or table. ($\epsilon_t \geq 0.004$.) Determine ϕ. Select $\overline{k}$ if it is to be used.
6. Calculate ϕM_n from $\phi N_C Z$ or $\phi N_T Z$, where

$$N_C = 0.85 f'_c ab$$

$$N_T = A_s f_y$$

$$a = \frac{A_s f_y}{0.85 f'_c b}$$

$$Z = d - \frac{a}{2}$$

or
calculate ϕM_n from $\phi M_n = \phi b d^2 \overline{k}$.

A flow diagram of this procedure is presented in Appendix B.

2-12 SLABS: INTRODUCTION

Slabs constitute a specialized category of bending members and are used in both structural steel and reinforced concrete structures. Probably the most basic and common type of slab is the *one-way slab*. A one-way slab may be described as a structural reinforced concrete slab supported on two opposite sides so that the bending occurs in one direction only—that is, perpendicular to the supported edges.

If a slab is supported along all four edges, it may be designated as a *two-way slab*, with bending occurring in two directions perpendicular to each other. If the ratio of the lengths of the two perpendicular sides is in excess of 2, however, the slab may be assumed to act as a one-way slab with bending primarily occurring in the short direction.

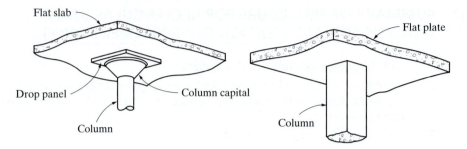

FIGURE 2-16   Reinforced concrete slabs.

A specific type of two-way slab is categorized as a *flat slab*. A flat slab may be defined as a concrete slab reinforced in two or more directions, generally without beams or girders to transfer the loads to supporting members. The slab, then, could be considered to be supported on a grid of shallow beams, which are themselves integral with and have the same depth as the slab. The columns tend to punch upward through the slab, resulting in a high shearing stress along with inclined slab cracking. (This "punching shear" is considered later in the discussion on column footings.) Thus it is common to both thicken the slab in the vicinity of the column, utilizing a *drop panel*, and at the same time enlarge the top of the column in the shape of an inverted frustum called a *column capital* (see Figure 2-16). Another type of two-way slab is a *flat plate*. This is similar to the flat slab, without the drop panels and column capitals; hence it is a slab of constant thickness supported directly on columns. Generally, the flat plate is used where spans are smaller and loads lighter than those requiring a flat slab design (see Figure 2-16).

2-13 ONE-WAY SLABS: ANALYSIS FOR MOMENT

In our discussion of slabs, we will be primarily concerned with one-way slabs. Such a slab is assumed to be a rectangular beam with a width $b = 12$ in. When loaded with a uniformly distributed load, the slab deflects so that it has curvature, and therefore bending moment, in only one direction. Hence the slab is analyzed and designed as though it were composed of 12-in.-wide segments placed side by side with a total depth equal to the slab thickness. With a width of 12 in., the uniformly distributed load, generally specified in pounds per square foot (psf) for buildings, automatically becomes the load per linear foot (lb/ft) for the design of the slab (see Figure 2-17).

In the one-way slab, the main reinforcement for bending is placed perpendicular to the supports. Because analysis and design will be done for a typical 12-in.-wide segment, it will be necessary to specify the amount of steel in that segment. Reinforcing steel in slabs is normally specified by bar size and center-to-center spacing, and the amount of steel considered to exist in the 12-in.-wide typical segment is an *average amount*. Table A-4 is provided to facilitate this determination. For example, if a slab is reinforced with No. 7 bars spaced 15 in. apart (center to center), a typical

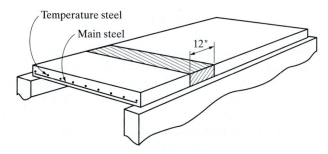

FIGURE 2-17   One-way slab.

12-in.-wide segment of the slab would contain an average steel area of 0.48 in.^{2} This is denoted 0.48 in.^{2}/ft.

In addition, the ACI Code stipulates that reinforcement for shrinkage and temperature stresses normal to the principal reinforcement must be provided in structural floor and roof slabs where the principal reinforcement extends in one direction only. The ACI Code (Section 7.12.2) further states that for grade 40 or 50 deformed bars, the minimum area of such steel must be $A_s = 0.0020bh$, and for grade 60 deformed bars, the minimum area must be $A_s = 0.0018bh$, where b = width of member (12 in. for slabs) and h = total slab thickness. The ACI Code also stipulates that in structural slabs of uniform thickness, the minimum amount of reinforcement in the direction of the span (principal reinforcement) must not be less than that required for shrinkage and temperature reinforcement (ACI Code, Section 10.5.4). Further, according to ACI Code, Section 7.6.5, the principal reinforcement shall not be spaced farther apart than three times the slab thickness nor more than 18 in. ACI Code, Section 7.12.2.2, requires that shrinkage and temperature reinforcement shall not be spaced farther apart than five times the slab thickness nor more than 18 in.

The required thickness of a one-way slab may depend on the bending, deflection, or shear strength requirements. The ACI Code imposes span/depth criteria in an effort to prevent excessive deflection, which might adversely affect the strength or performance of the structure at service loads. Table 9.5a of the ACI Code establishes minimum thicknesses for beams and one-way slabs in terms of fractions of the span length. These may be used for members not supporting or attached to construction likely to be damaged by large deflections. If the member supports such construction, deflections must be calculated. For members of lesser thickness than that indicated in the table, deflection should be computed, and if satisfactory, the member may be used. The tabular values are for use with non-prestressed reinforced concrete members made with normal-weight concrete and grade 60 reinforcement. If a different grade of reinforcement is used, the tabular values must be multiplied by a factor equal to

$$0.4 + \frac{f_y}{100,000}$$

where f_y is in units of psi. As an example, for simply supported solid one-way slabs of normal-weight concrete and grade 60 steel, the minimum thickness required when deflections are not computed equals $\ell/20$, where ℓ is the span length of the slab. Deflections are discussed further in Chapter 7.

The ACI Code, Section 7.7.1, discusses *cover*, concrete protection for the reinforcement against weather and other effects, which is measured from the surface of the steel to the nearest surface of the concrete. The cover for reinforcement in slabs must be not less than ¾ in. for surfaces not exposed directly to the weather or in contact with the ground. This is applicable for No. 11 and smaller bars. For surfaces exposed to the weather or in contact with the ground, the minimum cover for reinforcement is 2 in. for No. 6 through No. 18 bars and 1½ in. for No. 5 and smaller bars. If a slab is cast against and permanently exposed to the ground, the minimum concrete cover for all reinforcement is 3 in.

Example 2-4

A one-way structural interior slab having the cross section shown spans 12 ft. The steel is A615 grade 40. The concrete strength is 3000 psi, and the cover is ¾ in. Determine the service live load (psf) that the slab can support.

Solution:

From Figure 2-18, the bars in this slab are No. 5 bars spaced 7 in. apart. This is sometimes denoted "7 in. o.c." (7 in. on center), meaning 7 in. center-to-center distance. They are perpendicular to the supports.

1.   Given:

$$A_s = 0.53 \text{ in.}^2/\text{ft} \qquad \text{(Table A-4)}$$

$$f_c' = 3000 \text{ psi}, \quad f_y = 40,000 \text{ psi}$$

$$b = 12 \text{ in.}$$

$$d = 6.5 - 0.75 - \frac{0.625}{2} = 5.44 \text{ in.}$$

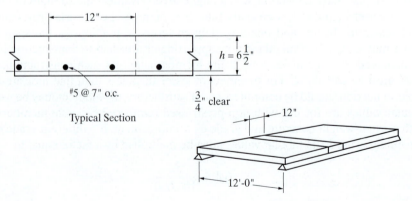

FIGURE 2-18   Sketch for Example 2-4.

2.   Find ϕM_n and the permissible service live load.

3.
$$\rho = \frac{A_s}{bd} = \frac{0.53}{12(5.44)} = 0.0081$$

4.   The minimum flexural reinforcement is slabs is that required for shrinkage and temperature steel:

$$A_{s,min} = 0.0020bh = 0.0020(12)(6.5) = 0.16 \text{ in.}^2/\text{ft}$$

$$0.53 \text{ in.}^2 > 0.16 \text{ in.}^2 \qquad\qquad\qquad\qquad \text{(O.K.)}$$

5.   From Table A-7, for $\rho = 0.0081$, ϵ_t is not tabulated. Therefore $\epsilon_t > 0.005$, this section is tension controlled, and $\phi = 0.90$. Also from Table A-7, $\bar{k} = 0.3034$ ksi.

6.   Calculate a, Z, and ϕM_n:

$$a = \frac{A_s f_y}{0.85f_c'b} = \frac{0.53(40)}{0.85(3)(12)} = 0.693 \text{ in.}$$

$$Z = d - \frac{a}{2} = 5.44 - \frac{0.693}{2} = 5.09 \text{ in.}$$

$$\phi M_n = \phi A_s f_y Z = 0.9(0.53)(40)(5.09) = 97.1 \text{ in.-kips (per foot of slab width)}$$

$$\phi M_n = \frac{97.1 \text{ in.-kips}}{12 \text{ in./ft}} = 8.09 \text{ ft-kips}$$

or

from Table A-7, $\bar{k} = 0.3034$ ksi, from which

$$\phi M_n = \phi bd^2 \bar{k}$$

$$= \frac{0.90(12 \text{ in.})(5.44 \text{ in.})^2(0.3034 \text{ kips/in.}^2)}{12 \text{ in./ft}} = 8.08 \text{ ft-kips}$$

The service live load that the slab can support will be found next. (The notation M_u is used to denote moment resulting from *factored* applied loads.) As

$$M_u = \frac{w_u \ell^2}{8}$$

the total factored design load that can be supported by the slab is

$$w_u = \frac{8M_u}{\ell^2}$$

Because, as a limit, $M_u = \phi M_n$,

$$w_u = \frac{8\phi M_n}{\ell^2} = \frac{8(8.09)}{12^2} = 0.449 \text{ kip/ft}$$

The slab weight is

$$w_{\mathrm{DL}} = \frac{6.5(12)}{144}(0.150) = 0.0813 \text{ kip/ft}$$

As the total factored design load is

$$w_u = 1.2w_{\mathrm{DL}} + 1.6w_{\mathrm{LL}}$$

w_{LL} may be found from

$$w_{\mathrm{LL}} = \frac{w_u - 1.2w_{\mathrm{DL}}}{1.6} = \frac{0.449 - 1.2(0.0813)}{1.6}$$

$$= 0.220 \text{ kip/ft}$$

Because the segment is 12 in. wide, this is equivalent to 220 psf. It will be noted that the procedure for finding ϕM_n for a one-way slab is almost identical to that for a beam. See Section 2-11 for a summary of this procedure.

2-14   RECTANGULAR BEAM DESIGN FOR MOMENT (TENSION REINFORCEMENT ONLY)

In the design of rectangular sections for moment, with f_c' and f_y usually prescribed, *three basic quantities* are to be determined: beam width, beam depth, and steel area. It should be recognized that there is a large multitude of combinations of these three quantities that will satisfy the moment strength required in a particular application. Theoretically, a wide, shallow beam may have the same ϕM_n as a narrow, deep beam. It must also be recognized that practical considerations and code restraints will affect the final choices of these quantities. There is no easy way to determine the *best* cross section, because economy depends on much more than simply the volume of concrete and amount of steel in a beam.

We have previously developed the analysis expression for the resisting moment of a rectangular beam with tension reinforcement only:

$$\phi M_n = \phi N_C Z = \phi N_T Z$$

We subsequently modified the equation for ϕM_n for the use of tables (Tables A-7 through A-11):

$$\phi M_n = \phi b d^2 \overline{k}$$

This equation will now be used for the design of rectangular reinforced concrete sections. The first example is one where the cross-section width b and overall depth h are known by either practical or architectural considerations, leaving the selection of the reinforcing bars as the only unknown.

Material properties, sizes, and availability of reinforcing steel in the form of bars and welded wire fabric were discussed in Section 1-7. In our discussion we will be concerned with bars only and the ACI Code recommendations as to details governing minimum clearance and cover requirements for steel reinforcing bars. Clearance details are governed by the requirement for concrete to pass through a layer of bars without undue segregation of the aggregates. Cover details are governed by the necessity that the concrete must provide protection against corrosion for the bars. Required minimums of both spacing and cover also play a role in preventing the splitting of the concrete in the proximity of highly stressed tension bars.

Spacing requirements in the ACI Code indicate that the clear space between bars in a single layer shall be not less than

1.   The bar diameter, but not less than 1 in. (ACI Code, Section 7.6.1).
2. $1\frac{1}{3} \times$ maximum aggregate size (ACI Code, Section 3.3.2).

Also, should multiple layers of bars be necessary, a 1-in. minimum clear distance is required between layers (ACI Code, Section 7.6.2) and bars in the upper layers shall be placed directly above bars in the bottom layer. When multiple layers of steel are required, short transverse *spacer bars* may be used to separate the layers and support the upper layers. A No. 8 spacer bar will provide the minimum 1-in. separation between layers. This detail is illustrated in Figure 3-6. In general, in this text, clear distance between layers of steel is noted with a dimension.

Cover requirements for cast-in-place concrete are stated in the ACI Code, Section 7.7.1. This listing is extensive. For beams, girders, and columns not exposed to the weather or in contact with the ground, however, the minimum concrete cover on any steel is $1\frac{1}{2}$ in. Cover requirements for slabs were discussed in Section 2-13 of this text.

Table A-3 combines spacing and cover requirements into a tabulation of minimum beam widths for multiples of various bars. The assumptions are stated. It should be noted that No. 3 stirrups are assumed. Stirrups are a special form of reinforcement that primarily resist shear forces and will be discussed in Chapter 4. Stirrups are common in rectangular beams and will be assumed to exist in all further rectangular beam examples and problems in this text. One type of stirrup, called a *loop stirrup*, can be observed in Figure 2-19, in Example 2-5.

Example 2-5

Design a rectangular reinforced concrete beam to carry a service dead load moment of 100 ft-kips (which includes the moment due to the weight of the beam) and a service live load moment of 75 ft-kips. Architectural considerations require the beam width to be 10 in. and the total depth (h) to be 25 in. Use $f'_c = 3000$ psi and $f_y = 60,000$ psi.

Solution:

Of the three basic quantities to be found, two are specified in this example, and the solution for the required steel area is direct. The procedure outlined here is summarized in Section 2-15.

1. The total design moment is

$$M_u = 1.2M_{DL} + 1.6M_{LL}$$
$$= 1.2(100) + 1.6(75)$$
$$= 240 \text{ ft-kips}$$

2. Estimate d to be equal to $h - 3$ in. This is conservative for a single layer of bars. The effective depth d in the resulting section will be a bit larger.

$$d = 25 - 3 = 22 \text{ in.}$$

Because $\phi M_n = \phi bd^2\overline{k}$ and because ϕM_n must equal M_u as a lower limit, the expression may be written as $M_u = \phi bd^2\overline{k}$ and, assuming $\phi = 0.90$, subject to later check,

$$\text{required } \overline{k} = \frac{M_u}{\phi bd^2} = \frac{240(12)}{0.9(10)(22)^2} = 0.6612 \text{ ksi}$$

3. From Table A-8, $\overline{k}$ of 0.6649 ksi will be provided if the steel ratio $\rho = 0.0131$. Therefore required $\rho = 0.0131$. Also from Table A-8, because ϵ_t is not tabulated, $\epsilon_t > 0.005$, this is a tension-controlled section, and $\phi = 0.90$.

4. $$\text{required } A_s = \rho bd$$
$$= 0.0131(10)(22) = 2.88 \text{ in.}^2$$

Check $A_{s,\min}$. From Table A-5,

$$A_{s,\min} = 0.0033b_w d$$
$$= 0.0033(10)(22) = 0.73 \text{ in.}^2$$

5. Select the bars. Theoretically, any bar or combination of bars that provides at least 2.88 in.² of steel area will satisfy the design requirements. Preferably, no fewer than two bars should be used. The bars should be of the same diameter and placed in one layer whenever possible. We next consider the bar selection based on the foregoing required A_s (2.88 in.²). The following combinations may be considered:

$$2 \text{ No. 11 bars: } A_s = 3.12 \text{ in.}^2$$
$$3 \text{ No. 9 bars: } A_s = 3.00 \text{ in.}^2$$
$$4 \text{ No. 8 bars: } A_s = 3.16 \text{ in.}^2$$
$$5 \text{ No. 7 bars: } A_s = 3.00 \text{ in.}^2$$

A review of Table A-3 indicates that the most acceptable combination is three No. 9 bars. The minimum width of beam required for three No. 9 bars is 9½ in., which is satisfactory.

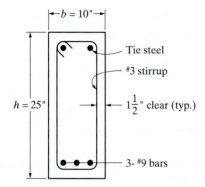

FIGURE 2-19   Design sketch for Example 2-5.

At this point we should check the actual effective depth d and compare it with the estimated d:

$$\text{actual } d = h - \text{cover} - \text{stirrup} - d_b/2$$

$$= 25 - 1.5 - 0.38 - \frac{1.128}{2} = 22.6 \text{ in.}$$

This is slightly in excess of the estimated d and is therefore conservative (on the safe side). Because of the small difference, no revision is either suggested or required.

6.   Generally, concrete dimensions should be to an increment no smaller than $\frac{1}{2}$ in. In this example, whole inch dimensions are given. The final design sketch should show the following (see Figure 2-19):

a.   Beam width
b.   Total beam depth
c.   Main reinforcement size and number of bars
d.   Cover on reinforcement
e.   Stirrup size

As a final note for Example 2-5, the reader has probably recognized that there is a direct solution that avoids the use of Tables A-7 through A-11. A quadratic equation results when the design expression is written as

$$M_u = \phi M_n = \phi A_s f_y \left(d - \frac{a}{2} \right)$$

where
$$a = \frac{A_s f_y}{0.85 f'_c b}$$

With known quantities of M_u, f'_c, f_y, b, and d, the quadratic equation can be solved for the required steel area A_s. Generally, the use of the tables results in a much faster solution.

A rule of thumb or approximate equation commonly used in design practice to determine the approximate area of flexural reinforcement is [2]

$$A_s = \frac{M_u}{4d}$$

where

M_u = the factored moment in ft-kips
d = the effective depth in inches
A_s = reinforcement area, in.^{2}

This equation is derived from the equation

$$M_u = \phi M_n = \phi A_s f_y \left(d - \frac{a}{2} \right)$$

by substituting f_y of 60 ksi, assuming tension-controlled section or $\phi = 0.9$, and assuming the internal moment arm

$$\left(d - \frac{a}{2} \right) \approx 0.9d$$

A second type of design problem is presented in Example 2-6. This may be categorized as a *free design* because of the *three* unknown variables: beam width, beam depth, and area of reinforcing steel. There is, therefore, a large number of combinations of these variables that will theoretically solve the problem. We do have some idea as to the required or desired relationships between these unknowns, however.

Whenever possible, flexural members should be proportioned so they are tension-controlled sections ($\epsilon_t \geq 0.005$.) This allows the strength-reduction factor ϕ to be taken at its maximum value of 0.90. The reinforcement ratio ρ corresponding to $\epsilon_t = 0.005$ may be determined from Tables A-7 through A-11. Additionally, we know that the steel area must not be less than $A_{s,\min}$ (from Table A-5). This in effect establishes the range of the acceptable amount of steel. Table A-5 contains recommended values of ρ and associated values of $\overline{k}$ to use for design purposes. These are recommended maximum values. Try not to use larger ρ values. If larger values are used, a smaller concrete section will result, with potential deflection problems.

These ρ values are based on a previous ACI Code, which stipulated that

where $\rho > \dfrac{0.18f_c}{f_y}$, deflection must be checked

where $\rho < \dfrac{0.18f_c}{f_y}$, deflection need not be checked

This stipulation was deleted in more current ACI Codes. Nevertheless, it remains a valid guide for selecting a preliminary value for the reinforcement ratio. The assumption of a value for reinforcement ratio will reduce from three to two the number of unknown quantities to be determined (yet to be found are b and d).

Experience and judgment developed over the years have also established a range of acceptable and economical depth/width ratios for rectangular beams. Although there is no code requirement for the d/b ratio to be within a given range, rectangular beams commonly have d/b ratios between 1 and 3. *Desirable d/b* ratios lie between 1.5 and 2.2. There are situations in which d/b ratios outside this range have applications, however; therefore it is the designer's choice.

Example 2-6

Design a simply supported rectangular reinforced concrete beam with tension steel only to carry a service dead load of 1.35 kips/ft and a service live load of 1.90 kips/ft. (The dead load does not include the weight of the beam.) The span is 18 ft. Assume No. 3 stirrups. Use $f_c' = 4000$ psi and $f_y = 60,000$ psi. See Figure 2-20.

Solution:

1. Find the applied design moment M_u, temporarily neglecting the beam weight, which will be included at a later step:

$$w_u = 1.2w_{DL} + 1.6w_{LL}$$

$$= 1.2(1.35) + 1.6(1.90) = 4.66 \text{ kips/ft}$$

Note that this is the factored design load,

$$M_u = \frac{w_u \ell^2}{8} = \frac{4.66(18)^2}{8} = 189 \text{ ft-kips}$$

2. Assume a value for ρ. Use $\rho = 0.0120$ (see Table A-5).
3. From Table A-5, the associated $\bar{k}$ value is 0.6438 ksi (alternatively, Table A-10 can be used).
4. At this point we have two unknowns, b and d, which can be established using two different approaches. One approach is to assume b and then solve for d. This is logical as practical and architectural considerations

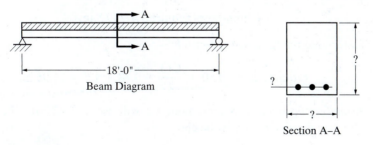

Beam Diagram

Section A–A

FIGURE 2-20 Sketch for Example 2-6.

often establish b within narrow limits. Assume $\phi = 0.90$, subject to later check.

Assuming that $b = 11$ in. and utilizing the relationship $M_u = \phi bd^2\overline{k}$,

$$\text{required } d = \sqrt{\frac{M_u}{\phi b \overline{k}}} = \sqrt{\frac{189(12)}{0.9(11)(0.6438)}} = 18.9 \text{ in.}$$

A check of the d/b ratio gives $18.9/11 = 1.72$, which is a reasonable value.

5. At this point, the beam weight may be estimated. Realizing that the total design moment will increase, the final beam size may be estimated to be about 11 in. $\times$ 23 in. for purposes of calculating its weight. Thus

$$\text{beam dead load} = \frac{11(23)}{144}(0.150) = 0.264 \text{ kip/ft}$$

6. The additional M_u due to beam weight is

$$M_u = 1.2\left(\frac{0.264(18)^2}{8}\right) = 12.8 \text{ ft-kips}$$

$$\text{total } M_u = 189 + 12.8 = 202 \text{ ft-kips}$$

7. Using the same $\rho, \overline{k}$, and b as previously, compute the new required d:

$$\text{required } d = \sqrt{\frac{M_u}{\phi b \overline{k}}} = \sqrt{\frac{202(12)}{0.9(11)(0.6438)}} = 19.5 \text{ in.}$$

A check of the d/b ratio gives $19.50/11 = 1.77$, which is reasonable.

8. $$\text{required } A_s = \rho bd$$

$$= 0.0120(11)(19.50) = 2.57 \text{ in.}^2$$

Check $A_{s,\text{min}}$. From Table A-5,

$$A_{s,\text{min}} = 0.0033b_w d$$

$$= 0.0033(11)(19.50) = 0.71 \text{ in.}^2 \qquad \text{(O.K.)}$$

9. From Table A-2, select three No. 9 bars. Therefore $A_s = 3.00$ in.^{2}. From Table A-3,

$$\text{minimum required } b = 9.5 \text{ in.} \qquad \text{(O.K.)}$$

10. Determine the total beam depth h:

$$\text{required } h = 19.50 + \frac{1.13}{2} + 0.38 + 1.5 = 21.95 \text{ in.}$$

Rounding up by a $\frac{1}{2}$ in. increment, we will use $h = 22.0$ in. The actual effective depth d may now be checked:

$$d = 22.0 - 1.5 - 0.38 - \frac{1.13}{2} = 19.55 \text{ in.} > 19.50 \text{ in.} \qquad \text{(O.K.)}$$

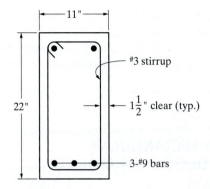

FIGURE 2-21   Design sketch for Example 2-6.

11.   A design sketch is shown in Figure 2-21. This design is not necessarily the best. The final ρ is

$$\frac{3.0}{11(19.55)} = 0.01395$$

Check ϵ_t: from Table A-10, $\epsilon_t > 0.005$. Therefore, the assumed ϕ is O.K. The final d/b ratio is

$$\frac{19.55}{11} = 1.78 \qquad\qquad\text{(O.K.)}$$

The design moment was based on a section 11 in. × 23 in., where the designed beam turned out to be 11 in. × 22 in. Small modifications in proportions could very well make more efficient use of the steel and concrete provided.

The percentage of overdesign could be determined (not necessary, however) by analyzing the designed cross section and comparing moment strength with applied moment.

There are several alternative approaches to this problem. At step 4, we could establish a desired d/b ratio and then mathematically solve for b and d. For example, if we establish a desirable d/b ratio of 2.0, then $d = 2b$. Again using the relationship $M_u = \phi b d^2 \bar{k}$,

$$\text{required } bd^2 = \frac{M_u}{\phi \bar{k}} = \frac{189(12)}{0.9(0.6438)} = 3914 \text{ in.}^3$$

We can substitute for d as follows:

$$\text{required } bd^2 = 3914 \text{ in.}^3$$
$$b(2b)^2 = 3914 \text{ in.}^3$$
$$b^3 = \frac{3914}{4} = 979 \text{ in.}^3$$
$$\text{required } b = \sqrt[3]{979} = 9.93 \text{ in.}$$

Hence, assuming that $b = 10$ in., the required effective depth d can then be calculated in a manner similar to that previously used. Also, at step 7, with a width b and a depth h established, along with the related design moment M_u, we could revert to the procedure for design of a rectangular section where the area is known (see Example 2-5, step 2.)

2-15   SUMMARY OF PROCEDURE FOR RECTANGULAR REINFORCED CONCRETE BEAM DESIGN FOR MOMENT (TENSION REINFORCEMENT ONLY)

A.   Cross Section (b and h) Known; Find the Required A_s

1.   Convert the service loads or moments to design M_u (include the beam weight).

2.   Based on knowing h, estimate d by using the relationship $d = h - 3$ in. (conservative for bars in a single layer). Calculate the required $\bar{k}$ using an assumed ϕ value of 0.90, subject to later check.

$$\text{required } \bar{k} = \frac{M_u}{\phi b d^2}$$

3.   From Tables A-7 through A-11, find the required steel ratio ρ and ensure that $\epsilon_t \geq 0.005$. If ϵ_t is within the range $0.004 \leq \epsilon_t \leq 0.005$, then ϕ will have to be reduced.

4.   Compute the required A_s:

$$\text{required } A_s = \rho b d$$

Check $A_{s,\text{min}}$. Use Table A-5.

5.   Select the bars. Check to see if the bars can fit into the beam in one layer (preferable). Check the actual effective depth and compare with the assumed effective depth. If the actual effective depth is slightly in excess of the assumed effective depth, the design will be slightly conservative (on the safe side). If the actual effective depth is less than the assumed effective depth, the design is on the unconservative side and should be revised.

6.   Sketch the design. (See Example 2-5, step 6, for a discussion.)

B.   Design for Cross Section and Required A_s

1.   Convert the service loads or moments to design moment M_u. An estimated beam weight may be included in the dead load if desired. Be sure to apply the load factor to this additional dead load.

2.   Select a desired steel ratio ρ. (See Table A-5 for recommended values. Use the ρ values from Table A-5 unless a small cross section or decreased steel is desired.)

3. From Table A-5 (or from Tables A-7 through A-11), find $\overline{k}$.

4. Assume b and compute the d required:

$$\text{required } d = \sqrt{\frac{M_u}{\phi b \overline{k}}}$$

   If the d/b ratio is reasonable (1.5 to 2.2), use these values for the beam. If the d/b ratio is not reasonable, increase or decrease b and compute the new required d.

5. Estimate h and compute the beam weight. Compare this with the estimated beam weight if an estimated beam weight was included.

6. Revise design M_u to include the moment due to the beam's own weight using the latest weight determined. Note that at this point, one could go to step 2 in design procedure A, where the cross section is known.

7. Using b and $\overline{k}$ previously determined along with the new total design M_u, find the new required d:

$$\text{required } d = \sqrt{\frac{M_u}{\phi b \overline{k}}}$$

   Check to see if the d/b ratio is reasonable.

8. Find the required A_s:

$$\text{required } A_s = \rho b d$$

   Check $A_{s,\text{min}}$. Use Table A-5.

9. Select the bars and check to see if the bars can fit into a beam of width b in one layer (preferable).

10. Establish the final h, rounding this upward to the next ½ in. This will make the actual effective depth greater than the design effective depth, and the design will be slightly conservative (on the safe side).

11. Check ϵ_t. Check the ϕ assumption. Sketch the design. (See Example 2-5, step 6, for a discussion.)

A flow diagram of this procedure is presented in Appendix B.

2-16   DESIGN OF ONE-WAY SLABS FOR MOMENT (TENSION REINFORCEMENT ONLY)

As higher-strength steel and concrete have become available for use in reinforced concrete members, the sizes of the members have decreased. *Deflections* of members are affected very little by material strength but are affected greatly by the size of a cross section and its related moment of inertia. Therefore deflections will be larger for a member of high-strength materials than would be the deflections for the same member fabricated from lower-strength materials, because the latter member will be essentially larger in cross-sectional area. Deflections are discussed in detail in Chapter 7. As discussed previously, one method that the ACI Code allows to limit adverse deflections is the use of a minimum thickness (see Section 2-13). A slab that meets the minimum thickness requirement must still be designed for flexure. Deflections need not be calculated or checked unless the slab supports or is attached to construction likely to be damaged by large deflections, however.

Example 2-7 illustrates the use of the ACI minimum thickness for one-way slabs. With respect to the span length to be used for design, the ACI Code, Section 8.9.1, recommends for beams and slabs not integral with supports

$$\text{span length} = \text{clear span} + \text{depth of member}$$

but not to exceed the distance between centers of supports. For design purposes, we will use the distance between centers of supports, as the slab thickness is not yet determined.

Example 2-7

Design a simple-span one-way slab to carry a uniformly distributed live load of 400 psf. The span is 10 ft (center-to-center of supports). Use $f_c' = 4000$ psi and $f_y = 60,000$ psi. Select the thickness to be not less than the ACI minimum thickness requirement.

Solution:

Determine the required minimum h and use this to estimate the slab dead weight.

1. From ACI Table 9-5(a), for a simply supported, solid, one-way slab,

$$\text{minimum } h = \frac{\ell}{20} = \frac{10(12)}{20} = 6.0 \text{ in.}$$

Try $h = 6$ in. and design a 12-in.-wide segment.

2. Determine the slab weight dead load:

$$\frac{6(12)}{144}(0.150) = 0.075 \text{ kip/ft}$$

The total design load is

$$w_u = 1.2w_{DL} + 1.6w_{LL}$$
$$= 1.2(0.075) + 1.6(0.400)$$
$$= 0.730 \text{ kip/ft}$$

3. Determine the design moment:

$$M_u = \frac{w_u \ell^2}{8} = \frac{0.730(10)^2}{8} = 9.13 \text{ ft-kips}$$

4. Establish the approximate d. Assuming No. 6 bars and minimum concrete cover on the bars of $\frac{3}{4}$ in.,

$$\text{assumed } d = 6.0 - 0.75 - 0.375 = 4.88 \text{ in.}$$

5. Determine the required $\bar{k}$ assuming $\phi = 0.90$:

$$\text{required } \bar{k} = \frac{M_u}{\phi b d^2}$$

$$= \frac{9.13(12)}{0.90(12)(4.88)^2} = 0.4260 \text{ ksi}$$

6. From Table A-10, for a required $\bar{k} = 0.4260$, the required $\rho = 0.0077$. (Note that the required ρ selected is the next *higher* value from Table A-10.) Also note that ϵ_t is not tabulated. Therefore, $\epsilon_t > 0.005$, this is a tension-controlled section, and $\phi = 0.90$.
    Use $\rho = 0.0077$.

7. $$\text{required } A_s = \rho b d = 0.0077(12)(4.88) = 0.45 \text{ in.}^2/\text{ft}$$

8. Select the main steel (from Table A-4). Select No. 5 bars at 8 in. o.c. ($A_s = 0.46$ in.^{2}). The assumption on bar size was satisfactory (actual $d >$ assumed d). The code requirements for maximum spacing have been discussed in Section 2-13. Minimum spacing of bars in slabs, practically, should not be less than 4 in., although the ACI Code allows bars to be placed closer together, as discussed in Example 2-5. Check the maximum spacing (ACI Code, Section 7.6.5):

$$\text{maximum spacing} = 3h \text{ or } 18 \text{ in.}$$

$$3h = 3(6) = 18 \text{ in.} \qquad\qquad \text{(O.K.)}$$

$$8 \text{ in.} < 18 \text{ in.}$$

Therefore use No. 5 bars at 8 in. o.c.

9. Select shrinkage and temperature reinforcement (ACI Code, Section 7.12):

$$\text{required } A_s = 0.0018bh$$

$$= 0.0018(12)(6) = 0.13 \text{ in.}^2/\text{ft}$$

Select No. 3 bars at 10 in. o.c. ($A_s = 0.13$ in.^{2}) or No. 4 bars at 18 in. o.c. ($A_s = 0.13$ in.^{2}):

$$\text{maximum spacing} = 5h \text{ or } 18 \text{ in.}$$

Use No. 3 bars at 10 in. o.c.

10. The main steel area must exceed the area required for shrinkage and temperature steel (ACI Code, Section 10.5.4):

$$0.46 \text{ in.}^2 > 0.13 \text{ in.}^2 \qquad\qquad \text{(O.K.)}$$

11. A design sketch is shown in Figure 2-22.

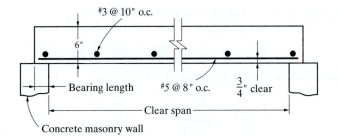

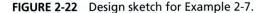

FIGURE 2-22   Design sketch for Example 2-7.

2-17   SUMMARY OF PROCEDURE FOR DESIGN OF ONE-WAY SLABS FOR MOMENT (TO SATISFY ACI MINIMUM *h*)

1. Compute the minimum *h* based on the ACI Code, Table 9.5(a). The slab thickness *h* can be rounded to the next higher ¼ in. for slabs up to 6 in. thickness and to the next higher ½ in. for slabs thicker than 6 in.
2. Compute the slab weight and compute w_u (total design load).
3. Compute the design moment M_u.
4. Calculate an assumed effective depth *d* (assuming No. 6 bars and ¾-in. cover) by using the relationship

$$d = h - 1.12 \text{ in.}$$

5. Calculate the required $\bar{k}$ assuming $\phi = 0.90$:

$$\text{required } \bar{k} = \frac{M_u}{\phi bd^2}$$

6. From Tables A-7 through A-11, find the required steel ratio ρ. Check ϵ_t to verify the ϕ assumption. If $\epsilon_t < 0.005$, the slab must be made thicker.
7. Compute the required A_s:

$$\text{required } A_s = \rho bd$$

8. Select the main steel (Table A-4). Check with maximum spacing of 3*h* or 18 in. Check the assumption of step 4.
9. Select shrinkage and temperature steel as per the ACI Code:

$$\text{required } A_s = 0.0020bh \quad \text{(grade 40 and 50 steel)}$$

$$\text{required } A_s = 0.0018bh \quad \text{(grade 60 steel)}$$

Check with maximum spacing of 5*h* or 18 in.

10. The main steel area *cannot* be less than the area of steel required for shrinkage and temperature.
11. Sketch the design.

2-18   SLABS ON GROUND

The previous discussion primarily concerns itself with a structural slab. Another category of slabs, generally used as a floor, may be termed a *slab on grade* or a *slab on ground*. As the name implies, it is a slab that is supported throughout its entire area by some form of subgrade. The design of such a slab is significantly different from the design of a structural slab.

A theoretical approach to determine the required floor slab thickness must consider the following factors:

1.   Strength of subgrade and subbase
2.   Strength of concrete
3.   Magnitude and type of load (including contact area of loads)

The subgrade is the natural ground, graded and compacted, on which the floor is built. The pressure on the subgrade is generally low due to the rigidity of the concrete floor slab. The floors do not necessarily require strong support from the subgrade. It is important that the subgrade support be reasonably uniform without abrupt changes, horizontally, from hard to soft, however. The upper portion of the subgrade should be of uniform material and density.

The subbase is usually a thin layer of material placed on top of the prepared subgrade. It is generally used when a uniform subgrade cannot be developed by grading and compaction. It will serve to equalize minor surface defects as well as provide a capillary break and a working platform for construction activities. A 4-in. minimum thickness, compacted to a high density, is suggested.

The concrete strength for industrial and commercial slabs on ground should not be less than 4000 psi at 28 days. This furnishes satisfactory wear resistance in addition to strength. Generally, in residential construction the slabs on ground will have concrete strengths of 2500 psi or 3000 psi. A minimum of 3000 psi is recommended. The required slab thickness will be a function of the type of loading, magnitude of loading, and contact area of the load.

Reinforcing steel, in the form of welded wire reinforcement or deformed bars in both directions, is usually placed in the slab for a number of reasons. The reinforcement may add to the strength of the slab, particularly when the slab spans soft spots in the subgrade. It also acts as crack control by minimizing the width of the cracks that may develop between joints. In addition, the use of reinforcement will allow an increased joint spacing. Despite several other relatively minor advantages that result from the use of steel, questions have been raised as to whether reinforcement is always necessary, particularly with uniform support of the slab and short joint spacings. The steel does not prevent cracking nor does it add significantly to the load carrying capacity of the slab. It is also usually more economical to obtain increased strength in concrete slabs on ground by increasing the thickness of the slab.

Design aids and procedures (based primarily on research done for highway and airport pavements) have been developed and are available in specialized publications. (See references [3] through [8].)

REFERENCES

[1]   Charles S. Whitney. "Plastic Theory of Reinforced Concrete Design." *Trans. ASCE*, Vol. 68, 1942.

[2]   "Rule-of-Thumb for Flexural Steel Area." Concrete Q & A, Concrete International, October 2007, page 91.

[3]   *Guide for Concrete Floor and Slab Construction*. ACI 302.1R-04, American Concrete Institute, P.O. Box 9094, Farmington Hills, MI 48333-9094, 2004.

[4]   *Design of Slabs on Grade*. ACI 360R-06, American Concrete Institute, P.O. Box 9094, Farmington Hills, MI 48333-9094, 2006.

[5]   *Slabs on Grade*. Concrete Craftsman Series 1, American Concrete Institute, P.O. Box 9094, Farmington Hills, MI 48333-9094, 1994.

[6]   Robert G. Packard. *Slab Thickness Design for Industrial Concrete Floors on Grade*. Portland Cement Association, 5420 Old Orchard Road, Skokie, IL 60077, 1996.

[7]   J. A. Farny and S. M. Tarr. *Concrete Floors on Ground*. Portland Cement Association, 5420 Old Orchard Road, Skokie, IL 60077, 2008.

[8]   Robert B. Anderson. *Innovative Ways to Reinforce Slabs-on-Ground*. Tech Facts, Wire Reinforcement Institute, Inc., 942 Main Street, Suite 300, Hartford, CT 06103, 1996.

PROBLEMS

In the following problems, consider moment only and tension reinforcing only. For beams, assume $1^1/_2$-in. cover and No. 3 stirrups. For slabs, assume $3/_4$-in. cover. Unless noted otherwise, given loads are superimposed service loads and do not include weights of the members.

2-1. **(a)** The beam of the cross section shown has $f'_c = 3000$ psi and $f_y = 60,000$ psi. Neglect the tensile strain check and $A_{s,min}$ check. Calculate M_n.

   (b) Same cross section as part (a) but the steel is changed to four No. 10 bars. Calculate M_n. Calculate the percent increase in M_n and A_s.

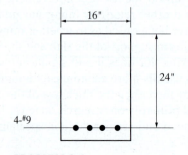

PROBLEM 2-1

(c) Same cross section as part (a) but the depth d is increased to 28 in. Calculate M_n. Calculate the percent increase in M_n and d.

(d) Same cross section as part (a) but f_c' is increased to 4000 psi. Calculate M_n. Calculate the percent increase in M_n and f_c'.

2-2. For the cross section of Problem 2-1(a), verify that the tension steel yields.

2-3. (a) Calculate the practical moment strength ϕM_n for a rectangular reinforced concrete cross section having a width b of 13 in. and an effective depth d of 24 in. The tension reinforcing steel is four No. 8 bars. $f_c' = 4000$ psi and $f_y = 40,000$ psi.

(b) Same as part (a) but $f_y = 60,000$ psi. Also calculate the percent increase in ϕM_n and f_y.

2-4. Determine ϕM_n for a reinforced concrete beam 16 in. wide by 32 in. deep reinforced with seven No. 10 bars (placed in two layers: five in the bottom layer, two in the top layer, with 1 in. clear between layers). Use $f_c' = 4000$ psi and $f_y = 60,000$ psi.

2-5. A reinforced concrete beam having the cross section shown is on a simple span of 28 ft. It carries uniform service loads of 3.60 kips/ft live load and 2.20 kips/ft dead load. Check the adequacy of the beam with respect to moment. Use $f_c' = 3000$ psi and $f_y = 40,000$ psi.

(a) Reinforcing is six No. 10 bars.

(b) Reinforcing is six No. 11 bars.

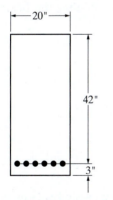

PROBLEM 2-5

2-6. Develop a spreadsheet application that will allow a user to input the basic information for the analysis of a rectangular R/C cross section and that will then calculate the practical moment strength ϕM_n. Set up the spreadsheet to be "user friendly" and fully label the output.

2-7. A 12-in.-wide by 20-in.-deep concrete beam is reinforced with three No. 8 bars. The beam supports a service live load of 2.5 kips/ft and a service dead load of 0.7 kip/ft on a simple span of 16 ft. Use $f_c' = 4000$ psi and $f_y = 60,000$ psi. Check the adequacy of the beam with respect to moment.

2-8. A reinforced concrete beam having the cross section shown is on a simple span of 26.5 ft. It supports uniformly distributed service loads of 3.20 kips/ft live load and 1.80 kips/ft dead load (excluding the beam weight). Reinforcing is as shown. Check the adequacy of the beam with respect to moment. Use $f_c' = 3000$ psi and $f_y = 60,000$ psi.

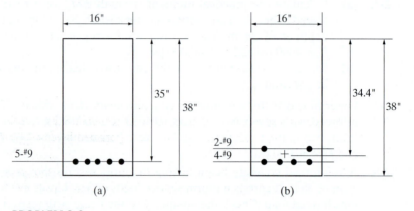

(a)   (b)

PROBLEM 2-8

2-9. A rectangular reinforced concrete beam carries *service loads* on a span of 20 ft as shown. Use $f_c' = 3000$ psi and $f_y = 60,000$ psi; $b = 14.5$ in., $h = 26$ in., and reinforcing is three No. 10 bars. Determine whether the beam is adequate with respect to moment.

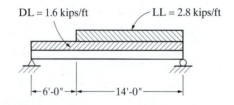

DL = 1.6 kips/ft   LL = 2.8 kips/ft

|←6'-0"→|←——14'-0"——→|

PROBLEM 2-9

2-10. A rectangular reinforced concrete beam 14 in. wide by 24 in. deep is to support a service dead load of 0.6 kip/ft and a service live load of 1.4 kips/ft. Reinforcing is four A615 grade 60 No. 9 bars. Use $f_c' = 4000$ psi. Determine the maximum simple span length on which this beam may be utilized.

2-11. A 10-in.-thick one-way slab supports a superimposed service live load of 600 psf on a simple span of 16 ft. Reinforcement is No. 7 at 6 in. on center. Check the adequacy of the slab with respect to moment. Use $f_c' = 3000$ psi and $f_y = 60,000$ psi.

2-12. The one-way slab shown spans 12 ft from center of support to center of support. Calculate ϕM_n and determine the *service live load* (psf) that the slab may carry. (Assume that the only dead load is the weight of the slab.) Use $f'_c = 3000$ psi and $f_y = 40{,}000$ psi.

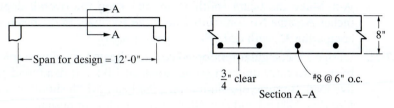

Span for design = 12'-0"

$\frac{3}{4}$" clear   #8 @ 6" o.c.

Section A–A

PROBLEM 2-12

2-13. An $8\frac{1}{2}$-in.-thick one-way reinforced concrete slab overhangs a simple support. The span of the overhang is 8 ft. Drawings called for the reinforcement to be placed with top cover of 1 in. The steel was misplaced, however, and later was found to be as much as $3\frac{1}{2}$ in. below the top of the concrete. Find ϕM_n for the slab as designed and as built and the percent of reduction in flexural strength. Use $f'_c = 4000$ psi and $f_y = 60{,}000$ psi. Bars are No. 7 at 11 in. o.c.

2-14. Design a rectangular reinforced concrete beam to resist a total design moment M_u of 133 ft-kips. (This includes the moment due to beam weight.) Architectural considerations require that the width (b) be $11\frac{1}{2}$ in. and the overall depth (h) be 23 in. Use $f'_c = 3000$ psi and $f_y = 60{,}000$ psi. Sketch your design.

2-15. Rework Problem 2-14 with $M_u = 400$ ft-kips, $b = 16$ in., $h = 28$ in., $f'_c = 4000$ psi, and $f_y = 60{,}000$ psi.

2-16. For the beam designed in Problem 2-15, if the main reinforcement were incorrectly placed so that the actual effective depth were 24 in., would the beam be adequate? Check by comparing M_u with the ϕM_n resulting from the beam using actual steel, actual b, and $d = 24$ in.

2-17. Design a rectangular reinforced concrete beam (tension steel only) for a simple span of 32 ft. Uniform service loads are 0.85 kip/ft dead load and 1.0 kip/ft live load. The beam is to be $11\frac{1}{2}$ in. wide and 26 in. deep overall (form reuse consideration). Use $f'_c = 4000$ psi and $f_y = 60{,}000$ psi. Calculate ϕM_n for the beam designed.

2-18. Design a rectangular reinforced concrete beam (tension steel only) for a simple span of 30 ft. There is no superimposed dead load (other than the weight of the beam) and the superimposed live load is 1.35 kips/ft. The beam is to be 12 in. wide and 27 in. deep overall. Use $f'_c = 5000$ psi and A615 grade 60 steel. As a check, calculate ϕM_n for the beam *designed*.

2-19. Rework Problem 2-18 assuming that the superimposed live load has increased to 1.75 kips/ft and there is now a 1.0 kip/ft superimposed dead load (in addition to the beam weight). As before, check ϕM_n for the beam designed.

2-20. Design a simply supported rectangular reinforced concrete beam to span 22 ft and to carry uniform service loads of 1.6 kips/ft dead load and 1.4 kips/ft live load. The assumed dead load includes an estimated beam weight. Use A615 grade 60 steel and f'_c = 3000 psi. Use the recommended ρ from Table A-5. Make the beam width 15 in. and keep the overall depth (h) to full inches. Assume No. 3 stirrups. Check the adequacy of the beam you design by comparing M_u with ϕM_n. Sketch your design.

2-21. Design a rectangular reinforced concrete beam for a simple span of 30 ft. The beam is to carry uniform *service* loads of 1.0 kip/ft dead load and 2.0 kips/ft live load. Because of column sizes, the beam width should not exceed 16 in. Use f'_c = 3000 psi and f_y = 60,000 psi. Sketch your design.

2-22. Rework Problem 2-21 assuming that the total depth h is not to exceed 30 in. and that there is no limitation on the width b.

2-23. Design a rectangular reinforced concrete beam for a simple span of 32 ft. Uniform service loads are 1.5 kips/ft dead load and 2.0 kips/ft live load. The width of the beam is limited to 18 in. Use f'_c = 3000 psi and f_y = 60,000 psi. Sketch your design.

2-24. Rework Problem 2-23 assuming that the total depth h is not to exceed 32 in. and that there is no limitation on the width b.

2-25. Design a rectangular reinforced concrete beam for a simple span of 40 ft. Uniform service loads are 0.8 kip/ft dead load and 1.4 kips/ft live load. Use f'_c = 4000 psi and f_y = 60,000 psi. Sketch your design.

2-26. Design a rectangular reinforced concrete beam (tension steel only) for the span and *superimposed* service loads shown. Use f'_c = 5000 psi and A615 grade 60 steel.

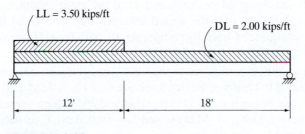

LL = 3.50 kips/ft   DL = 2.00 kips/ft

12'   18'

PROBLEM 2-26

2-27. Design a 28-ft, simple-span, rectangular reinforced concrete beam to support a uniform live load of 0.8 kip/ft and concentrated loads at midspan of 10 kips dead load and 14 kips live load. Use f'_c = 5000 psi and f_y = 60,000 psi.

2-28. Design a simply supported rectangular reinforced concrete beam for the span and *service loads* shown. Use f'_c = 3000 psi and f_y = 60,000 psi.

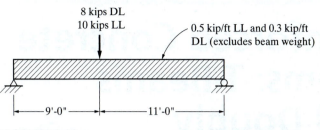

8 kips DL
10 kips LL

0.5 kip/ft LL and 0.3 kip/ft
DL (excludes beam weight)

|——9'-0"——|——11'-0"——|

PROBLEM 2-28

2-29. Rework Problem 2-28 assuming that the beam is to be extended to overhang the right support by 10 ft. The uniform load also extends to the end of the overhang.

2-30. Design a simply supported one-way reinforced concrete floor slab to span 8 ft and carry a service live load of 300 psf. Use $f'_c = 3000$ psi and $f_y = 60{,}000$ psi. Sketch your design.

2-31. Design a simply supported one-way reinforced concrete floor slab to span 10 ft and carry a service live load of 175 psf and a service dead load of 25 psf. Use $f'_c = 3000$ psi and $f_y = 60{,}000$ psi. Make the slab thickness to $\frac{1}{2}$-in. increments.

    (a) Design the slab for the ACI Code minimum thickness.

    (b) Design the thinnest possible slab allowed by the ACI Code.

2-32. Design the simply supported one-way reinforced concrete slab as shown. The service live load is 200 psf. Use $f'_c = 3000$ psi and $f_y = 60{,}000$ psi. Sketch your design.

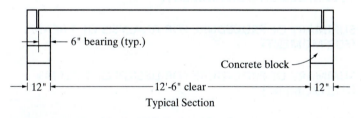

    6" bearing (typ.)

Concrete block

|→|12"|←|————12'-6" clear————|→|12"|←|

Typical Section

PROBLEM 2-32

Chapter 3

Reinforced Concrete Beams: T-Beams and Doubly Reinforced Beams

3-12   SUMMARY OF PROCEDURE FOR DESIGN OF DOUBLY REINFORCED BEAMS (FOR MOMENT)

3-13   ADDITIONAL CODE REQUIREMENTS FOR DOUBLY REINFORCED BEAMS

3-1   T-BEAMS: INTRODUCTION

Floors and roofs in reinforced concrete buildings may be composed of slabs that are supported so that loads are carried to columns and then to the building foundation. As previously discussed, these are termed *flat slabs* or *flat plates*. The span of such a slab cannot become very large before its own dead weight causes it to become un-economical. Many types of systems have been devised to allow greater spans with-out the problem of excessive weight.

One such system, called a *beam and girder system*, is composed of a slab on sup-porting reinforced concrete beams and girders. The beam and girder framework, in turn, is supported by columns. In such a system, the beams and girders are com-monly placed monolithically with the slab. Systems other than the monolithic sys-tem do exist, and these may make use of some precast and some cast-in-place concrete. These are generally of a proprietary nature. The typical monolithic system is shown in Figure 3-1. The beams are commonly spaced so that they intersect the girders at the midpoint, third points, or quarter points, as shown in Figure 3-2.

In the analysis and design of such floor and roof systems, it is common practice to assume that the monolithically placed slab and supporting beam interact as a unit in resisting *positive bending moment*. As shown in Figure 3-3, the slab becomes the compression flange, and the supporting beam becomes the web or stem. The inter-acting flange and stem produce the cross section having the typical T-shape from

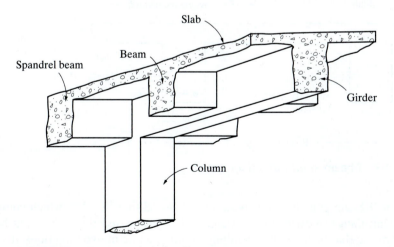

FIGURE 3-1   Beam and girder floor system.

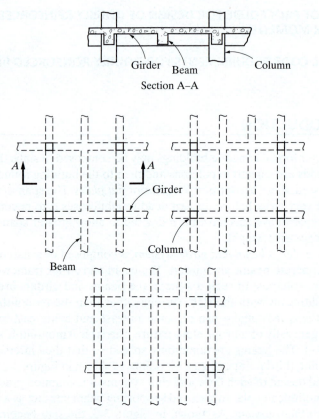

FIGURE 3-2   Common beam and girder layouts.

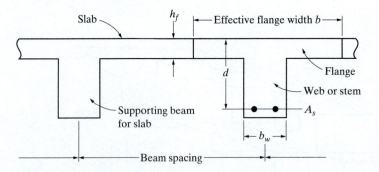

FIGURE 3-3   T-beam as part of a floor system.

which the T-beam gets its name. It should be noted that the slab, which comprises the T-beam flange, must itself be designed to span across the supporting beams. Therefore the slab behaves as a bending member acting in two directions. It should also be noted that should the T-beam cross section be subjected to *negative bending moment*, the slab at the top of the stem will be in tension while the bottom of the stem is in compression. It will be seen that this situation will occur at interior supports of continuous beams, which are discussed later.

To simplify the complex two-way behavior of the flange, the ACI Code, for design and analysis purposes, has established criteria whereby the flange, when acting together with the web, will have a limited width that may be considered effective in resisting applied moment. This effective flange width for symmetrical shapes will always be equal to or less than the beam spacing (see Figure 3-3).

3-2   T-BEAM ANALYSIS

For purposes of analysis and design, the ACI Code, Section 8-12, has established limits on the effective flange width as follows:

1. The effective flange width must not exceed one-fourth of the span length of the beam, and the effective overhanging flange width on each side of the web must not exceed eight times the thickness of the slab nor one-half of the clear distance to the next beam. In other words, the effective flange width must not exceed
    a. One-fourth of the span length.
    b. $b_w + 16h_f$.
    c. Center-to-center spacing of beams.
    The smallest of the three values will control.
2. For beams having a flange on one side only, the effective overhanging flange width must not exceed one-twelfth of the span length of the beam, nor six times the slab thickness, nor one-half of the clear distance to the next beam.
3. For isolated beams in which the T-shape is used only for the purpose of providing additional compressive area, the flange thickness must not be less than one-half of the width of the web, and the total flange width must not be more than four times the web width.

The ductility requirements for T-beams are similar to those for rectangular beams. To ensure ductile behavior, ACI Code, Section 10.3.5, requires a net tensile strain $\epsilon_t \geq 0.004$ for flexural members. A section is tension controlled, that is, completely ductile when the net tensile strain $\epsilon_t \geq 0.005$. It is always desirable and more efficient in the design of flexural members to strive for a tension-controlled section. The T-shape can be a factor in the determination of net tensile strain for a T-beam.

The procedure for determining the minimum steel for a T-beam is the same as for a rectangular beam when the T-beam flange is in compression (positive moment). Where tensile reinforcement is required by analysis, the steel area, A_s, shall not be less than that given by

$$A_{s,\min} = \frac{3\sqrt{f_c'}}{f_y}b_w d \geq \frac{200}{f_y}b_w d$$

Note that for T-beams, b_w represents the width of the web. Also note that the first expression controls only if $f'_c > 4440$ psi. The above expressions for minimum steel also apply to continuous T-beams.

For negative moment (flange in tension) in statically determinate members,

$$A_{s,min} = \text{the smaller of } \frac{6\sqrt{f'_c}}{f_y}b_w d \quad \text{or} \quad \frac{3\sqrt{f'_c}}{f_y}bd$$

The minimum steel requirements need not be applied if, at every section along the member, at least 33% more steel than is required by analysis is provided.

Because of the relatively large compression area available in the flange of the T-beam, the moment strength is usually limited by the yielding of the tensile steel. Therefore it is usual to assume that the tensile steel will yield before the concrete reaches its ultimate strain and crushes. The total tensile force, N_T, at the ultimate condition may then be found by

$$N_T = A_s f_y$$

To proceed with the analysis, the shape of the compressive stress block must be defined. As in our previous analyses, the total compressive force N_C must be equal to the total tensile force N_T. The shape of the stress block must be compatible with the area in compression. Two conditions may exist: The stress block may be completely within the flange, or it may cover the flange and extend into the web. These two conditions will result in what we will term, respectively, a *rectangular T-beam* and a *true T-beam*. In addition to the shape of the stress block, the basic difference between the two is that the rectangular T-beam with effective flange width b is analyzed in the same way as is a rectangular beam of width b, whereas the analysis of the true T-beam must consider the T-shaped stress block.

Example 3-1 _____

The T-beam shown in Figure 3-4 is part of a floor system. Determine the practical moment strength ϕM_n if $f_y = 60{,}000$ psi (A615 grade 60) and $f'_c = 3000$ psi.

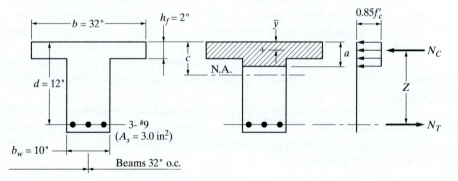

FIGURE 3-4   Sketch for Example 3-1.

Solution:

1. Because the span length is not given, determine the effective flange width in terms of the flange thickness and beam spacing:

$$b_w + 16h_f = 10 + 16(2) = 42 \text{ in.}$$

$$\text{beam spacing} = 32 \text{ in. o.c.}$$

$$\text{Use } b = 32 \text{ in.}$$

2. Check $A_{s,min}$. From Table A-5:

$$A_{s,min} = 0.0033b_w d$$

$$= 0.0033(10)(12) = 0.40 \text{ in.}^2$$

$$0.40 \text{ in.}^2 < 3.0 \text{ in.}^2 \hspace{2cm} \text{(O.K.)}$$

3. Assume that the steel yields and find N_T:

$$N_T = A_s f_y = 3.00(60{,}000) = 180{,}000 \text{ lb}$$

4. The flange alone, if fully stressed to $0.85f_c'$, would produce a total compressive force of

$$N_{Cf} = (0.85f_c')h_f b$$

$$= 0.85(3000)(2)(32) = 163{,}200 \text{ lb}$$

5. Because $180{,}000 > 163{,}200$, the stress block must extend below the flange far enough to provide the remaining compression:

$$180{,}000 - 163{,}200 = 16{,}800 \text{ lb}$$

Hence the stress block extends below the flange and the analysis is one for a true T-beam.

6. The remaining compression $(N_T - N_{Cf})$ may be obtained by the additional web area:

$$N_T - N_{Cf} = (0.85f_c')b_w(a - h_f)$$

Solving for a, we obtain

$$a = \frac{N_T - N_{Cf}}{(0.85f_c')b_w} + h_f = \frac{16{,}800}{0.85(3000)(10)} + 2$$

$$= 2.66 \text{ in.}$$

7. Determine net tensile strain ϵ_t (check ductility.) Using the relationship $a = \beta_1 c$, which is approximate for T-beams:

$$c = \frac{a}{\beta_1} = \frac{2.66 \text{ in.}}{0.85} = 3.13 \text{ in. (See Figure 3-4.)}$$

The distance d_t of the extreme tensile reinforcement from the compression face is 12 in. Therefore, the net tensile strain in the extreme tensile reinforcement is

$$\epsilon_t = 0.003 \frac{(d_t - c)}{c} = 0.003 \frac{(12 - 3.13)}{3.13} = 0.0085$$

8. Determine the strength reduction factor ϕ: because $0.0085 > 0.005$, this is a tension-controlled section and $\phi = 0.90$ (see Section 2-9).

9. To calculate the magnitude of the internal couple, it is necessary to know the lever-arm distance between N_C and N_T. The location of N_T is assumed at the centroid of the steel area, and we will locate N_C at the centroid of the T-shaped compression area (see Figure 3-5a). Using a reference axis at the top of the section, the centroid may be located a distance $\bar{y}$ below the reference axis, as follows:

$$\bar{y} = \frac{\Sigma(Ay)}{\Sigma A}$$

$$A_1 = 32(2) = 64 \text{ in.}^2, \quad A_2 = 10(0.66) = 6.6 \text{ in.}^2$$

$$\bar{y} = \frac{64(1) + 6.6(2 + 0.33)}{64 + 6.6} = 1.12 \text{ in.}$$

This locates N_C. Therefore

$$Z = d - \bar{y}$$

$$= 12 - 1.12 = 10.88 \text{ in.}$$

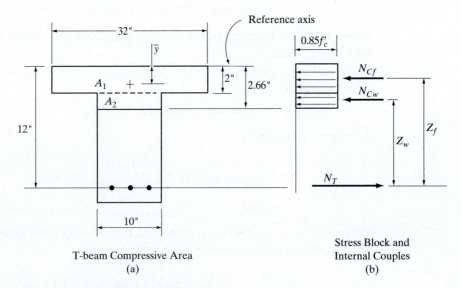

T-beam Compressive Area
(a)

Stress Block and
Internal Couples
(b)

FIGURE 3-5   Sketch for Example 3-1.

The nominal (or ideal) moment strength may be found:

$$M_n = N_T Z = \frac{180,000(10.88)}{12,000} = 163 \text{ ft-kips}$$

from which the practical moment strength is

$$\phi M_n = 0.9(163) = 147 \text{ ft-kips}$$

In the solution of Example 3-1, step 9 may be accomplished in a slightly different way. If the total internal couple M_n is assumed to be composed of two component couples, a flange couple (using compressive force N_{Cf} in Figure 3-5b) and a web couple (using compressive force N_{Cw} in Figure 3-5b), then its magnitude can be calculated from

$$M_n = \text{flange couple} + \text{web couple}$$

$$= N_{Cf}Z_f + N_{Cw}Z_w$$

$$= N_{Cf}\left(d - \frac{h_f}{2}\right) + (N_T - N_{Cf})\left[d - h_f - \left(\frac{a - h_f}{2}\right)\right]$$

This avoids the calculation of the centroid location and results in the same moment strength. The concept of two component couples is used again in the design of true T-beams.

Example 3-2 _____

For the T-beam shown in Figure 3-6, determine the practical moment strength ϕM_n if $f'_c = 3000$ psi and $f_y = 60,000$ psi. The beam span length is 24 ft.

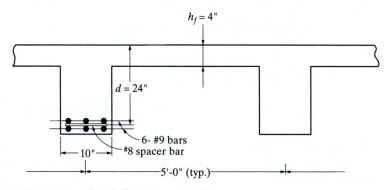

FIGURE 3-6   Sketch for Example 3-2.

Solution:

1. Find the effective flange width:

$$\frac{1}{4} \text{ span length} = \frac{24(12)}{4} = 72 \text{ in.}$$

$$b_w + 16h_f = 10 + 16(4) = 74 \text{ in.}$$

$$\text{beam spacing} = 60 \text{ in.}$$

$$\text{Use } b = 60 \text{ in.}$$

2. Check $A_{s,min}$. From Table A-5:

$$A_{s,min} = 0.0033b_w d$$

$$= 0.0033(10)(24) = 0.79 \text{ in.}^2$$

$$0.79 \text{ in.}^2 < 6.00 \text{ in.}^2 \qquad\qquad\qquad\qquad \text{(O.K.)}$$

3. Assume the steel yields and find N_T.

$$N_T = A_s f_y = 6.0(60,000) = 360,000 \text{ lb}$$

4. The flange itself is capable of furnishing a compression force of

$$N_{Cf} = (0.85f_c')bh_f = 0.85(3000)(60.0)(4) = 612,000 \text{ lb}$$

5. Because $612,000 > 360,000$, the flange furnishes sufficient compression area and the stress block lies entirely in the flange. Therefore analyze the T-beam as a rectangular T-beam of width $b = 60$ in.

6. Solve for the depth of the stress block.

$$a = \frac{A_s f_y}{0.85f_c'b} = \frac{6.00(60,000)}{0.85(3000)(60)} = 2.35 \text{ in.}$$

7. Determine the net tensile strain (check ductility.) The depth to the extreme tension steel d_t is determined using a #8 spacer bar between the two layers of #9 bars, as shown in Figure 3-6.

$$d_t = 24 + \frac{1.00}{2} + \frac{1.125}{2} = 25.1 \text{ in.}$$

$$c = \frac{a}{\beta_1} = \frac{2.35}{0.85} = 2.76 \text{ in.}$$

$$\epsilon_t = \frac{0.003(d_t - c)}{c} = \frac{0.003(25.1 - 2.76)}{2.76} = 0.0243$$

(Or use Eq. [2-1] from Section 2-9.)

Because $0.0243 > 0.005$, ductility is ensured, and this is a tension-controlled section.

8. Find ϕ. Because $\epsilon_t > 0.005$, $\phi = 0.90$.

9.   Find ϕM_n.

$$\phi M_n = \phi A_s f_y \left(d - \frac{a}{2} \right)$$

$$= \frac{0.90(6.00)(60)\left(24 - \dfrac{2.35}{2} \right)}{12} = 616 \text{ ft-kips}$$

This T-beam behaves like a rectangular beam having a width b of 60 in., so the expression for ϕM_n developed in Section 2-10 could be used at step 9:

$$\phi M_n = \phi b d^2 \overline{k}$$

$$\rho = \frac{A_s}{bd} = \frac{6.00}{60(24)} = 0.0042$$

From Table A-8, $\overline{k} = 0.2396$ ksi and $\epsilon_t > 0.005$. Therefore, $\phi = 0.90$. Then

$$\phi M_n = \frac{0.90(60)(24)^2(0.2396)}{12} = 621 \text{ ft-kips}$$

3-3   ANALYSIS OF BEAMS HAVING IRREGULAR CROSS SECTIONS

Beams having other than rectangular and T-shaped cross sections are common, particularly in structures using precast elements. The approach for the analysis of such beams is to use the internal couple in the normal way, taking into account any variation in the shape of the compressive stress block. The method is similar to that used for true T-beam analysis.

Example 3-3

The cross section shown in Figure 3-7 is sometimes referred to as an inverted T-girder. Find the practical moment strength ϕM_n. (The ledges in the beam cross section will possibly be used for support of precast slabs.) Use $f_y = 60,000$ psi (A615 grade 60) and $f'_c = 3000$ psi.

Solution:

1.   The effective flange width may be considered to be 7 in.
2.   Check $A_{s,\text{min}}$. From Table A-5:

$$A_{s,\text{min}} = 0.0033 b_w d$$

$$= 0.0033(17)(24) = 1.35 \text{ in.}^2$$

$$1.35 \text{ in.}^2 < 4.00 \text{ in.}^2 \qquad\qquad \text{(O.K.)}$$

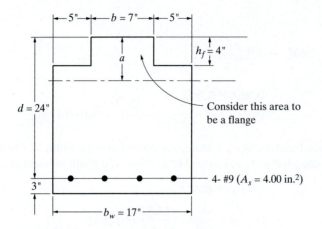

FIGURE 3-7   Sketch for Example 3-3.

3.   Find N_T:

$$N_T = A_s f_y = 4.00(60,000) = 240,000 \text{ lb}$$

4.   Determine the amount of compression that the 7 in. × 4 in. area is capable of furnishing. (This is the area we are considering to be the flange.)

$$N_{Cf} = (0.85 f_c')h_f b$$
$$= 0.85(3000)(4)(7) = 71,400 \text{ lb}$$

5.   Because 240,000 lb > 71,400 lb, the compressive stress block must extend below the ledges to provide the remaining compression ($a > h_f$):

$$N_T - N_{Cf} = 240,000 - 71,400 = 168,600 \text{ lb}$$

6.   The remaining compression will be furnished by additional beam area below the ledges. Referring to Figure 3-7,

$$a = \frac{N_T - N_{Cf}}{(0.85 f_c')b_w} + h_f$$

$$= \frac{168,600}{0.85(3000)(17)} + 4$$

$$= 7.89 \text{ in. from top of beam}$$

7.   Check ductility.

$$c = \frac{a}{\beta_1} = \frac{7.89 \text{ in.}}{0.85} = 9.28 \text{ in.}$$

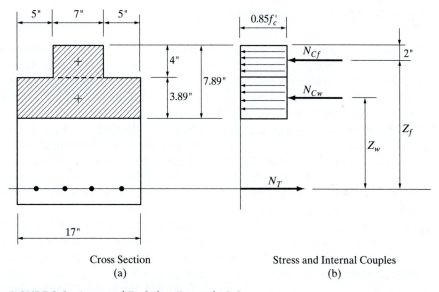

FIGURE 3-8 Inverted T-girder, Example 3-3.

The distance of the extreme tensile reinforcement from the compression face d_t is 24 in. Therefore, the net tensile strain in the extreme tensile reinforcement is

$$\epsilon_t = 0.003\frac{(d_t - c)}{c} = 0.003\frac{(24 - 9.28)}{9.28} = 0.00476$$

8. Therefore, this is a transition section ($0.004 < \epsilon_t < 0.005$), and the corresponding strength-reduction factor is

$$\phi = 0.65 + (\epsilon_t - 0.002)\left(\frac{250}{3}\right) = 0.65 + (0.00476 - 0.002)\left(\frac{250}{3}\right) = 0.875$$

Note that $0.65 < 0.875 < 0.90$.

9. ϕM_n will be calculated considering two component internal couples, a flange couple and a web couple. Refer to Figure 3-8.

$$\phi M_n = \phi(N_{Cf}Z_f + N_{Cw}Z_w)$$

$$= \phi\left\{N_{Cf}\left(d - \frac{h_f}{2}\right) + (N_T - N_{Cf})\left[d - h_f - \left(\frac{a - h_f}{2}\right)\right]\right\}$$

$$= \frac{0.875\left\{71.4\left(24 - \frac{4}{2}\right) + 168.6\left[24 - 4 - \left(\frac{3.89}{2}\right)\right]\right\}}{12}$$

$$= 337 \text{ ft-kips}$$

3-4   T-BEAM DESIGN (FOR MOMENT)

The design of the T-sections involves the dimensions of the flange and web and the area of the tension steel, a total of five unknowns. In the normal progression of a design, the flange thickness is determined by the design of the slab, and the web size is determined by the shear and moment requirements at the end supports of a beam in continuous construction. Practical considerations, such as column sizes and forming, may also dictate web width. Therefore, when the T-section is designed for positive moment, most of the five unknowns have been previously determined.

As indicated previously, the ACI Code dictates permissible effective flange width *b*. The flange itself generally provides more than sufficient compression area; therefore the stress block usually lies completely in the flange. Thus most T-beams are wide rectangular beams with respect to flexural behavior.

The recommended method for the design of T-beams will depend on whether the T-beam behaves as a rectangular T-beam or a true T-beam. The first step will be to answer this question. If the T-beam is determined to be a rectangular T-beam, the design procedure is the same as for the tensile reinforced rectangular beam where the size of the cross section is known (see Section 2-15). If the T-beam is determined to be a true T-beam, the design proceeds by designing a flange component and a web component and combining the two.

Example 3-4 _____

Design the T-beam for the floor system shown in Figure 3-9. The floor has a 4-in. slab supported by 22-ft-span-length beams cast monolithically with the slab. Beams are 8 ft-0 in. on center and have a web width of 12 in. and a total depth = 22 in.; f'_c = 3000 psi and f_y = 60,000 psi (A615 grade 60). Service loads are 0.125-ksf live load and 0.200-ksf dead load. The given dead load does not include the weight of the floor system.

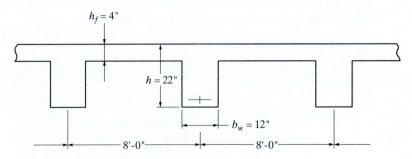

FIGURE 3-9   Typical floor section, Example 3-4.

Solution:

1. Establish the design moment:

$$\text{slab weight} = \frac{96(4)}{144}(0.150) = 0.400 \text{ kip/ft}$$

$$\text{stem weight} = \frac{12(18)}{144}(0.150) = \underline{0.225}$$

$$\text{total} = 0.625 \text{ kip/ft}$$

$$\text{service DL} = 8(0.200) = 1.60 \text{ kips/ft}$$

$$\text{service LL} = 8(0.125) = 1.00 \text{ kip/ft}$$

Calculate the factored load and moment:

$$w_u = 1.2(0.625 + 1.60) + 1.6(1.00) = 4.27 \text{ kips/ft}$$

$$M_u = \frac{w_u \ell^2}{8} = \frac{4.27(22)^2}{8} = 258 \text{ ft-kips}$$

2. Assume an effective depth $d = h - 3$ in.:

$$d = 22 - 3 = 19 \text{ in.}$$

3. Determine the effective flange width:

$$\frac{1}{4} \text{ span length} = 0.25(22)(12) = 66 \text{ in.}$$

$$b_w + 16h_f = 12 + 16(4) = 76 \text{ in.}$$

$$\text{beam spacing} = 96 \text{ in.}$$

Use an effective flange width $b = 66$ in.

4. Assume a tension-controlled section—that is, the net tensile strain $\epsilon_t \geq 0.005$; this assumption will be checked later. The net tensile strain value of 0.005 gives a strength-reduction factor $\phi = 0.90$.

5. Determine whether the beam behaves as a true T-beam or as a rectangular beam by computing the practical moment strength ϕM_{nf} with the full effective flange assumed to be in compression. This assumes that the bottom of the compressive stress block coincides with the bottom of the flange, as shown in Figure 3-10. Thus

$$\phi M_{nf} = \phi(0.85f_c')bh_f\left(d - \frac{h_f}{2}\right)$$

$$= \frac{0.9(0.85)(3)(66)(4)(19 - 4/2)}{12} = 858 \text{ ft-kips}$$

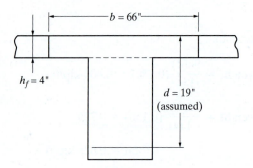

FIGURE 3-10   T-beam for Example 3-4.

6. Because 858 ft-kips > 258 ft-kips, the total effective flange need not be completely utilized in compression (i.e., $a < h_f$), and the T-beam behaves as a wide rectangular beam with a width b of 66 in.

7. Design as a rectangular beam with b and d as known values (see Section 2-15):

$$\text{required } \bar{k} = \frac{M_u}{\phi b d^2} = \frac{258(12)}{0.9(66)(19)^2} = 0.1444 \text{ ksi}$$

8. From Table A-8, select the required steel ratio to provide a $\bar{k}$ of 0.1444 ksi:

$$\text{required } \rho = 0.0025$$

9. Calculate the required steel area:

$$\text{required } A_s = \rho b d$$
$$= 0.0025(66)(19) = 3.14 \text{ in.}^2$$

10. Select the steel bars. Use four No. 8 bars ($A_s = 3.16 \text{ in.}^2$):
    From Table A-3,
$$\text{minimum } b_w = 11 \text{ in.} \qquad \text{(O.K.)}$$

    Check the effective depth d. Assume a No. 3 stirrup and 1½-in. cover, as shown in Figure 3-11.

$$d = 22 - 1.5 - 0.38 - \frac{1.00}{2} = 19.62 \text{ in.}$$

$$19.62 \text{ in.} > 19 \text{ in.} \qquad \text{(O.K.)}$$

11. Check $A_{s,\text{min}}$. From Table A-5,

$$A_{s,\text{min}} = 0.0033 b_w d$$
$$= 0.0033(12)(19.62) = 0.78 \text{ in.}^2$$
$$0.78 \text{ in.}^2 < 3.16 \text{ in.}^2 \qquad \text{(O.K.)}$$

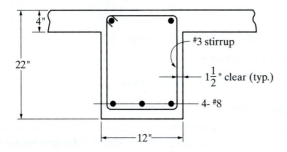

FIGURE 3-11 Design sketch for Example 3-4.

12. Check ϵ_t to ensure a tension-controlled section ($\epsilon_t \geq 0.005$.) From Section 2-10, for a rectangular section:

$$\epsilon_t = \frac{0.00255 f_c' \beta_1}{\rho f_y} - 0.003$$

$$= \frac{0.00255(3)(0.85)}{0.0025(60)} - 0.003 = 0.0404$$

Therefore, the net tensile strain is much larger than 0.005; this is a tension-controlled section, and $\phi = 0.90$, as assumed. (Note that $\epsilon_t > 0.005$ could also be confirmed from Table A-8.)

13. Sketch the design (see Figure 3-11).

Example 3-5 _____

Design a T-beam having a cross section as shown in Figure 3-12. Assume that the effective flange width given is acceptable. The T-beam will carry a total design moment M_u of 340 ft-kips. Use $f_c' = 3000$ psi and $f_y = 60,000$ psi. Use 1½-in. cover and No. 3 stirrups.

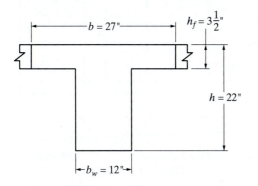

FIGURE 3-12 Sketch for Example 3-5.

Solution:

1. The design moment $M_u = 340$ ft-kips (given).
2. Assume an effective depth of

$$d = 22 - 3 = 19 \text{ in.}$$

3. The effective flange width = 27 in. (given).
4. Assume a tension-controlled section—that is, the net tensile strain $\epsilon_t \geq 0.005$; this assumption will be checked later. The net tensile strain value of 0.005 gives a strength reduction factor $\phi = 0.90$.
5. Determine ϕM_{nf} assuming the effective flange to be in compression over its full depth:

$$\phi M_{nf} = \phi(0.85f_c')bh_f\left(d - \frac{h_f}{2}\right)$$

$$= \frac{0.9(0.85)(3)(27)(3.5)(19 - 3.5/2)}{12} = 312 \text{ ft-kips}$$

6. $\phi M_{nf} < M_u$; therefore, the beam must behave as a true T-beam.
7. Two component couples will be designed, a flange couple (subscript f) and a web couple (subscript w). Refer to Figure 3-13.

   Calculate the required steel area A_{sf} for the flange couple:

$$\text{estimated } d_f = h - 3 \text{ in.} = 22 - 3 = 19 \text{ in.}$$

$$\text{estimated } Z_f = d_f - h_f/2 = 19 - 3.5/2 = 17.25 \text{ in.}$$

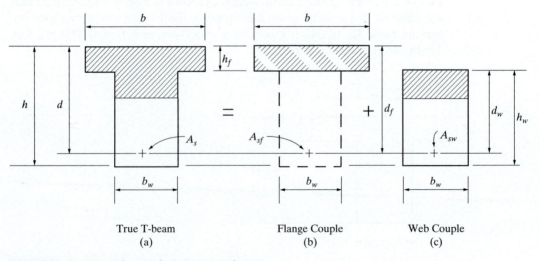

True T-beam   Flange Couple   Web Couple
(a)   (b)   (c)

FIGURE 3-13   True T-beam design, Example 3-5.

$$\text{required } A_{sf} = \frac{\phi M_{nf}}{\phi f_y Z_f}$$

$$= \frac{312(12)}{0.9(60)(17.25)}$$

$$= 4.02 \text{ in.}^2$$

8. The web couple will be designed for the remaining applied moment $(M_u - \phi M_{nf})$. The design is for a rectangular reinforced concrete beam having a depth $h_w = h - h_f$ and a width of b_w.

$$h_w = 22 - 3.5 = 18.5 \text{ in.}$$

$$\text{estimated } d_w = h_w - 3 \text{ in.} = 18.5 - 3 = 15.5 \text{ in.}$$

$$\text{required } \overline{k} = \frac{M_u - \phi M_{nf}}{\phi b_w d_w^2} = \frac{(340 - 312)(12)}{0.9(12)(15.5)^2} = 0.1295 \text{ ksi}$$

From Table A-8, the required $\rho = 0.0023$, from which we calculate

$$\text{required } A_{sw} = \rho b_w d_w = 0.0023(12)(15.5) = 0.43 \text{ in.}^2$$

9. Total required

$$A_s = A_{sf} + A_{sw} = 4.02 + 0.43 = 4.45 \text{ in.}^2$$

10. From Table A-2, select three No. 11 bars. $A_s = 4.68 \text{ in.}^2$ and minimum b_w is 11.0 in. Check d assuming No. 3 stirrups and 1½ in. cover:

$$d = 22 - 1.5 - 0.38 - \frac{1.41}{2} = 19.42 \text{ in.}$$

$$19.42 \text{ in.} > \text{estimated } d_f = 19.0 \text{ in.} \qquad\qquad \text{(O.K.)}$$

11. Check $A_{s,\min}$. From Table A-5,

$$A_{s,\min} = 0.0033 b_w d$$

$$= 0.0033(12)(19.42) = 0.77 \text{ in.}^2$$

$$0.77 \text{ in.}^2 < 4.68 \text{ in.}^2 \qquad\qquad \text{(O.K.)}$$

12. Check ϵ_t to ensure a tension-controlled section $(\epsilon_t \geq 0.005)$:

$$d_t = d = 19.42 \text{ in.}$$

$$N_T = A_s f_y = 4.68(60) = 281 \text{ kips}$$

$$N_{Cf} = 0.85 f_c' b h_f = 0.85(3)(27)(3.5) = 241 \text{ kips}$$

$$a = \frac{N_T - N_{Cf}}{0.85 f_c' b_w} + h_f = \frac{281 - 241}{0.85(3)(12)} + 3.5 = 4.81 \text{ in.}$$

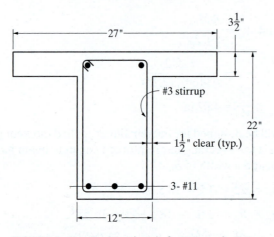

FIGURE 3-14 Design sketch for Example 3-5.

From Section 2-9, Equation (2-1):

$$\epsilon_t = \frac{0.003\beta_1 d_t}{a} - 0.003$$

$$= \frac{0.003(0.85)(19.42)}{4.81} - 0.003 = 0.00730$$

0.00730 > 0.005; therefore, this is a tension-controlled section and $\phi = 0.90$, as assumed. (O.K.)

13. Sketch the design. See Figure 3-14.

3-5 SUMMARY OF PROCEDURE FOR ANALYSIS OF T-BEAMS (FOR MOMENT)

1. Establish the effective flange width based on ACI criteria.
2. Check $A_{s,\text{min}}$. Use Table A-5.
3. To ensure ductility, assume a net tensile strain greater than or equal to 0.004; this assumption will be checked later. Compute the total tension in the steel:

$$N_T = A_s f_y$$

4. Compute the magnitude of the compression that the flange itself is capable of furnishing:

$$N_{Cf} = 0.85 f_c' b h_f$$

5. If $N_T > N_{Cf}$, the beam will behave as a true T-beam and the remaining compression, which equals $N_T - N_{Cf}$, will be furnished by additional web area. If $N_T < N_{Cf}$, the beam will behave as a rectangular beam of width b.

Rectangular T-Beam

6. Solve for the depth of the stress block:

$$a = \frac{A_s f_y}{0.85 f'_c b}$$

7. Check ductility; find ϵ_t.
8. Find ϕ ($0.65 \leq \phi \leq 0.90$).
9. Calculate ϕM_n.

$$\phi M_n = \phi A_s f_y \left(d - \frac{a}{2} \right)$$

Or, in place of steps 6–9, calculate ρ, obtain $\bar{k}$, check ϵ_t, determine ϕ, and use

$$\phi M_n = \phi b d^2 \bar{k}$$

A flow diagram of this procedure is presented in Appendix B.

True T-Beam

6. Determine the depth of the compressive stress block:

$$a = \frac{N_T - N_{Cf}}{0.85 f'_c b_w} + h_f$$

7. Check ductility; find ϵ_t.
8. Find ϕ ($0.65 \leq \phi \leq 0.90$).
9. a. Locate the centroid of the total compressive area referenced to the top of the flange using the relationship

$$\bar{y} = \frac{\Sigma(Ay)}{\Sigma A}$$

    from which

$$Z = d - \bar{y}$$

    Compute the practical moment strength ϕM_n:

$$\phi M_n = \phi N_C Z \quad \text{or} \quad \phi N_T Z$$

    or

    b. Calculate ϕM_n using a summation of internal couples contributed by the flange and the web:

$$\phi M_n = \phi \left\{ N_{cf} \left(d - \frac{h_f}{2} \right) + (N_T - N_{Cf}) \left[d - h_f - \left(\frac{a - h_f}{2} \right) \right] \right\}$$

3-6   SUMMARY OF PROCEDURE FOR DESIGN OF T-BEAMS (FOR MOMENT)

1. Compute the design moment M_u.
2. Assume that the effective depth $d = h - 3$ in.
3. Establish the effective flange width based on ACI criteria.
4. Assume a net tensile strain $\epsilon_t \geq 0.005$; this will give a strength reduction factor $\phi = 0.90$. This assumption will be checked later.
5. Compute the practical moment strength ϕM_{nf} assuming that the total effective flange is in compression:

$$\phi M_{nf} = \phi(0.85f_c')bh_f\left(d - \frac{h_f}{2}\right)$$

6. If $\phi M_{nf} > M_u$, the beam will behave as a rectangular T-beam of width b. If $\phi M_{nf} < M_u$, the beam will behave as a true T-beam.

Rectangular T-Beam

7. Design as a rectangular beam with b and d as known values. Compute the required $\bar{k}$:

$$\text{required } \bar{k} = \frac{M_u}{\phi bd^2}$$

8. From the tables in Appendix A, determine the required ρ for the required $\bar{k}$ of step 7.
9. Compute the required A_s:

$$\text{required } A_s = \rho bd$$

10. Select bars and check the beam width. Check the actual d and compare it with the assumed d. If the actual d is slightly in excess of the assumed d, the design will be slightly conservative (on the safe side). If the actual d is less than the assumed d, the design may be on the nonconservative side (depending on the steel provided) and should be more closely investigated for possible revision.
11. Check $A_{s,min}$. Use Table A-5.
12. Check ductility. Find ϵ_t. If $d = d_t$, Tables A-7 through A-11 can be used, or use Equation (2-2) from Section 2-10. If $d \neq d_t$, use basic principles or Equation (2-1) from Section 2-9. Check the assumed value of ϕ.
13. Sketch the design.

True T-Beam

7. Using an estimated $d_f = h - 3''$ and $Z_f = d_f - h_f/2$, determine the steel area A_{sf} required for the flange couple:

$$\text{required } A_{sf} = \frac{\phi M_{nf}}{\phi f_y Z_f}$$

8. Design the web couple as a rectangular reinforced concrete beam having a total depth $h_w = h - h_f$, using an estimated $d_w = h_w - 3''$ and a beam width of b_w. Design for an applied moment of $M_u - \phi M_{nf}$. Determine required $\bar{k}$, required ρ, and required A_{sw}.

9. Total required $A_s = A_{sf} + A_{sw}$.

10. Select the bars. (Bars must fit into beam width b_w.) Check d as in step 10 of the rectangular T-beam design.

11. Check $A_{s,\min}$. Use Table A-5.

12. Check ductility. Find ϵ_t. Determine the stress block depth a and use the expression for ϵ_t at the end of Section 2-9.

13. Sketch the design.

Flowcharts of these procedures are presented in Appendix B.

3-7   DOUBLY REINFORCED BEAMS: INTRODUCTION

As indicated in Chapter 2, the practical moment strength of a rectangular reinforced concrete beam reinforced with only tensile steel may be determined by the expression $\phi M_n = \phi b d^2 \bar{k}$, where $\bar{k}$ is a function of the steel ratio ρ, f'_c, and f_y.

If we assume a given rectangular section with tension-only reinforcing that is tension-controlled, the upper limit of tension steel area can be established using the reinforcement ratio ρ associated with net tensile strain ϵ_t of 0.005. The maximum practical moment strength ϕM_n for the section may then be calculated using the associated value of $\bar{k}$.

Occasionally, practical and architectural considerations may dictate and limit beam sizes, whereby it becomes necessary to develop more moment strength from a given cross section. When this situation occurs, the ACI Code, Section 10.3.5.1, permits the addition of tensile steel over and above the code maximum provided that compression steel is also added in the compression zone of the cross section. The result constitutes a combined tensile and compressive reinforced beam commonly called a *doubly reinforced beam*.

Where beams span more than two supports (continuous construction), practical considerations are sometimes the reason for the existence of main steel in compression zones. In Figure 3-15, positive moments exist at A and C; therefore, the main tensile reinforcement would be placed in the bottom of the beam. At B, however, a negative moment exists and the bottom of the beam is in compression. The tensile reinforcement must be placed near the top of the beam. It is general practice that at least some of the tension steel in each of these cases will be extended the length of the beam and will pass through compression zones. In this case the compression steel may sometimes be used for additional strength. In Chapter 7 we will see that compression reinforcement aids significantly in reducing long-term deflections. In fact, the use of compression steel to increase the bending strength of a reinforced concrete beam is an inefficient way to utilize steel. More commonly, deflection control will be the reason for the presence of compression steel.

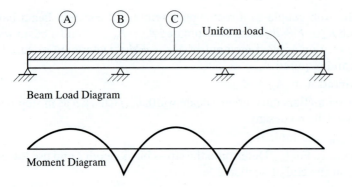

Beam Load Diagram

Moment Diagram

FIGURE 3-15   Continuous beam.

3-8   DOUBLY REINFORCED BEAM ANALYSIS FOR MOMENT (CONDITION I)

The basic assumptions for the analysis of doubly reinforced beams are similar to those for tensile reinforced beams. One additional significant assumption is that the *compression steel stress* (f_s') is a function of the strain at the level of the centroid of the compression steel. As discussed previously, the steel will behave elastically up to the point where the strain exceeds the yield strain ϵ_y. In other words, as a limit, $f_s' = f_y$ when the compression steel strain $\epsilon_s' \geq \epsilon_y$. If $\epsilon_s' < \epsilon_y$, the compression steel stress $f_s' = \epsilon_s' E_s$, where E_s is the modulus of elasticity of the steel.

With two different materials, concrete and steel, resisting the compressive force N_C, the total compression will now consist of two forces: N_{C1}, the compression resisted by the concrete, and N_{C2}, the compression resisted by the compressive steel. For analysis, the total resisting moment of the beam will be assumed to consist of two parts or two internal couples: the part due to the resistance of the compressive concrete and tensile steel, and the part due to the compressive steel and additional tensile steel. The two internal couples are illustrated in Figure 3-16.

Notation for doubly reinforced beams is as follows:

A_s' = total compression steel cross-sectional area

d = effective depth of tension steel

d' = depth to centroid of compression steel from compression face of beam

A_{s1} = amount of tension steel used by the concrete–steel couple

A_{s2} = amount of tension steel used by the steel–steel couple

A_s = total tension steel cross-sectional area $(A_s = A_{s1} + A_{s2})$

M_{n1} = nominal moment strength of the concrete–steel couple

M_{n2} = nominal moment strength of the steel–steel couple

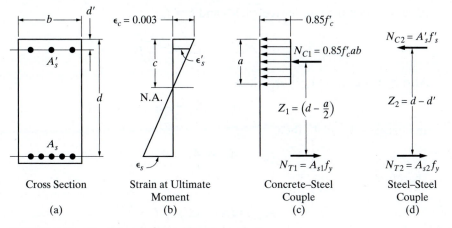

Cross Section	Strain at Ultimate Moment	Concrete–Steel Couple	Steel–Steel Couple
(a)	(b)	(c)	(d)

FIGURE 3-16 Doubly reinforced beam analysis.

M_n = nominal moment strength of the (doubly reinforced) beam

ϵ_s = unit strain at the centroid of the tension steel

ϵ_s' = unit strain at the centroid of the compression steel

The total nominal moment strength may be developed as the sum of the two internal couples, neglecting the concrete displaced by the compression steel.

The strength of the steel–steel couple is evaluated as follows:

$$M_{n2} = N_{T2}Z_2$$

Assuming that $f_s = f_y$ (tensile steel yields),

$$M_{n2} = A_{s2}f_y(d - d')$$

Also, because $\Sigma H_F = 0$ and $N_{C2} = N_{T2}$,

$$A_s'f_s' = A_{s2}f_y$$

If we assume that the compression steel yields and that $f_s' = f_y$, then

$$A_s'f_y = A_{s2}f_y$$

from which

$$A_s' = A_{s2}$$

Therefore

$$M_{n2} = A_s'f_y(d - d')$$

The strength of the concrete–steel couple is evaluated as follows:

$$M_{n1} = N_{T1}Z_1$$

Assuming that the tensile steel yields and $f_s = f_y$,

$$M_{n1} = A_{s1}f_y\left(d - \frac{a}{2}\right)$$

Also, because $A_s = A_{s1} + A_{s2}$, then

$$A_{s1} = A_s - A_{s2}$$

and because $A_{s2} = A_s'$, then

$$A_{s1} = A_s - A_s'$$

Therefore

$$M_{n1} = (A_s - A_s')f_y\left(d - \frac{a}{2}\right)$$

Summing the two couples, we arrive at the nominal moment strength of a doubly reinforced beam:

$$M_n = M_{n1} + M_{n2}$$

$$= (A_s - A_s')f_y\left(d - \frac{a}{2}\right) + A_s'f_y(d - d')$$

The practical moment strength of ϕM_n may then be calculated.

The foregoing expressions are based on the assumption that both tension and compression steels yield prior to concrete strain reaching 0.003. This may be checked by determining the strains that exist at the nominal moment, which depend on the location of the neutral axis. The neutral axis may be located, as previously, by the depth of the compressive stress block and the relationship $a = \beta_1 c$. Thus

$$N_T = N_{C1} + N_{C2}$$

$$A_s f_y = (0.85f_c')ab + A_s'f_y$$

$$a = \frac{(A_s - A_s')f_y}{(0.85f_c')b}$$

which may also be expressed as

$$a = \frac{A_{s1}f_y}{(0.85f_c')b}$$

With the calculation of the a distance, the neutral axis location c may be determined and the assumptions checked.

Check the net tensile strain ϵ_t in the extreme tensile reinforcement and ensure that $\epsilon_t \geq 0.004$ to satisfy ACI 318-05, Section 10.3.5.

$$\epsilon_t = 0.003\frac{(d_t - c)}{c}$$

where d_t = the distance of the extreme tensile reinforcement from the compression face. The corresponding strenth-reduction factor is then calculated as described in Section 2-9.

The case where both tensile and compressive steels yield prior to the concrete strain reaching 0.003 will be categorized as *condition I* (see Example 3-6). The case where the tensile steel yields but the compressive steel does not yield prior to the concrete strain reaching 0.003 is categorized as *condition II* (see Example 3-7). The maximum area of steel in the beam permitted by ACI Code is that area of steel that would result in a net tensile strain of 0.004.

Example 3-6 _____

Compute the practical moment strength ϕM_n for a beam having a cross section as shown in Figure 3-17. Use $f_c' = 3000$ psi and $f_y = 60,000$ psi.

Solution:

1. Assume that all the steel yields:

$$f_s' = f_y \quad \text{and} \quad f_s = f_y$$

Therefore $A_{s2} = A_s'$ (see Figure 3-16d).

2. $$A_s = A_{s1} + A_{s2}$$

$$A_{s1} = A_s - A_s'$$

$$= 6.00 - 2.54 = 3.46 \text{ in.}^2$$

From the concrete–steel couple, the stress block depth can be found:

$$a = \frac{A_{s1}f_y}{(0.85f_c')b} = \frac{3.46(60)}{0.85(3.0)(11)} = 7.40 \text{ in.}$$

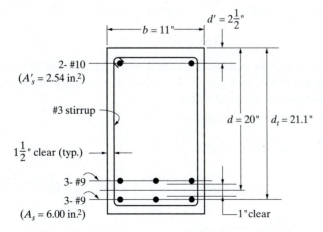

FIGURE 3-17 Sketch for Example 3-6.

3.   Assuming that the same relationship ($a = \beta_1 c$) exists between the depth of the stress block and the beam's neutral axis as existed in singly reinforced beams, the neutral axis may now be located for purposes of checking the steel strains. From Figure 3-16 at the nominal moment,

$$c = \frac{a}{\beta_1} = \frac{a}{0.85} = \frac{7.40}{0.85} = 8.71 \text{ in.}$$

This value of c is based on the assumption in step 1 and will be verified in step 4.

4.   Check the strains to determine whether the assumptions are valid and that both steels yield before the concrete crushes (see Figure 3-16b). The strains calculated exist at the nominal moment:

$$\epsilon'_s = \frac{c - d'}{c}(0.003)$$

$$= \frac{8.71 - 2.5}{8.71}(0.003) = 0.00214$$

Therefore, $\epsilon'_s > \epsilon_y = 0.00207$ (from Table A-1).

Check the ductility of the beam by calculating the net tensile strain in the extreme tensile reinforcement:

$$\epsilon_t = 0.003\frac{(d_t - c)}{c} = 0.003\frac{(21.2 - 8.71)}{8.71} = 0.0043$$

$$0.0043 > 0.004 \qquad \text{(O.K.)}$$

Therefore, the tensile steel yields, and ductility is assured. Because $0.004 \le \epsilon_t \le 0.005$, this is a transition section, and the strength reduction factor ϕ, is calculated as follows:

$$\phi = 0.65 + (\epsilon_t - 0.002)\left(\frac{250}{3}\right) = 0.65 + (0.0043 - 0.002)\left(\frac{250}{3}\right) = 0.84$$

$$0.65 < 0.84 < 0.90 \qquad \text{(O.K.)}$$

Because $\epsilon'_s > \epsilon_y$, the compression steel will yield before the concrete strain reaches 0.003, and $f'_s = f_y$. Therefore, the assumption concerning the compression steel stress is O.K.

5.   From the concrete–steel couple:

$$M_{n1} = A_{s1}f_y\left(d - \frac{a}{2}\right)$$

$$= 3.46(60)\left(20 - \frac{7.40}{2}\right) = 3384 \text{ in.-kips}$$

$$\frac{3384}{12} = 282 \text{ ft-kips}$$

From the steel–steel couple:

$$M_{n2} = A'_s f_y (d - d') = 2.54(60)(20 - 2.5)$$

$$= 2667 \text{ in-kips}$$

$$\frac{2667}{12} = 222 \text{ ft-kips}$$

$$M_n = M_{n1} + M_{n2}$$

$$= 282 + 222 = 504 \text{ ft-kips}$$

6.
$$\phi M_n = 0.84(504)$$

$$= 423 \text{ ft-kips}$$

In this example, because both the compressive steel and tensile steel yield prior to the concrete reaching a compressive strain of 0.003 and because the net tensile strain ϵ_t is greater than 0.004, a ductile failure mode is assured.

3-9   DOUBLY REINFORCED BEAM ANALYSIS FOR MOMENT (CONDITION II)

As has been pointed out, usually the compression steel (A'_s) will reach its yield stress before the concrete reaches a strain of 0.003. This may not occur in shallow beams reinforced with the higher-strength steels, however. Referring to Figure 3-16b, if the neutral axis is located relatively high in the cross section, it is possible that $\epsilon'_s < \epsilon_y$ at the nominal moment. The magnitude of ϵ'_s (and therefore f'_s) depends on the location of the neutral axis. The depth of the compressive stress block a also depends on c, because $a = \beta_1 c$.

The total compressive force must be equal to the total tensile force $A_s f_y$, and an equilibrium equation can be written to solve for the exact required value of c. This turns out to be a quadratic equation. This situation and its solution may be observed in the following example.

Example 3-7 _____

Compute the practical moment strength ϕM_n for a beam having a cross section as shown in Figure 3-18. Use $f'_c = 5000$ psi and $f_y = 60,000$ psi.

Solution:

1.   Assume that all the steel yields. This results in

$$A_{s2} = A'_s$$

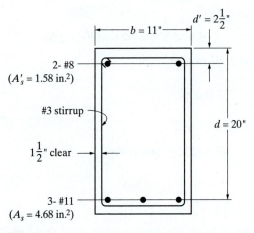

FIGURE 3-18   Sketch for Example 3-7.

2.   With reference to Figure 3-19b,

$$a = \frac{(A_s - A'_s)f_y}{(0.85f'_c)b}$$

$$= \frac{3.10(60)}{0.85(5)(11)}$$

$$= 3.98 \text{ in.}$$

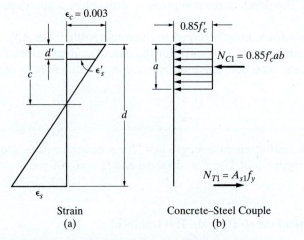

FIGURE 3-19   Compatibility check, Example 3-7.

3.   Locate the neutral axis:

$$a = \beta_1 c, \quad \beta_1 = 0.80 \quad \text{(reference, ACI Code, Section 10.2.7.3)}$$

$$c = \frac{a}{\beta_1} = \frac{3.98}{0.80} = 4.98 \text{ in.}$$

This value of c is based on the assumption in step 1 and will be verified in step 4.

4.   By similar triangles of Figure 3-19a, check the steel strains:

$$\text{compressive steel: } \epsilon'_s = \frac{0.003(c - d')}{c}$$

$$\epsilon'_s = \frac{2.48}{4.98}(0.003) = 0.0015$$

Calculate the net tensile strain in the extreme reinforcement based on the depth of the neutral axis obtained in step 3. Note that $d_t = d$ and, therefore, $\epsilon_s = \epsilon_t$.

$$\epsilon_t = 0.003\frac{(d_t - c)}{c} = 0.003\frac{(20 - 4.98)}{4.98} = 0.009$$

$$0.009 > 0.004 \qquad\qquad\qquad\qquad \text{(O.K.)}$$

For grade 60 steel, $\epsilon_y = 0.00207$ (from Table A-1). Because $\epsilon_s > \epsilon_y < \epsilon'_s$, the tensile steel *has* yielded and the compression steel *has not* yielded. Therefore the assumptions of step 1 are incorrect.

5.   With the original assumptions incorrect, a solution for the location of the neutral axis must be established. With reference to Figure 3-16, c will be determined by using the condition that horizontal equilibrium exists. That is, $\Sigma H_F = 0$. Thus

$$N_T = N_{C1} + N_{C2}$$

$$A_s f_y = (0.85f'_c)ba + f'_s A'_s$$

But

$$a = \beta_1 c$$

and

$$f'_s = \epsilon'_e E_s = \left[\frac{c - d'}{c}(0.003)\right]E_s$$

Then, by substitution,

$$A_s f_y = (0.85f'_c)b\beta_1 c + \left[\frac{c - d'}{c}(0.003)\right]E_s A'_s$$

Multiplying by c and expanding, we obtain

$$A_s f_y c = (0.85 f_c')b\beta_1 c^2 + c(0.003)E_s A_s' - d'(0.003)E_s A_s'$$

Rearranging yields

$$(0.85 f_c' b\beta_1)c^2 + (0.003 E_s A_s' - A_s f_y)c - d'(0.003)E_s A_s' = 0$$

With $E_s = 29{,}000$ ksi, the expression becomes

$$(0.85 f_c' b\beta_1)c^2 + (87 A_s' - A_s f_y)c - 87 d' A_s' = 0$$

where

$A_s = 4.68$ in.^{2}

$f_y = 60$ ksi

$f_c' = 5$ ksi

$b = 11$ in.

$\beta_1 = 0.80$

$A_s' = 1.58$ in.^{2}

$d' = 2.5$ in.

Substitution yields

$$[0.85(5)(11)(0.80)]c^2 + [87(1.58) - 4.68(60)]c - 87(2.5)(1.58) = 0$$

$$37.4c^2 - 143.34c - 343.65 = 0$$

$$c^2 - 3.83c - 9.19 = 0$$

This may be solved using the usual formula for the roots of a quadratic equation:

$$\frac{-b \pm \sqrt{b^2 - 4ac}}{2a}$$

where the coefficients are

$$a = 1.0$$
$$b = -3.83$$
$$c = -9.19$$

Or the square may be completed as follows:

$$c^2 - 3.83c = 9.19$$

$$c^2 - 3.83c + \left(\frac{-3.83}{2}\right)^2 = 9.19 + \left(\frac{-3.83}{2}\right)^2$$

$$c^2 - 3.83c + 3.67 = 9.19 + 3.67 = 12.86$$

$$(c - 1.92)^2 = 12.86$$

$$c - 1.92 = \sqrt{12.86} = 3.59$$

$$c = 5.51 \text{ in.}$$

The solution of the quadratic equation for c may be simplified as follows:

$$c = \pm\sqrt{Q + R^2} - R$$

where

$$R = \frac{87A_s' - A_s f_y}{1.7f_c'b\beta_1}$$

$$Q = \frac{87d'A_s'}{0.85f_c'b\beta_1}$$

Note that basic units are kips and inches, so the value of f_y, for instance, must be in ksi, not in psi.

6. With this value of c, all the remaining unknowns may be found:

$$f_s' = \frac{c - d'}{c}(87) = \frac{5.51 - 2.50}{5.51}(87)$$

$$= 47.5 \text{ ksi} < 60 \text{ ksi} \quad \text{(as expected)}$$

7. $$a = \beta_1 c = 0.80(5.51) = 4.41 \text{ in.}$$

The actual net tensile strain is calculated as

$$\epsilon_t = 0.003\frac{(d_t - c)}{c} = 0.003\frac{(20 - 5.51)}{5.51} = 0.0079$$

$$0.0079 > 0.004 \qquad\qquad \text{(O.K.)}$$

Thus, the beam is ductile as assumed and because $\epsilon_t > 0.005$, the strength-reduction factor $\phi = 0.90$, as discussed in Section 2-9.

8. $$N_{C1} = (0.85f_c')ab = 0.85(5)(4.41)(11.0) = 206.2 \text{ kips} \quad \text{(O.K.)}$$

$$N_{C2} = A_s'f_s' = 47.5(1.58) = 75.1 \text{ kips}$$

$$N_C = 281.3 \text{ kips}$$

$$\text{Check: } N_T = A_s f_y = 4.68(60) = 281 \text{ kips}$$

$$N_T \approx N_C$$

9. $$M_{n1} = N_{C1}Z_1 = N_{C1}\left(d - \frac{a}{2}\right) = 206.2\left(20 - \frac{4.41}{2}\right) = 3670 \text{ in.-kips}$$

$$M_{n2} = N_{C2}Z_2 = N_{C2}(d - d') = 75.1(20 - 2.5) = 1314 \text{ in.-kips}$$

$$M_n = 4984 \text{ in.-kips} = 415.3 \text{ ft-kips}$$

10. $$\phi M_n = 0.9(415.3)$$

$$= 373.8 \text{ ft-kips}$$

3-10 SUMMARY OF PROCEDURE FOR ANALYSIS OF DOUBLY REINFORCED BEAMS (FOR MOMENT)

1. Assume that all the steel yields, $f_s = f_s' = f_y$. Therefore

$$A_{s2} = A_s'$$

2. Using the concrete–steel couple and $A_{s1} = A_s - A_s'$, compute the depth of the compression stress block:

$$a = \frac{A_{s1}f_y}{(0.85f_c')b} = \frac{(A_s - A_s')f_y}{(0.85f_c')b}$$

3. Compute the location of the neutral axis:

$$c = \frac{a}{\beta_1}$$

This value of c is based on the assumption in step 1 and will be verified in step 4.

4. Using the strain diagram, check the strain in the compression reinforcement and the net tensile strain in the extreme tensile reinforcement to determine whether the assumption in step 1 is valid:

$$\epsilon_s' = \frac{0.003(c - d')}{c}$$

$$\epsilon_t = \frac{0.003(d_t - c)}{c}$$

It is required that $\epsilon_t \geq 0.004$. Therefore, the tensile steel has yielded ($0.004 > \epsilon_y$). The following two conditions may exist. In each of the two cases, the strength-reduction factor ϕ must be determined as discussed in Section 2-9.

   a. Condition I: $\epsilon_s' \geq \epsilon_y$. This indicates that the assumption of step 1 is correct and the compression steel has yielded.

   b. Condition II: $\epsilon_s' < \epsilon_y$. This indicates that the assumption of step 1 is incorrect and the compression steel has not yielded.

Condition I

5. If ϵ_s' and ϵ_s both exceed ϵ_y, compute the nominal moment strengths M_{n1} and M_{n2}. For a steel–steel couple:

$$M_{n2} = A_s'f_y(d - d')$$

For a concrete–steel couple:

$$M_{n1} = A_{s1}f_y\left(d - \frac{a}{2}\right)$$

and

$$M_n = M_{n1} + M_{n2}$$

6. Practical moment strength $= \phi M_n$.

Condition II

5. If ϵ_s' is less than ϵ_y and $\epsilon_s \geq \epsilon_y$, compute c using the following formula:

$$(0.85 f_c' b \beta_1) c^2 + (87 A_s' - A_s f_y) c - 87 d' A_s' = 0$$

and solve the quadratic equation for c, or use the simplified formula approach from Example 3-7, step 5. Note that the basic units are kips and inches.

6. Compute the compressive steel stress (to be less than f_y):

$$f_s' = \frac{c - d'}{c}(87)$$

7. Solve for a using

$$a = \beta_1 c$$

To check ductility, recalculate the net tensile strain,

$$\epsilon_t = 0.003 \frac{(d_t - c)}{c}$$

The strength-reduction factor ϕ is determined as discussed in Section 2-9.

8. Compute the compressive forces:

$$N_{C1} = (0.85 f_c') ba$$

$$N_{C2} = A_s' f_s'$$

Check these by computing the tensile force:

$$N_T = A_s f_y$$

Note that N_T should equal $N_{C1} + N_{C2}$.

9. Compute the ideal resisting moment strengths of the individual couples:

$$M_{n1} = N_{C1}\left(d - \frac{a}{2}\right)$$

and

$$M_{n2} = N_{C2}(d - d')$$

$$M_n = M_{n1} + M_{n2}$$

10. Practical moment strength $= \phi M_n$.

3-11 DOUBLY REINFORCED BEAM DESIGN FOR MOMENT

If a check shows that a singly reinforced rectangular section is inadequate and the size of the beam cannot be increased, a doubly reinforced section may be designed using a procedure that consists of the separate design of the two component couples such that their summation will result in a beam of the required strength.

Example 3-8

Design a rectangular reinforced concrete beam to carry a design moment M_u of 697 ft-kips. Physical limitations require that $b = 14$ in. and $h = 30$ in. If compression steel is needed, $d' = 3$ in. Use $f'_c = 3000$ psi and $f_y = 60,000$ psi. The beam cross section is shown in Figure 3-20.

Solution:

1. Assume that $d = h - 4 = 26$ in. (because of the probability of two rows of steel).
2. The design moment M_u is given; $M_u = 697$ ft-kips.
3. Determine if a singly reinforced beam will work. From Table A-8, maximum ρ for $\epsilon_t = 0.005$ is 0.01355. This is if $d = d_t$. For $d_t > d$, the maximum ρ for moment-strength calculation can be found by proportion. Assuming a #8 bar, #3 stirrups, and 1½-in. cover,

$$d_t = 30 - 1.5 - 0.38 - \frac{1.00}{2} = 27.6 \text{ in.}$$

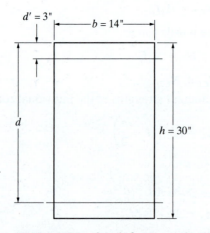

FIGURE 3-20 Sketch for Example 3-8.

Then

$$\rho_{max} = \frac{d_t}{d}(\rho_{max} \text{ from Table A-8})$$

$$= \frac{27.6}{26}(0.01355) = 0.01438$$

The associated $\bar{k}$ (Table A-8) is 0.7177 ksi. Assuming a tension-controlled section ($\epsilon_t \geq 0.005$), ϕ will be 0.90. This assumption will be checked later.

$$\text{maximum } \phi M_n = \phi b d^2 \bar{k}$$

$$= \frac{0.90(14)(26^2)(0.7177)}{12} = 509 \text{ ft-kips}$$

4. 509 ft-kips $<$ 697 ft-kips. Therefore, a doubly reinforced beam is required.
5. Provide a concrete–steel couple having ϕM_{n1} of 509 ft-kips. Therefore, $\rho = 0.01438$ and $\bar{k} = 0.7177$ ksi:

$$\text{required } A_{s1} = \rho b d = 0.01438(14)(26) = 5.23 \text{ in.}^2$$

6. The steel–steel couple must be proportioned to have moment strength equal to the remainder of the design moment:

$$\text{required } \phi M_{n2} = M_u - \phi M_{n1} = 697 - 509 = 188 \text{ ft-kips}$$

7. Considering the steel–steel couple, we have

$$\phi M_{n2} = \phi N_{C2}(d - d')$$

$$N_{C2} = \frac{\phi M_{n2}}{\phi(d - d')} = \frac{188(12)}{0.90(26 - 3)} = 109.0 \text{ kips}$$

8. Because $N_{C2} = A_s' f_s'$, compute f_s' using the neutral axis location of the concrete–steel couple and check the strain ϵ_s' in the compression steel (see Figure 3-21).

$$a = \frac{A_{s1} f_y}{0.85 f_c' b} = \frac{5.23(60)}{0.85(3)(14)} = 8.79 \text{ in.}$$

$$c = \frac{a}{\beta_1} = \frac{8.79}{0.85} = 10.34 \text{ in.}$$

$$\epsilon_s' = \frac{(c - d')}{c}(0.003) = \frac{(10.34 - 3.00)}{10.34}(0.003) = 0.00213$$

$\epsilon_y = 0.00207$ from Table A-1. Because $\epsilon_s' > \epsilon_y$, the compression steel will yield before the concrete strain reaches 0.003, and $f_s' = f_y$.

9. Determine the required compression steel:

$$\text{required } A_s' = \frac{N_{C2}}{f_s'} = \frac{N_{C2}}{f_y} = \frac{109.0}{60} = 1.82 \text{ in.}^2$$

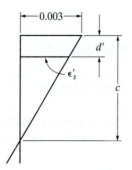

FIGURE 3-21   Concrete strain diagram for Example 3-8.

10. Because $f_s' = f_y$, required A_{S2} = required A_s' = 1.82 in.^{2}
11. The total required tension steel is

$$A_s = A_{s1} + A_{s2} = 5.23 + 1.82 = 7.05 \text{ in.}^2$$

12. Select the compression steel. Two No. 9 bars will provide $A_s' = 2.00$ in.^{2}
13. Select the tension steel. Six No. 10 bars will provide $A_s = 7.62$ in.^{2} Place the tension steel in two layers of three bars each with 1 in. clear between layers. Minimum beam width for three #10 bars is 10.5 in., from Table A-3.
14. Assume No. 3 stirrups and determine the actual depth to the centroid of the bar group by considering the total depth (30 in.), required cover (1½ in.), stirrup size (No. 3), tension bar size (No. 10), and required 1-in. minimum clear space between layers:

$$\text{actual } d = 30 - 1.5 - 0.38 - 1.27 - 0.5 = 26.35 \text{ in.}$$

The assumed d was 26 in.

$$26.35 \text{ in.} > 26 \text{ in.} \hspace{3cm} \text{(O.K.)}$$

15. Check d_t.

$$d_t = 30 - 1.5 - 0.38 - \frac{1.27}{2} = 27.5 \text{ in.} \approx 27.6 \text{ in.} \hspace{1cm} \text{(Say O.K.)}$$

Recalculating the design is an option. The differences will be very small.

16. Check the assumption of step 3 ($\phi = 0.90$). We will use the cross section designed (see the design sketch: Figure 3-22).

$$A_s = 7.62 \text{ in.}^2, A_s' = 2.00 \text{ in.}^2$$

$$d = 26.4 \text{ in. (step 14)}, d_t = 27.5 \text{ in. (step 15)}$$

$$d' = 1.5 + 0.38 + 1.13/2 = 2.45 \text{ in.}$$

$$\text{Assume } f_s = f_s' = f_y; A_{s2} = A_s'$$

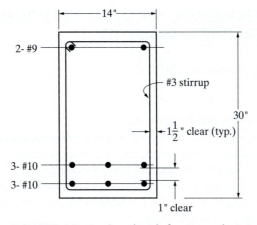

2- #9

#3 stirrup

30"

$1\frac{1}{2}$ " clear (typ.)

3- #10

3- #10

1" clear

14"

FIGURE 3-22 Design sketch for Example 3-8.

$$a = \frac{(A_s - A_s')f_y}{0.85f_c'b} = \frac{(7.62 - 2.00)(60)}{0.85(3)(14)} = 9.45 \text{ in.}$$

$$c = \frac{a}{\beta_1} = \frac{9.45}{0.85} = 11.12 \text{ in.}$$

$$\epsilon_s' = \frac{0.003(c - d')}{c} = \frac{0.003(11.12 - 2.45)}{11.12} = 0.00234 > 0.00207 \ (\text{O.K.})$$

$$\epsilon_t = \frac{0.003(d_t - c)}{c} = \frac{0.003(27.5 - 11.12)}{11.12} = 0.00442 > 0.004$$

Therefore, this is in the transition zone and ϕ must be reduced below 0.90:

$$\phi = 0.65 + (\epsilon_t - 0.002)\left(\frac{250}{3}\right) = 0.65 + (0.00442 - 0.002)\left(\frac{250}{3}\right) = 0.852$$

$$M_{n1} = A_{s1}f_y\left(d - \frac{a}{2}\right) = \frac{(7.62 - 2.00)(60)\left(26.4 - \frac{9.45}{2}\right)}{12} = 609 \text{ ft-kips}$$

$$M_{n2} = A_s'f_y(d - d') = \frac{2.00(60)(26.4 - 2.45)}{12} = 240 \text{ ft-kips}$$

$$\phi M_n = \phi(M_{n1} + M_{n2}) = 0.852(609 + 240) = 723 \text{ ft-kips}$$

$$723 \text{ ft-kips} > 697 \text{ ft-kips} \qquad (\text{O.K.})$$

17. Figure 3-22 is a sketch of the design.

3-12   SUMMARY OF PROCEDURE FOR DESIGN OF DOUBLY REINFORCED BEAMS (FOR MOMENT)

The size of the beam cross section is fixed.

1. Assume that $d = h - 4$ in. Estimate d_t.
2. Establish the total design moment M_u.
3. Check to see if a doubly reinforced beam is necessary. Compute maximum ϕM_n for a singly reinforced beam. Assume a tension-controlled section (net tensile strain $\epsilon_t \geq 0.005$) and use the corresponding strength-reduction factor, $\phi = 0.90$; use maximum $\overline{k}$ from the tables in Appendix A and the corresponding maximum steel ratio ρ_{max} for $\epsilon_t = 0.005$. If $d_t > d$, adjust ρ_{max} for $\epsilon_t = 0.005$ and select maximum $\overline{k}$ accordingly.

$$\text{maximum } \phi M_n = \phi b d^2 \overline{k}$$

4. If $\phi M_n < M_u$, design the beam as a doubly reinforced beam. If $\phi M_n \geq M_u$, the beam can be designed as a beam reinforced with tension steel only.

For a Doubly Reinforced Beam

5. Provide a concrete–steel couple having the maximum ϕM_n from step 3. This is ϕM_{n1}.
    Using ρ from step 3, find the steel required for the concrete–steel couple:

$$\text{required } A_{s1} = \rho b d$$

6. Find the remaining moment that must be resisted by the steel–steel couple:

$$\text{required } \phi M_{n2} = M_u - \phi M_{n1}$$

7. Considering the steel–steel couple, find the required compressive force in the steel (assume that $d' = 3$ in.):

$$N_{C2} = \frac{\phi M_{n2}}{\phi(d - d')}$$

8. Because $N_{C2} = A'_s f'_s$, compute f'_s so that A'_s may eventually be determined. This can be accomplished by using the neutral-axis location of the concrete–steel couple and checking the strain ϵ'_s in the compression steel with ϵ_y from Table A-1. Thus

$$a = \frac{A_{s1} f_y}{(0.85 f'_c) b}$$

$$c = \frac{a}{\beta_1}$$

$$\epsilon'_s = \frac{0.003(c - d')}{c}$$

If $\epsilon'_s \geq \epsilon_y$, the compressive steel has yielded at the nominal moment and $f'_s = f_y$. If $\epsilon'_s < \epsilon_y$, then calculate $f'_s = \epsilon'_s E_s$ and use this stress in the following steps.

9. Because $N_{C2} = A'_s f'_s$,

$$\text{required } A'_s = \frac{N_{C2}}{f'_s}$$

10. Determine the required A_{s2}:

$$A_{s2} = \frac{f'_s A'_s}{f_y}$$

11. Find the total tension steel required:

$$A_s = A_{s1} + A_{s2}$$

12. Select the compressive steel (A'_s).
13. Select the tensile steel (A_s). Check the required beam width. Preferably, place the bars in one layer.
14. Check the actual d and compare it with the assumed d. If the actual d is slightly in excess of the assumed d, the design will be slightly conservative (on the safe side). If the actual d is less than the assumed d, the design may be on the unconservative side and an analysis and possible revision should be considered.
15. Check d_t and compare with the assumed d_t.
16. Check the ϕ value assumption.
17. Sketch the design.

3-13 ADDITIONAL CODE REQUIREMENTS FOR DOUBLY REINFORCED BEAMS

The compression steel in beams, whether it is in place to increase flexural strength or to control deflections, will act similarly to all typical compression members in that it will tend to buckle, as shown in Figure 3-23. Should this buckling occur, it will naturally be accompanied by spalling of the concrete cover. To help guard against this type

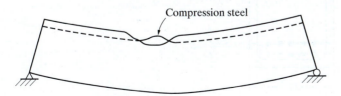

Compression steel

FIGURE 3-23   Possible failure mode for compression steel.

of failure, the ACI Code requires that the compression bars be tied into the beam in a manner similar to that used for reinforced concrete columns (discussed in Chapter 9). Compression reinforcement in beams or girders must be enclosed by ties or stirrups. The size of the ties or stirrups is to be at least No. 3 for No. 10 longitudinal bars or smaller and No. 4 for No. 11 longitudinal bars or larger. The spacing of the ties or stirrups is not to exceed the smaller of 16 longitudinal bar diameters, 48 tie (or stirrup) bar diameters, or the least dimension of the beam. Alternatively, welded wire fabric of equivalent area may be used. The ties or stirrups are to be used throughout the area where compression reinforcement is required (see the ACI Code, Section 7.11).

PROBLEMS

In the following problems, unless otherwise noted, assume No. 3 stirrups, 1½-in. cover for beams, and 1-in. clear space between layers of bars. In all problems, check the net tensile strain ϵ_t in the extreme tension steel to ensure that it is within the allowable limits. Unless otherwise noted or shown, $d_t = d$.

3-1. Find ϕM_n for the following T-beam: $b = 36$ in., $b_w = 12$ in., $h_f = 4$ in., $d = 22$ in., $f'_c = 4000$ psi, and $f_y = 60,000$ psi. The reinforcement is four No. 8 bars.

3-2. Rework Problem 3-1 with $b = 48$ in.

3-3. Rework Problem 3-1 with reinforcement of three No. 11 bars.

3-4. Rework Problem 3-1 with $f'_c = 5000$ psi.

3-5. The simple-span T-beam shown is part of a floor system of span length 20 ft-0 in. and beam spacing 45 in. o.c. Use $b = 45$ in., $f'_c = 3000$ psi and $f_y = 60,000$ psi. The bars are placed with 1-in. clear space between layers.

    (a) Find the practical moment strength ϕM_n.

    (b) How much steel would be required in this beam cross section to make the compressive stress block just completely cover the flange?

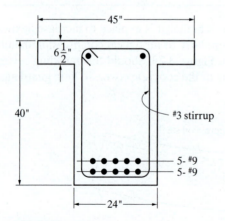

PROBLEM 3-5

3-6. Find ϕM_n for a typical T-beam in the floor system shown. The beams span 24 ft and are spaced 6'-0" on center. Use $f_c' = 3000$ psi and $f_y = 60,000$ psi.

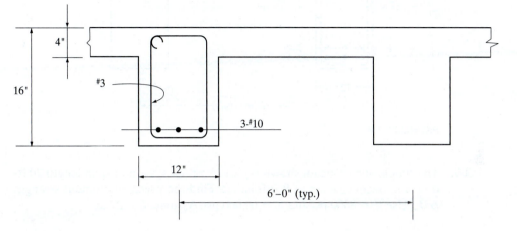

PROBLEM 3-6

3-7. Find the practical moment strength ϕM_n for the T-beam in the floor system shown. The beam span is 31 ft-6 in. Use $f_c' = 4000$ psi and $f_y = 60,000$ psi.

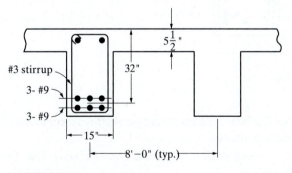

PROBLEM 3-7

3-8. The T-beam shown is on a simple span of 30 ft. Use $f_c' = 3000$ psi and $f_y = 60,000$ psi. No dead load exists other than the weight of the floor system. Assume that the slab design is adequate.

(a) Find the practical moment strength ϕM_n.

(b) Compute the permissible service live load that can be placed on the floor (psf).

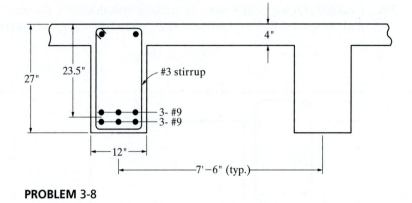

PROBLEM 3-8

3-9. The simple-span T-beam shown is part of a floor system of span length 20 ft-0 in. and beam spacing of 8 ft-0 in. o.c. Find the practical moment strength ϕM_n. Use $f'_c = 3000$ psi and $f_y = 60,000$ psi. Assume $d_t = 27$ in.

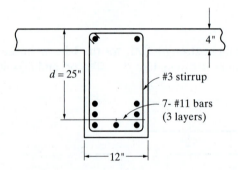

PROBLEM 3-9

3-10. Find ϕM_n for the beams of cross section shown. Assume that the physical dimensions are acceptable.

    (a) Spandrel beam with a flange on one side only. Use $f'_c = 4000$ psi and $f_y = 60,000$ psi.

    (b) Box beam: $f'_c = 3000$ psi and $f_y = 60,000$ psi.

3-11. Determine ϕM_n for the cross section that has a rectangular duct cast in it, as shown. Use $f'_c = 3000$ psi, $f_y = 60,000$ psi, 1-in. clear space between bar layers, $1\frac{1}{2}$-in. cover, and eight No. 7 bars, as shown.

3-12. Design a typical interior tension-reinforced T-beam to resist positive moment. A cross section of the floor system is shown. The service loads are 50 psf dead load (this does not include the weight of the beam and slab) and 325 psf live load. The beam is on a simple span of 18 ft. Use $f'_c = 4000$ psi and $f_y = 60,000$ psi.

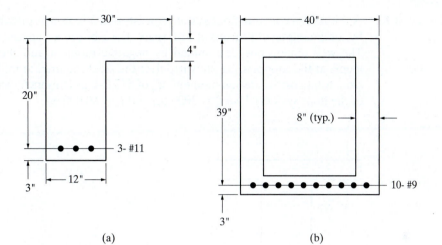

(a)   (b)

PROBLEM 3-10

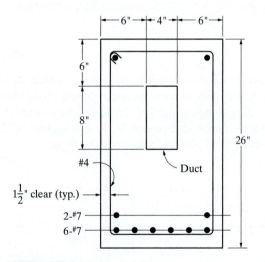

PROBLEM 3-11

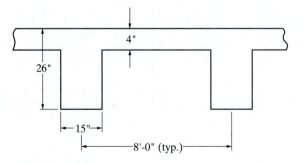

PROBLEM 3-12

3-13. A reinforced concrete floor system consists of a 3-in. concrete slab supported by continuous-span T-beams of 24-ft spans. The T-beams are spaced 4 ft-8 in. o.c. The web dimensions, determined by negative moment and shear requirements at the supports, are shown. Select the steel required at midspan to resist a total positive design moment M_u of 575 ft-kips (this includes the weight of the floor system). Use $f'_c = 3000$ psi and $f_y = 60,000$ psi.

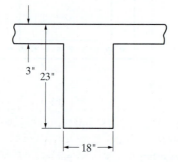

PROBLEM 3-13

3-14. A reinforced concrete floor system is to have a 4-in.-thick slab supported on 16-in.-wide beams, as shown. At one location, penetrations through the floor slab limit the effective flange width for the supporting beam to 20 in. (Use $b = 20$ in.) The positive factored moment M_u at this section is 320 ft-kips. Design the T-beam using $f'_c = 3000$ psi and $f_y = 60,000$ psi.

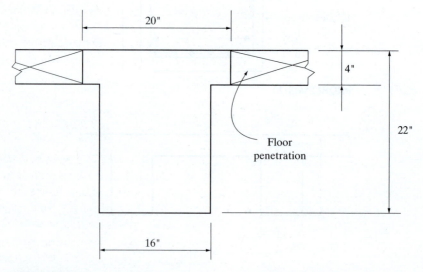

PROBLEM 3-14

3-15. Select steel for the beams of cross section shown. The positive service moments are 160 ft-kips live load and 100 ft-kips dead load (this includes beam weight). Use $f_c' = 3000$ psi and $f_y = 60,000$ psi.

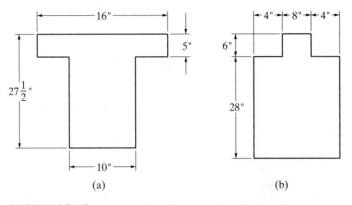

(a)   (b)

PROBLEM 3-15

3-16. Find ϕM_n for the beam of cross section shown. Use $f_c' = 4000$ psi and $f_y = 60,000$ psi.

3-17. The beam of cross section shown is to span 28 ft on simple supports. The uniform load on the beam (in addition to its own weight) will be composed of equal service dead load and live load. Use $f_c' = 4000$ psi and $f_y = 60,000$ psi.

(a) Find the service loads that the beam can carry (in addition to its own weight).

(b) Compare the practical moment strength ϕM_n of the beam as shown with ϕM_n of a beam of similar size reinforced with the area of steel associated with (1) $\epsilon_t = 0.005$ and (2) $\epsilon_t = 0.004$.

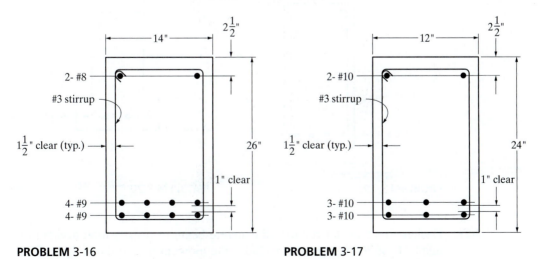

PROBLEM 3-16   **PROBLEM** 3-17

3-18. Find ϕM_n for the beam cross section shown. Use $f_c' = 4000$ psi and $f_y = 60,000$ psi.

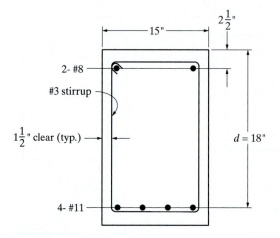

PROBLEM 3-18

3-19. Compute the practical moment strength ϕM_n for the beam of cross section shown. How much can ϕM_n be increased if four No. 8 bars are added to the top of the beam? Use $d' = 2\frac{1}{2}$ in., $f_c' = 3000$ psi, and $f_y = 60,000$ psi.

3-20. Compute the practical moment strength ϕM_n for the simply supported precast inverted T-girder shown. Use $f_c' = 3000$ psi and $f_y = 60,000$ psi.

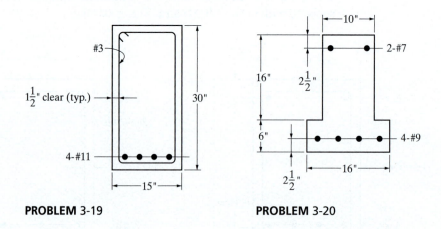

PROBLEM 3-19   **PROBLEM 3-20**

3-21. Design a rectangular reinforced concrete beam to resist a total design moment M_u of 765 ft-kips (this includes the moment due to the weight of the beam). The beam size is limited to 15 in. maximum width and 30 in. maximum overall depth. Use $f_c' = 3000$ psi and $f_y = 60,000$ psi. If compression steel is required, make $d' = 2\frac{1}{2}$ in.

3-22. Design a rectangular reinforced concrete beam to resist service moments of 150 ft-kips dead load (includes moment due to weight of beam) and 160 ft-kips due to live load. Architectural considerations require that width be limited to 11 in. and overall depth be limited to 23 in. Use $f_c' = 3000$ psi and $f_y = 60,000$ psi.

3-23. Design a rectangular reinforced concrete beam to carry service loads of 1.25 kips/ft dead load (includes beam weight) and 2.60 kips/ft live load. The beam is a simple span and has a span length of 18 ft. The overall dimensions are limited to width of 10 in. and overall depth of 20 in. Use $f_c' = 3000$ psi and $f_y = 60,000$ psi.

3-24. Redesign the beam of Problem 3-23 for tension reinforcing only and increased width. Keep an overall depth of 20 in.

3-25. A simply supported precast inverted T-girder having the cross section shown is subjected to a total positive design moment M_u of 280 ft-kips. Select the required reinforcement (both tensile and compressive). Assume a No. 3 stirrup and $1\frac{1}{2}$ in. cover on the tension side of the beam. Use $f_c' = 4000$ psi and $f_y = 60,000$ psi.

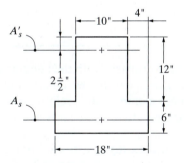

PROBLEM 3-25

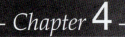

Shear in Beams

4-1   INTRODUCTION

In prior chapters we have been concerned primarily with the bending strength of reinforced concrete beams and slabs. The shear forces create additional tensile stresses that must be considered. In these members steel reinforcing must be added specifically to provide additional shear strength if the shear is in excess of the shear strength of the concrete itself.

The concepts of bending stresses and shearing stresses in homogeneous elastic beams are generally discussed at great length in most strength-of-materials texts. The accepted expressions are

$$f = \frac{Mc}{I} \quad \text{and} \quad v = \frac{VQ}{Ib} \tag{4-1}$$

where f, M, c, and I are as defined in Chapter 1; v is the shear stress; V is the external shear; Q is the statical moment of area about the neutral axis; and b is the width of the cross section.

All points in the length of the beam, where the shear and bending moment are not equal to zero, and at locations other than the extreme fiber or neutral axis, are subject to both shearing stresses and bending stresses. The combination of these stresses is of such a nature that maximum normal and shearing stresses at a point in a beam exist on planes that are inclined with respect to the axis of the beam. It can be shown that maximum and minimum normal stresses exist on two perpendicular planes. These planes are commonly called the *principal planes*, and the stresses that act on them are called *principal stresses*. The principal stresses in a beam subjected to shear and bending may be calculated using the following formula:

$$f_{pr} = \frac{f}{2} \pm \sqrt{\frac{f^2}{4} + v^2} \tag{4-2}$$

where f_{pr} is the principal stress and f and v are the bending and shear stresses, respectively, calculated from Equation (4-1).

The *orientation* of the principal planes may be calculated using the following formula:

$$\tan 2\alpha = \frac{2v}{f} \tag{4-3}$$

where α is the angle measured from the horizontal.

The magnitudes of the shearing stresses and bending stresses vary along the length of the beam and with distance from the neutral axis. It then follows that the inclination of the principal planes as well as the magnitude of the principal stresses will also vary. At the neutral axis the principal stresses will occur at a 45° angle. This may be verified by Equation (4-3), substituting $f = 0$, from which $\tan 2\alpha = \infty$ and $\alpha = 45°$.

In Figure 4-1 we isolate a small, square unit element from the neutral axis of a beam (where $f = 0$). The vertical shear stresses are equal and opposite on the two vertical faces by reason of equilibrium. If these were the only two stresses present, the element would rotate. Therefore there must exist equal and opposite horizontal shear stresses on the horizontal faces and of the same magnitude as the vertical shear stresses. (The concept of horizontal shear stresses equal in magnitude to the vertical shear stresses at any point in a beam can also be found in almost any strength-of-materials text.)

If we consider a set of orthogonal planes that are inclined at 45° with respect to the original element and resolve the shear stresses into components that are parallel and perpendicular to these planes, the effect will be as shown in Figure 4-2. Note

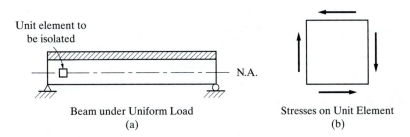

Unit element to
be isolated

N.A.

Beam under Uniform Load
(a)

Stresses on Unit Element
(b)

FIGURE 4-1 Shear stress relationship.

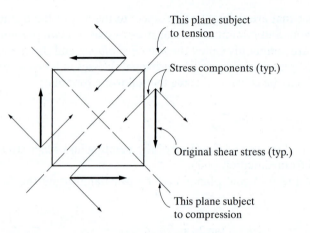

This plane subject
to tension

Stress components (typ.)

Original shear stress (typ.)

This plane subject
to compression

FIGURE 4-2   Effect of shear stresses on inclined planes.

that the components combine so that one of the inclined planes is in compression
while the other is in tension. Concrete is strong in compression but weak in tension,
and there is a tendency for the concrete to crack on the plane subject to tension
should the stress become large enough. The tensile force resulting from the tensile
stress acting on a diagonal plane has historically been designated as *diagonal tension*.
When it becomes large enough, it will necessitate that shear reinforcing be provided.

As stated previously, tensile stresses of various inclinations and magnitudes, resulting from either shear alone or the combined action of shear and bending, exist in
all parts of a beam and must be taken into consideration in both analysis and design.

The preceding discussion is a fairly accurate conceptual description of what occurs in a plain concrete beam. In the beams with which we are concerned, where the
length over which a shear failure could occur (the *shear span*) is in excess of approximately three times the effective depth, the diagonal tension failure would be the
mode of failure in shear. Such a failure is shown in Figure 4-3. For shorter spans, the

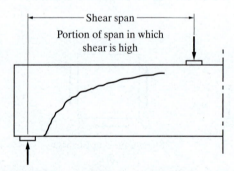

Shear span

Portion of span in which
shear is high

FIGURE 4-3   Typical diagonal tension failure.

failure mode would actually be some combination of shear, crushing, and splitting. For the longer shear spans in plain concrete beams, cracks due to flexural tensile stresses would occur long before cracks due to the diagonal tension. The earlier flexural cracks would initiate the failure, and shear would be of little consequence. In a concrete beam reinforced for flexure (moment) where tensile strength is furnished by steel, however, tensile stresses due to flexure and shear will continue to increase with increasing load. The steel placed in the beam to reinforce for moment is not located where the large diagonal tension stresses (due to shear) occur. The problem then becomes one of furnishing additional reinforcing steel to resist the diagonal tension stresses.

Considerable research over the years has attempted to establish the exact distribution of the shear stresses over the depth of the beam cross section. Despite extensive studies and ongoing research, the precise shear-failure mechanism is still not fully understood.

As with several previous codes, ACI 318-08 furnishes design guidelines for shear reinforcement based on the vertical shear force V_u that develops at any given cross section of a member. Although it is really the diagonal tension for which shear reinforcing must be provided, diagonal tensile forces (or stresses) are not calculated. Historically, vertical shear force (and in older codes, vertical shear stress) has been taken to be a good indicator of the diagonal tension present.

4-2   SHEAR REINFORCEMENT DESIGN REQUIREMENTS

ACI 318-08, Chapter 11, addresses shear and torsion design provisions for both non-prestressed and prestressed concrete members. Our discussion in this chapter is limited to non-prestressed concrete members.

The design of bending members for shear is based on the assumption that the concrete resists part of the shear, and any excess over and above what the concrete is capable of resisting has to be resisted by shear reinforcement. The basic rationale for the design of the shear reinforcement, or *web reinforcement* as it is usually called in beams, is to provide steel to cross the diagonal tension cracks and subsequently keep them from opening. Visualizing this basic rationale with reference to Figure 4-3, it is seen that the web reinforcement may take several forms.

The code allows vertical stirrups and welded wire reinforcement with wires located perpendicular to the axis of the member as well as spirals, circular ties, or hoops.

Additionally, for non-prestressed members, the code allows shear reinforcement to be composed of inclined or diagonal stirrups and main reinforcement bent to act as inclined stirrups.

The most common form of web reinforcement used is the vertical stirrup. The web reinforcement contributes very little to the shear resistance *prior* to the formation of the inclined cracks but appreciably increases the *ultimate* shear strength of a bending member.

For members of normal-weight concrete that are subject to shear and flexure only, the amount of shear force that the concrete alone, unreinforced for shear, can resist is V_c:

$$V_c = 2\lambda\sqrt{f_c'}b_w d \qquad\qquad \text{[ACI Eq. (11-3)]}$$

In the expression for V_c, the terms are as previously defined with units for f_c' in psi and units for b_w and d in inches. Lamda (λ) is described in Chapter 1 and for normal-weight concrete, $\lambda = 1.0$. In this chapter we will consider only normal-weight concrete and therefore λ will be omitted. V_c will be in units of pounds. For rectangular beams, b_w is equivalent to b. The nominal shear strength of the concrete will be reduced to a dependable shear strength by applying a strength-reduction factor ϕ of 0.75 (ACI Code, Section 9.3.2). Should members be subject to other effects of axial tension or compression, other expressions for V_c can be found in Section 11.2 of the code. Also, it is permitted to calculate V_c using a more detailed calculation (ACI 318-08, Section 11.2.2). For most designs, it is convenient and conservative to use ACI Equation (11-3).

The design shear force is denoted V_u and results from the application of factored loads. Values of V_u are most conveniently determined using a typical shear force diagram. Theoretically, no web reinforcement should be required if $V_u \leq \phi V_c$. The code, however, requires that a minimum area of shear reinforcement be provided in all reinforced concrete flexural members where V_u exceeds $\frac{1}{2}\phi V_c$ except as follows:

1. In slabs and footings
2. In concrete joist construction as defined by the ACI Code, Section 8.13
3. In beams with a total depth of not greater than 10 in., $2\frac{1}{2}$ times the flange thickness, or one-half the width of the web, whichever is greater

This provision of the code is primarily to guard against those cases where an unforeseen overload would cause failure of the member due to shear. Tests have shown the shear failure of a flexural member to be sudden and without warning. In cases where it is not practical to provide shear reinforcement (footings and slabs) and sufficient thickness is provided to resist V_u, the minimum area of web reinforcement is not required. In cases where shear reinforcement is required for strength or because $V_u >$ $\frac{1}{2}\phi V_c$, the minimum area of shear reinforcement shall be calculated from

$$A_v = 0.75\sqrt{f_c'}\frac{b_w s}{f_{yt}} \geq \frac{50 b_w s}{f_{yt}} \qquad\qquad \text{[ACI Eq. (11-13)]}$$

In the preceding equation and with reference to Figure 4-4,

A_v = total cross-sectional area of web reinforcement within a distance s; for single-loop stirrups, $A_v = 2A_s$, where A_s is the cross-sectional area of the stirrup bar (in.^{2})

b_w = web width = b for rectangular sections (in.)

s = center-to-center spacing of shear reinforcement in a direction parallel to the longitudinal reinforcement (in.)

f_{yt} = yield strength of web reinforcement steel (psi)

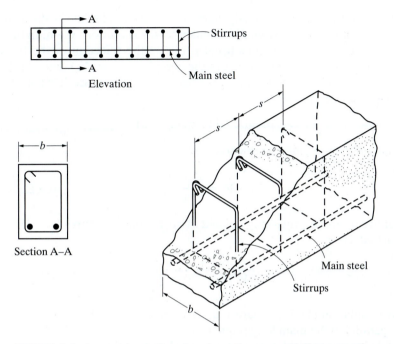

FIGURE 4-4   Isometric section showing stirrups partially exposed.

Note that for $f_c' \leq 4444$ psi, the minimum area of shear reinforcement will be controlled by

$$\frac{50 b_w s}{f_{yt}}$$

When determining the shear strength V_c of reinforced or prestressed concrete beams and concrete joist construction, the value of $\sqrt{f_c'}$ is limited to 100 psi unless minimum web reinforcement is provided.

In any span, that portion in which web reinforcement is theoretically necessary can be determined by using the shear (V_u) diagram. When the applied shear V_u exceeds the capacity of the concrete web ϕV_c, web reinforcement is required. In addition, according to the code, web reinforcement at least equal to the minimum required must be provided elsewhere in the span, where the applied shear is greater than one-half of ϕV_c. The ACI Code, Section 11.1.1, states that the basis for shear design must be

$$\phi V_n \geq V_u \qquad\qquad \text{[ACI Eq. (11-1)]}$$

where

$$V_n = V_c + V_s \qquad\qquad \text{[ACI Eq. (11-2)]}$$

from which

$$\phi V_c + \phi V_s \geq V_u$$

where V_u, ϕ, and V_c are as previously defined; V_n is the total nominal shear strength; and V_s is the nominal shear strength provided by shear reinforcement. In the design process, the design of the stirrups usually follows the selection of the beam size. Therefore, V_c can be determined, as can the complete shear (V_u) diagram. The stirrups to be designed will provide the shear strength V_s. Therefore, it is convenient to write the preceding expression as

$$\text{required } \phi V_s = V_u - \phi V_c$$

For vertical stirrups, V_s may be calculated from

$$V_s = \frac{A_v f_{yt} d}{s} \qquad\qquad \text{[ACI Eq. (11-15)]}$$

where all terms are as previously defined. Also, for inclined stirrups at 45°, V_s may be calculated using

$$V_s = \frac{1.414 A_v f_{yt} d}{s}$$

from ACI Equation (11-16), where s is the horizontal center-to-center distance of stirrups parallel to the main longitudinal steel.

It will be more practical if ACI Equations (11-15) and (11-16) are rearranged as expressions for spacing, because the stirrup bar size, strength, and beam effective depth are usually predetermined. The design is then for stirrup spacing. For vertical stirrups,

$$\text{required } s = \frac{A_v f_{yt} d}{\text{required } V_s}$$

Because it is the required ϕV_s that will be conveniently determined, the preceding expression is rewritten

$$\text{required } s = \frac{\phi A_v f_{yt} d}{\text{required } \phi V_s}$$

or

$$\text{required } s = \frac{\phi A_v f_{yt} d}{V_u - \phi V_c}$$

Similarly, for 45° stirrups,

$$\text{required } s = \frac{1.414 \, \phi A_v f_{yt} d}{V_u - \phi V_c}$$

Note that these equations give *maximum* spacing of stirrups based on *required strength*.

4-3   SHEAR ANALYSIS PROCEDURE

The shear analysis procedure involves checking the shear strength in an existing member and verifying that the various code requirements have been satisfied. The member may be reinforced or plain.

Example 4-1

A reinforced concrete beam of rectangular cross section shown in Figure 4-5 is reinforced for moment only (no shear reinforcement). Beam width $b = 18$ in., $d = 10.25$ in., and the reinforcing is five No. 4 bars. Calculate the maximum factored shear force V_u permitted on the member by the ACI Code. Use $f'_c = 4000$ psi and $f_y = 60,000$ psi.

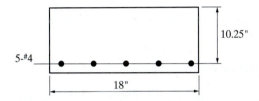

FIGURE 4-5   Cross section, Example 4-1.

Solution:

Because no shear reinforcement is provided, the ACI Code, Section 11.4.6.1, requires that V_u not exceed $0.5\,\phi V_c$:

$$\text{maximum } V_u = 0.5\,\phi V_c$$

$$= 0.5\,\phi(2\sqrt{f'_c}\,b_w d)$$

$$= 0.5(0.75)(2)(\sqrt{4000})(18)(10.25)$$

$$= 8750 \text{ lb}$$

Example 4-2

A reinforced concrete beam of rectangular cross section shown in Figure 4-6 is reinforced with seven No. 6 bars in a single layer. Beam width $b = 18$ in., $d = 33$ in., single-loop No. 3 stirrups are placed 12 in. on center, and typical cover is $1\frac{1}{2}$ in. Find V_c, V_s, and the maximum factored shear force V_u permitted on this member. Use $f'_c = 4000$ psi and $f_y = 60,000$ psi.

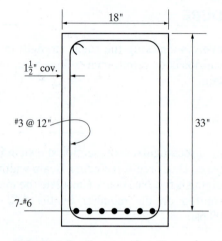

FIGURE 4-6   Cross section, Example 4-2.

Solution:

V_c and V_s will be expressed in units of kips.

$$V_c = 2\sqrt{f'_c}b_w d = \frac{2\sqrt{4000}(18)(33)}{1000} = 75.1\,\text{kips}$$

$$V_s = \frac{A_v f_{yt} d}{s} = \frac{2(0.11)(60)(33)}{12} = 36.3\,\text{kips}$$

$$\text{maximum } V_u = \phi V_c + \phi V_s = 0.75(75.1 + 36.3) = 83.6\,\text{kips}$$

In the general case of shear analysis, one must ensure that at all locations in the member, $\phi V_c + \phi V_s \geq V_u$. In addition, all other details of the reinforcement pattern must be checked to ensure that they comply with code provisions. (Refer to "Notes on Stirrup Design" in Section 4-4.)

4-4   STIRRUP DESIGN PROCEDURE

In the design of stirrups for shear reinforcement, the end result is a determination of stirrup size and spacing pattern. A *general procedure* may be adopted as follows:

1. Determine the shear values based on clear span and draw a shear (V_u) diagram.
2. Determine if stirrups are required.

3. Determine the length of span over which stirrups are required (assuming that stirrups *are* required).
4. On the V_u diagram, determine the area representing "required ϕV_s." This will display the required strength of the stirrups to be provided.
5. Select the size of the stirrup. See item 2a in "Notes on Stirrup Design." Find the spacing required at the critical section (a distance d from the face of the support). See "Notes on Stirrup Design" item 3b.
6. Establish the ACI Code maximum spacing requirements.
7. Determine the spacing requirements based on shear strength to be furnished by web reinforcing.
8. Establish the spacing pattern and show sketches.

Notes on Stirrup Design

1. Materials and maximum stresses
    a. To reduce excessive crack widths in beam webs subject to diagonal tension, the ACI Code, Section 11.4.2, limits the design yield strength of shear reinforcement to 60,000 psi. This increases to 80,000 psi for deformed welded wire reinforcing.
    b. The value of V_s must not exceed $8\sqrt{f_c'}b_w d$ irrespective of the amount of web reinforcement (ACI Code, Section 11.4.7.9).
2. Bar sizes for stirrups
    a. The most common stirrup size used is a No. 3 bar. Under span and loading conditions where the shear values are relatively large, it may be necessary to use a No. 4 bar. Rarely is anything larger than a No. 4 bar stirrup ever required, however. In large beams, multiple stirrup sets are sometimes provided in which a diagonal crack would be crossed by four or more vertical bars at one location of a beam. Single-loop stirrups, as shown in Figure 4-4, are generally satisfactory for $b \leq 24$ in.; double-loop stirrups are satisfactory for 24 in. $< b \leq 48$ in.; and triple-loop stirrups are satisfactory for $b > 48$ in.
    b. When conventional single-loop stirrups are used, the web area A_v provided by each stirrup is twice the cross-sectional area of the bar (No. 3 bars, $A_v = 0.22$ in.^{2}; No. 4 bars, $A_v = 0.40$ in.^{2}) because each stirrup crosses a diagonal crack twice.
    c. If possible, do not vary the stirrup bar sizes; use the same bar sizes unless all other alternatives are not reasonable. Spacing should generally be varied and size held constant.
3. Stirrup spacings
    a. When stirrups are required, the maximum spacing for vertical stirrups must not exceed $d/2$ or 24 in., whichever is smaller (ACI Code, Section 11.4.5.1). If V_s exceeds $4\sqrt{f_c'}b_w d$, the maximum spacing must not exceed $d/4$ or 12 in., whichever is smaller (ACI Code, Section 11.4.5.3). The

maximum spacing may also be governed by ACI Equation (11-13), which gives

$$s_{max} = \frac{A_v f_{yt}}{0.75\sqrt{f_c' b_w}} \leq \frac{A_v f_{yt}}{50 b_w}$$

b.   It is usually undesirable to space vertical stirrups closer than 4 in.

c.   It is generally economical and practical to compute the spacing required at several sections and to place stirrups accordingly in groups of varying spacing. Spacing values should be made to not less than 1-in. increments.

d.   The code permits (Section 11.1.3), when the support reaction introduces a vertical compression into the end region of a member, no concentrated load occurs between the face of the support and distance d from the face of the support, and the beam is loaded at or near the top, that the maximum shear to be considered is that at the section a distance d from the face of the support (except for brackets, short cantilevers, and special isolated conditions). For stirrup design, the section located a distance d from the face of the support will be called the *critical section*. Sections located less than a distance d from the face of the support may be designed for the same V_u as that at the critical section. Therefore, stirrup spacing should be constant from the critical section back to the face of the support based on the spacing requirements *at* the critical section. The first stirrup should be placed at a maximum distance of $s/2$ from the face of the support, where s equals the immediately adjacent required stirrup spacing (a distance of 2 in. is commonly used). For the balance of the span, the stirrup spacing is a function of the shear strength required to be provided by the stirrups or the maximum spacing limitations.

e.   The actual stirrup pattern used in the beam is the designer's choice. The choice will be governed by strength requirements and economy. Many patterns will satisfy the strength requirements. In most cases, the shear decreases from the support to the center of the span, indicating that the stirrup spacing could be continually increased from the critical section up to the maximum spacing allowed by the code. This would create tedious design, detailing, and bar placing operations, but would nevertheless result in the least steel used. This is not warranted economically or within the framework of the philosophy outlined in Appendix B. In the usual uniformly loaded beams, no more than two or three different spacings should be used within a pattern. Longer spans or concentrated loads may warrant more detailed spacing patterns.

Example 4-3 _____

A simply supported rectangular concrete beam shown in Figure 4-7 is 16 in. wide and has an effective depth of 25 in. The beam supports a total factored load (w_u) of 11.5 kips/ft on a clear span of 20 ft. The given load includes the

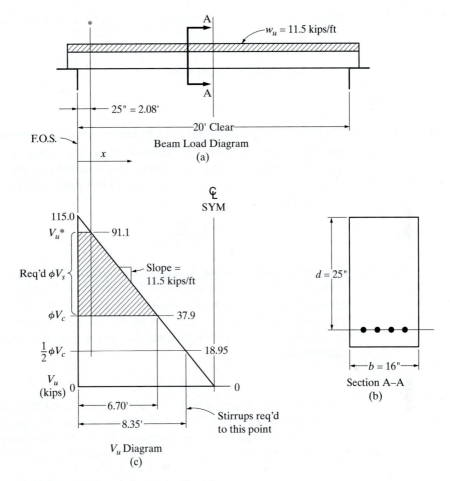

FIGURE 4-7   Sketch for Example 4-3.

weight of the beam. Design the web reinforcement. Use $f'_c = 4000$ psi and $f_y = 60,000$ psi.

Solution:

1.   Draw the shear force (V_u) diagram:

$$\text{maximum } V_u = \frac{w_u \ell}{2} = \frac{11.5(20)}{2} = 115.0 \text{ kips}$$

The quantities at the critical section (25 in. or 2.08 ft from the face of the support) are designated with an asterisk. Therefore

$$V_u^* = 115.0 - 2.08(11.5) = 91.1 \text{ kips}$$

2. Determine if stirrups are required. The ACI Code, Section 11.4.6.1, requires that stirrups be supplied if $V_u > 0.5\phi V_c$. Thus

$$\phi V_c = \phi(2\sqrt{f_c'}b_w d) = 0.75\left[\frac{2\sqrt{4000}(16)(25)}{1000}\right] = 37.9 \text{ kips}$$

$$0.5\phi V_c = 0.5(37.9) = 18.95 \text{ kips}$$

Therefore, because 91.1 kips > 18.95 kips, stirrups are required.

3. Find the length of span over which stirrups are required. As stirrups must be provided to the point where $V_u = 0.5\phi V_c = 18.95$ kips, find where this shear exists on the V_u diagram of Figure 4-7c. From the face of the support,

$$\frac{115.0 - 18.95}{11.5} = 8.35 \text{ ft}$$

Note this location on the V_u diagram as well as the location at 6.70 ft from the face of the support, where $V_u = \phi V_c = 37.9$ kips.

4. Designate as "required ϕV_s" the area enclosed by the ϕV_c line, the V_u^* line, and the sloping V_u line. This shows the required strength of the shear reinforcing at any point along the span and graphically depicts the relationship

$$\phi V_c + \phi V_s \geq V_u$$

At any location, the required ϕV_s can be determined from the diagram as the distance between the V_u^* (or V_u) line and the ϕV_c line (the height of the crosshatched area). For this particular V_u diagram, designating the slope (kips/ft) as m, taking x (ft) from the face of the support, and considering the range $2.08 \text{ ft} \leq x \leq 6.70$ ft, we have

$$\text{required } \phi V_s = \text{maximum } V_u - \phi V_s - mx$$

$$= 115.0 - 37.9 - 11.5x$$

$$= 77.1 - 11.5x$$

5. Assume a No. 3 vertical stirrup ($A_v = 0.22 \text{ in.}^2$) and establish the spacing requirement at the critical section based on the required ϕV_s. At this location the stirrups will be most closely spaced. From ACI Equation (11-15),

$$\text{required } s^* = \frac{\phi A_v f_{yt} d}{\text{required } \phi V_s^*} = \frac{0.75(0.22)(60)(25)}{91.1 - 37.9} = 4.65 \text{ in.}$$

(Note that the denominator in the preceding expression, required ϕV_s^*, is equal to $V_u^* = \phi V_c$.)

We will use a 4-in. spacing. This is the spacing used in the portion of the beam between the face of the support and the critical section, which lies a distance d from the face of the support, and it is based on the amount of shear strength that must be provided by the shear reinforcing. Had the required spacing in this case turned out to be less than 4 in. (see item 3b in "Notes on Stirrup Design" in Section 4-4), a larger bar would have been selected for the stirrups.

6. Establish ACI Code maximum spacing requirements. Recall that if V_s is less than $4\sqrt{f_c'}b_w d$, the maximum spacing is $d/2$ or 24 in., whichever is smaller. Therefore, compare V_s^* at the critical section with $4\sqrt{f_c'}b_w d$:

$$4\sqrt{f_c'}b_w d = \frac{4\sqrt{4000}(16)(25)}{1000} = 101.2 \text{ kips}$$

$$V_s^* = \frac{\phi V_s^*}{\phi} = \frac{91.1 - 37.9}{0.75} = 70.9 \text{ kips}$$

Because 70.9 kips < 101.2 kips, the maximum spacing should be the smaller of $d/2$ or 24 in.:

$$\frac{d}{2} = \frac{25}{2} = 12.5 \text{ in.}$$

$$\text{Use 12 in.}$$

A second criterion is based on the code minimum area requirement (ACI Code, Section 11.4.6.3). ACI Equation (11-13) may be rewritten in the form

$$s_{\max} = \frac{A_v f_{yt}}{0.75\sqrt{f_c'}b_w} \leq \frac{A_v f_{yt}}{50 b_w}$$

where the units of f_{yt} should be carefully noted as psi. Evaluate these two expressions:

$$\frac{A_v f_{yt}}{0.75\sqrt{f_c'}b_w} = \frac{0.22(60,000)}{0.75\sqrt{4000}(16)} = 17.39 \text{ in.}$$

$$\frac{A_v f_{yt}}{50 b_w} = \frac{0.22(60,000)}{50(16)} = 16.50 \text{ in.}$$

Therefore, of these two, 16.50 in. controls.

Of the foregoing two maximum spacing criteria, the smaller value will control. Therefore the 12-in. maximum spacing controls throughout the beam wherever stirrups are required.

7. Determine the spacing requirements based on shear strength to be furnished. At this point we know that the spacing required at the critical section is 4.65 in., that the maximum spacing allowed in this beam is 12 in. where stirrups are required, and that stirrups are required to 8.35 ft from F.O.S.

To establish a spacing pattern for the rest of the beam, the spacing required should be established at various distances from the face of support. This will permit the placing of stirrups in groups, with each group having a different spacing. The number of locations at which the required spacing should be determined is based on judgment and should be a function of the shape of the required ϕV_s portion of the V_u diagram.

To aid in the determination of an acceptable spacing pattern, a plot is developed using the formula

$$\text{required } s = \frac{\phi A_v f_{yt} d}{\text{required } \phi V_s}$$

where the denominator can be determined from the expression given in step 4.

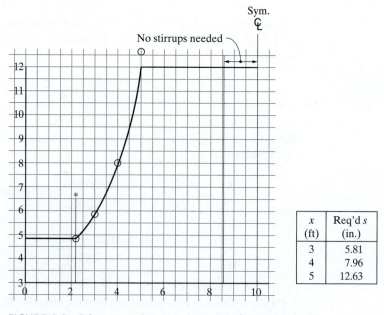

x (ft)	Req'd s (in.)
3	5.81
4	7.96
5	12.63

FIGURE 4-8 Stirrup spacing requirements for Example 4-3.

For plotting purposes, the required spacing will arbitrarily be found at 1-ft intervals beyond the critical section. At 3 ft from the face of the support,

$$\text{required } s = \frac{\phi A_v f_{yt} d}{\text{required } \phi V_s} = \frac{0.75(0.22)(60)(25)}{77.1 - 11.5(3)} = 5.81 \text{ in.}$$

Similarly, required spacing may be found at other points along the beam. The results of these calculations are tabulated and plotted in Figure 4-8. Note that required spacings need not be determined beyond the point where s_{max} (12 in.) has been exceeded.

8. The plot of Figure 4-8 may be readily used to aid in establishing a final pattern. For example, the 5-in. spacing could be started about 2.4 ft from the face of the support (F.O.S.), and the maximum spacing of 12 in. could be started at about 4.9 ft from the face of the support. The first stirrup will be placed away from the face of the support a distance equal to one-half the required spacing at the critical section. The rest of the stirrup pattern may be selected using an approach as shown in the following table.

Spacing (in.)	Theoretical stopping point (from F.O.S.)	Length required to cover (in.)	Number of spaces to use	Actual length covered (in.)	Actual stopping point (inches from F.O.S.)
2	—	—	1	2	2
4	2.4′ = 29″	27	7	28	30
5	4.9′ = 59″	29	6	30	60
12	120″	60	5	60	120

The spacings and theoretical stopping points for those spacings are determined from Figure 4-8. The 4-in. spacing must run from the end of the 2-in. spacing (2 in. from the face of the support) to the theoretical stopping for the 4-in. spacing (29 in.). Therefore, the 4-in. spacings must cover 27 in. For the 4-in. spacing, the required number of spaces is then calculated from $27/4 = 6.75$ spaces; use 7 spaces. This places the last stirrup of the 4-in. spacing group at 30 in. from the face of the support, where the 5-in. spacing will begin. The final pattern for No. 3 single-loop stirrups is: one at 2 in., seven at 4 in., six at 5 in., and five at 12 in. This places the last stirrup in the 12-in. spacing group on the beam centerline—thus, remembering symmetry about the centerline, providing stirrups across the full length of the beam. This is common practice and is conservative. The final stirrup pattern is shown in the design sketch of Figure 4-9.

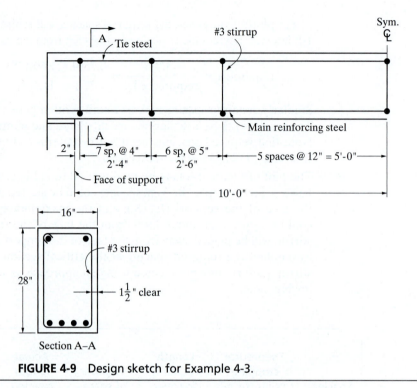

FIGURE 4-9   Design sketch for Example 4-3.

An alternative approach to stirrup spacing is to select a desired spacing (in this case between 4 in. and 12 in.) and to compute the distance from the face of the support at which that spacing may begin. This can be accomplished by rewriting the expression

$$\text{required } s = \frac{\phi A_v f_{yt} d}{\text{required } \phi V_s}$$

$$= \frac{\phi A_v f_{yt} d}{(\max V_u - \phi V_c - mx)}$$

in the form

$$x = \frac{\max V_u - \phi V_c - \dfrac{\phi A_v f_{yt} d}{s}}{m}$$

This expression will furnish the distance x at which any desired stirrup spacing s may commence. Returning to Example 4-3, we can compute where a 5-in. spacing may begin relative to the face of the support,

$$x = \frac{115.0 - 37.9 - \dfrac{0.75(0.22)(60)(25)}{5}}{11.5} = 2.40 \text{ ft}$$

and a 12-in. spacing may begin at

$$x = \frac{115.0 - 37.9 - \dfrac{0.75(0.22)(60)(25)}{12}}{11.5} = 4.91 \text{ ft}$$

Note that Figure 4-8 may be generated equally well with these data.

Example 4-4

A continuous reinforced concrete beam shown in Figure 4-10 is 15-in. wide and has an effective depth of 31 in. The factored loads are shown. (The factored uniform load includes the weight of the beam.) Design the web reinforcement using the V_u diagram shown in Figure 4-11. Use $f'_c = 4000$ psi and $f_y = 60{,}000$ psi.

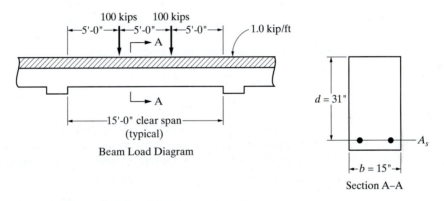

FIGURE 4-10   Sketch for Example 4-4.

Solution:

1. Note that, by reason of symmetry, it is necessary to show only half of the V_u diagram.
2. Determine if stirrups are required:

$$\phi V_c = \phi 2\sqrt{f'_c}\, b_w d = \frac{0.75(2\sqrt{4000})(15)(31)}{1000} = 44.1 \text{ kips}$$

$$0.5\phi V_c = 0.5(44.1) = 22.1 \text{ kips}$$

Therefore, because 104.9 kips > 22.1 kips, stirrups are required.

3. Find the length of span over which stirrups are required. Stirrups are required to the point where

$$V_u = 0.5\phi V_c = 22.1 \text{ kips}$$

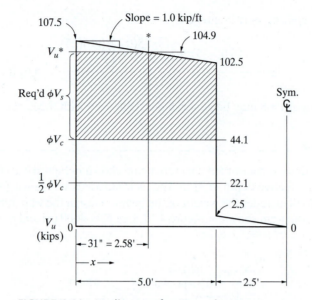

FIGURE 4-11 V_u diagram for Example 4-4.

From Figure 4-11, the point where V_u is equal to 22.1 kips may be determined by inspection to be at the concentrated load, 5 ft-0 in. from the face of the support. No stirrups are required between the two concentrated loads.

4.   Designate "required ϕV_s" on the V_u diagram:

$$\text{required } \phi V_s = \text{maximum } V_u = \phi V_c - mx$$

$$= 107.5 - 44.1 - 1.0x$$

$$= 63.4 - 1.0x \qquad \text{(applies for the range}$$
$$2.58 \text{ ft} \le x \le 5.0 \text{ ft)}$$

5.   Assume a No. 3 vertical stirrup ($A_v = 0.22$ in.^{2}):

$$\text{required } s^* = \frac{\phi A_v f_{yt} d}{\text{required } \phi V_s^*} = \frac{0.75(0.22)(60)(31)}{104.9 - 44.1} = 5.05 \text{ in.}$$

Use 5 in.

6.   Establish ACI Code maximum spacing requirements:

$$4\sqrt{f_c'} b_w = \frac{4\sqrt{4000}(15)(31)}{1000} = 117.6 \text{ kips}$$

$$V_s^* = \frac{f V_s^*}{f} = \frac{104.9 - 44.1}{0.75} = 81.1 \text{ kips}$$

Because 81.1 kips < 117.6 kips, the maximum spacing should be the smaller of $d/2$ or 24 in.:

$$\frac{d}{2} = \frac{31}{2} = 15.5 \, \text{in.}$$

Also check s_{max}. Because $f_c' < 4444 \, \text{psi}$,

$$s_{max} = \frac{A_v f_{yt}}{50 b_w} = \frac{0.22(60{,}000)}{50(15)} = 17.6 \, \text{in.}$$

Therefore, use a maximum spacing of 15 in.

7.   Determine the spacing requirements between the critical section and the concentrated load based on shear strength to be furnished. The denominator of the following formula for required spacing uses the expression for required ϕV_s from step 4:

$$\text{required } s = \frac{\phi A_v f_{yt} d}{\text{required } \phi V_s} = \frac{0.75(0.22)(60)(31)}{63.4 - 1.0x}$$

The results of these calculations are tabulated and plotted in Figure 4-12.

8.   With reference to Figure 4-12, no stirrups are required in the portion of the beam between the point load and center of the beam. A spacing of 5 in. will be used between the face of support and the point load. In the center portion of the beam, between the point loads, stirrups will be placed at a spacing slightly less than the maximum spacing as a conservative measure and to create a convenient spacing pattern. The design sketches are shown in Figure 4-13. Note the symmetry about the span centerline.

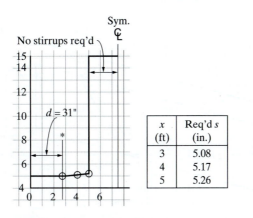

x (ft)	Req'd s (in.)
3	5.08
4	5.17
5	5.26

FIGURE 4-12   Stirrup spacing requirements for Example 4-4.

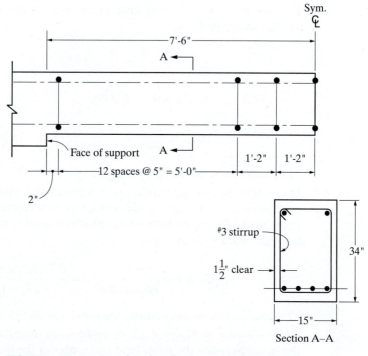

FIGURE 4-13   Design sketches for Example 4-4.

4-5   TORSION OF REINFORCED CONCRETE MEMBERS

The torsion or twisting of reinforced concrete members is caused by a torsional moment that acts about the longitudinal axis of the member due to unbalanced loads
applied to the member. The torsional moment usually acts in combination with
bending moment and shear force as shown in Figure 4-14.

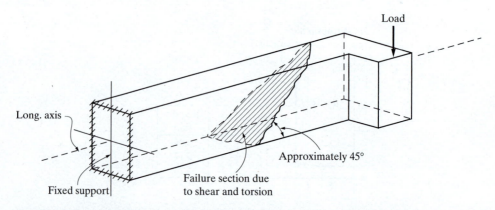

FIGURE 4-14   Cantilever beam subject to combined shear, moment, and torsion.

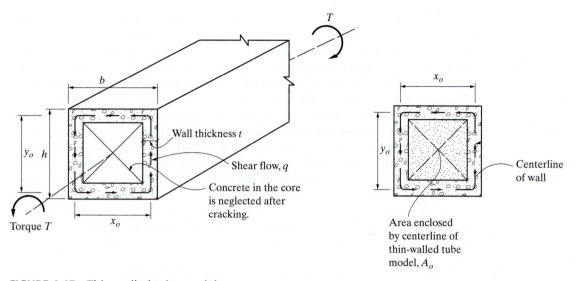

FIGURE 4-15   Thin-walled tube model.

A typical example of torsion in concrete members occurs in a rectangular beam supporting precast hollow-core slabs (or planks). The torsion may be due to unequal live loads on adjacent spans of the hollow-core planks or due to unequal adjacent spans of the hollow-core planks supported on the beam. Torsion in such beams could also be due to the construction sequence that has the hollow-core planks fully installed on one side of the beam before being installed on the adjacent side of the beam. Rectangular and L-beams are more susceptible to torsion than T-beams. In the ACI Code, the design for torsion in solid and hollow concrete beams is based on a thin-walled tube space truss model (see Figure 4-15 and Figure 4-16). In the thin-walled space truss model, the outer concrete cross section that is centered on the

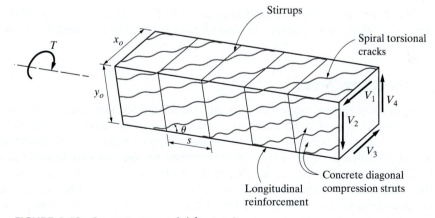

FIGURE 4-16   Space truss model for torsion.

stirrups is assumed to resist the torsion while the concrete in the core is neglected because after cracking, this core is relatively ineffective in resisting torsion. Torsional moments cause additional shear stresses that result in diagonal tension stresses in the concrete member. These diagonal tension stresses cause spiral inclined cracks to form around the surface of the concrete member, as shown in Figure 4-16. After cracking, the torsional resistance of a concrete member is provided by the outermost closed stirrups and the longitudinal reinforcement located near the surface of the beam, and this is modeled by the space truss shown in Figure 4-16, where the longitudinal reinforcement acts as the truss tension members, the stirrups act as the tension web members, and the inclined concrete struts between the diagonal cracks act as the compression web members of the space truss.

In the thin-walled tube model, the shear flow q, which is assumed to be constant around the perimeter of the beam, is equal to the product of the shear stress τ and the wall thickness, t. Using the thin-walled tube model, and summing the torques, the equilibrium of torsional moments yields

$$T - (q\,x_o)y_o - (q\,y_o)x_o = 0$$

or

$$T = 2q\,x_o y_o = 2q\,A_o$$

Therefore,

$$q = T/2A_o \qquad (4\text{-}4)$$

where

A_o = area enclosed by centerline of the shear flow path = $x_o y_o$

q = shear flow (i.e., force per unit length)

x_o and y_o are the width and height of the space truss model measured between the centerlines of the tube walls—that is, the centerlines of the longitudinal corner bars.

There are two conditions that may occur in the design of reinforced concrete members for torsion: primary or equilibrium torsion and secondary or compatibility torsion.

Compatibility Torsion

Compatibility torsion occurs in statically indeterminate structures, and the design torque, which cannot be obtained from statics alone, may be reduced due to redistribution of internal forces to maintain compatibility of deformations. Members subjected to compatibility torsion may be designed for the cracking torque multiplied by the resistance factor (i.e., ϕT_{cr}), but the redistribution of internal forces due to the reduction of the torque to ϕT_{cr} must be taken into account in the design of all the adjoining structural members. One example of compatibility torsion occurs in spandrel beams (see Figure 4-17), where the rotation of the slab is restrained by the

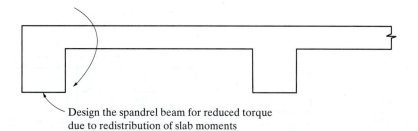

Design the spandrel beam for reduced torque
due to redistribution of slab moments

FIGURE 4-17 Compatibility torsion in spandrel beams.

spandrel beam. For compatibility, the restraining moment at the exterior end of the slab is equal to the uniform torsional moment per unit length on the spandrel beam. As the slab rotates and cracks, and the slab moments are redistributed, the torsional moment on the spandrel beam is reduced until it reaches the cracking torque of the spandrel beam, at which point a hinge is formed at the exterior end of the slab.

Equilibrium Torsion

For statically determinate structures, the design torque, which can be obtained from statics considerations alone, cannot be reduced because redistribution of internal moments and forces is not possible in such structures, and in order to maintain equilibrium, the full design torque has to be resisted by the beam. This is called *equilibrium torsion*. Examples of concrete members in equilibrium torsion are shown in Photo 4-1 and Figure 4-18.

PHOTO 4-1 Equilibrium torsion.

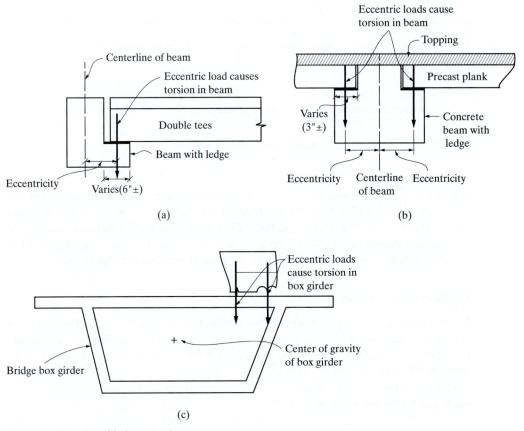

FIGURE 4-18   Equilibrium torsion.

Torsion Design of Reinforced Concrete Members (ACI Code Section, 11.5)

The ACI Code design approach for torsion follows a similar approach to the design for shear. Like shear, the critical section for torsion is located at a distance d from the face of a support, but where a concentrated torque occurs within a distance of d from the face of a support, the critical section for torsion shall be at the face of the support. Torsion can be neglected if the factored torque is less than or equal to one-quarter of the cracking torque of the beam section. That is, if

$$T_u \leq 0.25\phi T_{cr} \tag{4-5}$$

where the cracking torque for non-prestressed members not subject to axial tension or compression force is

$$T_{cr} = 4\lambda\sqrt{f_c'}\frac{(A_{cp})^2}{p_{cp}} \tag{4-6}$$

Thus, torsion may be neglected when

$$T_u \leq 0.25\phi\left(4\lambda\sqrt{f_c'}\frac{(A_{cp})^2}{p_{cp}}\right) = \lambda\phi\sqrt{f_c'}\frac{(A_{cp})^2}{p_{cp}} \qquad (4\text{-}7)$$

where

$\phi = 0.75$ (ACI Code, Section 9.3.2.3)

A_{cp} = area of outside perimeter of the cross section = bh (for rectangular beams not cast monolithic with a slab)

p_{cp} = outside perimeter of the cross section = $2(b + h)$

b and h = cross section width and depth (see Figure 4-15)

For isolated beams cast monolithic with a slab, the area, A_{cp}, can be determined from ACI, Sections 11.5.1.1 and 13.2.4.

λ is the lightweight aggregate factor discussed in Chapter 1, and for normal weight concrete, $\lambda = 1.0$. When the torque is small enough such that Equation (4-5) or (4-7) is satisfied, closed stirrups are not required.

Torsional reinforcement is required to resist the full applied torsional moment as specified in ACI, Section 11.5.2.1, when

$$T_u > \lambda\phi\sqrt{f_c'}\frac{(A_{cp})^2}{p_{cp}} \qquad (4\text{-}8)$$

To reduce unsightly cracks on the surface of the beam and to prevent crushing of the surface concrete from stresses in the inclined concrete compression struts (see Figure 4-16) due to combined shear and torsion, the ACI Code requires that *solid concrete beam* cross-sectional dimensions be such that ACI Code Equations (11-18) and (11-19) are satisfied. The size of the beam should be increased if these relationships are not satisfied.

$$\sqrt{\left(\frac{V_u}{b_w d}\right)^2 + \left(\frac{T_u p_h}{1.7(A_{oh})^2}\right)^2} \leq \phi\left(\frac{V_c}{b_w d} + 8\sqrt{f_c'}\right) \qquad \text{[ACI Eq. (11-18)]}$$

Similarly, the cross-sectional dimensions of *hollow concrete beams* are limited as follows:

$$\frac{V_u}{b_w d} + \frac{T_u p_h}{1.7(A_{oh})^2} \leq \phi\left(\frac{V_c}{b_w d} + 8\sqrt{f_c'}\right) \qquad \text{[ACI Eq. (11-19)]}$$

where

$V_c = 2\lambda\sqrt{f_c'}b_w d$

A_{oh} = area enclosed by the centerline of the outermost closed stirrup = $x_1 y_1$

$p_h = 2(x_1 + y_1)$

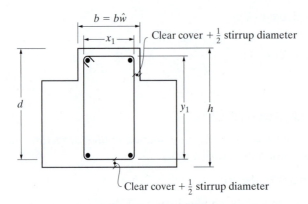

FIGURE 4-19   Definition of x_1 and y_1.

x_1 and y_1 are the width and height of the space truss model measured to the center-line of the outermost closed stirrup as shown in Figure 4-19. It should be noted that the nominal concrete shear strength, V_c, is assumed to be unaffected by torsion.

For hollow sections with varying wall thickness, ACI Equation (11-19) shall be evaluated at a location where the left-hand side of the equation is at its maximum value. If the wall thickness of the hollow concrete beam, t, is less than A_{oh}/p_h at the location where the torsional stresses are being determined, the second term in ACI Equation (11-19) shall be taken as $T_u/(1.7A_{oh}\,t)$.

Torsion Reinforcement

The reinforcement required for torsion shall be added to that required for other load effects that act in combination with the torsional moment, and the most restrictive spacing requirements must be satisfied (ACI Code, Section 11.5.3.8). The torsional reinforcement is determined using the space truss model shown in Figure 4-16. The yield strength of the torsional reinforcement shall not be greater than 60,000 psi per ACI Code, Section 11.5.3.4. The torsional reinforcement consists of closed stirrups and longitudinal reinforcement at the corners of the beam.

Vertical Equilibrium of Forces

Considering a free body diagram of the vertical forces acting on the front wall of the space truss model (Figure 4-16) as shown in Figure 4-20, the equilibrium of the vertical forces yields

$$V_2 = A_t f_{yt}\left(\frac{y_o \cot\theta}{s}\right) \tag{4-9}$$

From Equation (4-4) and noting that the shear flow is constant, V_2 can be calculated as

$$V_2 = q y_o = \left(\frac{T}{2A_o}\right) y_o \tag{4-10}$$

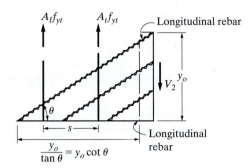

FIGURE 4-20   Free body diagram for vertical equilibrium of forces in the space truss model.

Substituting Equation (4-10) into (4-9) and rearranging yields, the ratio of the area of the torsional stirrup to the spacing of the stirrup as

$$\frac{A_t}{s} = \frac{T}{2A_o f_{yt} \cot \theta} \tag{4-11}$$

Horizontal Equilibrium of Forces

Considering a free body diagram of the horizontal forces acting on the front wall of the space truss model (Figure 4-16) as shown in Figure 4-21, the equilibrium of the horizontal forces yields

$$A_\ell f_{y\ell} = \Sigma V_i \cot \theta = \Sigma (q y_i) \cot \theta = q \cot \theta \, \Sigma y_i \tag{4-12}$$

Substituting Equation (4-4) into Equation (4-12) yields

$$A_\ell f_{y\ell} = \left(\frac{T}{2A_o}\right) \cot \theta \, \Sigma y_i = \left(\frac{T}{2A_o}\right) \cot \theta \, (x_o + x_o + y_o + y_o)$$

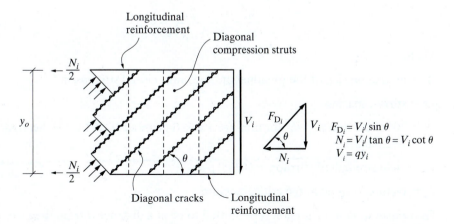

FIGURE 4-21   Free body diagram for horizontal equilibrium of forces in the space truss model.

Therefore,

$$A_\ell f_{y\ell} = \left(\frac{T}{2A_o}\right) \cot \theta [2(x_o + y_o)] \tag{4-13}$$

Substituting for the torque, T, in Equation (4-13) using Equation (4-11) gives

$$A_\ell f_{y\ell} = \left[\left(\frac{A_t}{s}\right) f_{yt} \cot \theta\right] \cot \theta [2(x_o + y_o)]$$

or

$$A_\ell = \left(\frac{A_t}{s}\right)\frac{f_{yt}}{f_{y\ell}} \cot^2 \theta [2(x_o + y_o)] \tag{4-14}$$

It should be recalled that x_o and y_o are the width and height measured to the center-lines of the tube wall in the thin-walled tube model, while x_1 and y_1 are the width and height measured to the centerlines of the outermost closed stirrup. After crack-ing, the shear flow path is defined more by the center-to-center dimensions between the outer most closed stirrup. Therefore, in subsequent equations, assuming that $x_o = x_1$ and $y_o = y_1$ yields the equation for the longitudinal torsion reinforcement as

$$A_\ell = \left(\frac{A_t}{s}\right)\frac{f_{yt}}{f_{y\ell}} \cot^2 \theta [2(x_1 + y_1)] \tag{4-15}$$

That is,

$$A_\ell = \left(\frac{A_t}{s}\right)\frac{f_{yt}}{f_{y\ell}} \cot^2 \theta (p_h) \qquad \text{[ACI Eq. (11-22)]}$$

Transverse Reinforcement Required for Torsion (Stirrups)

Using Equation (4-11) and the limit states design requirement that the design tor-sional strength, ϕT_n be greater than or equal to the factored torsional moment, T_u yields the required torsional stirrup area,

$$\frac{A_t}{s} = \frac{T_u}{2\phi A_o f_{yt} \cot \theta} \tag{4-16}$$

where

A_t = area of *one* leg of the torsional or outermost closed stirrups, in.[2]

s = stirrup spacing

θ = 30° to 60°; *use* θ = 45° for non-prestressed members (ACI Code, Section 11.5.3.6[a])

f_{yt} = yield strength of stirrups = 60,000 psi

T_u = factored torque at the critical section

The *critical section* for torsion is permitted to be at a distance d from the face of the beam support provided no concentrated torque occurs within a distance of d from the face of the support. If a concentrated torque occurs within a distance d from the face of support, the critical section will be at the face of the support (ACI Code, Section 11.5.2.4).

$$d = \text{effective depth of the beam}$$

$$A_o \approx 0.85\, A_{oh} = 0.85\, x_1 y_1$$

$$A_{oh} = \text{Area enclosed by the centerline of outermost closed stirrup}$$

The nominal torsional strength, T_n, is obtained from Equation (4-11) as

$$T_n = \frac{2 A_o A_t f_{yt}}{s} \cot\theta \qquad \text{[ACI Eq. (11-21)]}$$

The *total* equivalent transverse reinforcement or stirrups required for *combined shear plus torsion* is obtained from the ACI Code, Section 11.5.5.2, as

$$\frac{A_{vt}}{s} = \frac{A_v}{s} + \frac{2 A_t}{s} \geq \frac{50 b_w}{f_{yt}}$$

$$\geq 0.75 \sqrt{f_c'}\, \frac{b_w}{f_{yt}} \qquad\qquad (4\text{-}17)$$

where

A_{vt} = area of *two* legs of closed stirrups required for combined shear plus torsion

A_t = area of *one* leg of closed stirrups required for torsion

A_v = area of *two* legs of closed stirrups required for shear

f_{yt} = yield strength of stirrups = 60,000 psi

b_w = width of beam stem

s = spacing of stirrups $\leq p_h/8$ and 12 in. (ACI Code, Section 11.5.6.1)

Note that A_v/s is a slightly modified form for the stirrup requirement for shear from Section 4-2.

Additional Longitudinal Reinforcement Required for Torsion

The additional longitudinal reinforcement required to resist torsion, and to be added to the reinforcement required for bending, is obtained from ACI Equations (11-22) and (11-24) as

$$A_\ell = \left(\frac{A_t}{s}\right) p_h \frac{f_{yt}}{f_{y\ell}} \cot^2\theta \qquad \text{[ACI Eq. (11-22)]}$$

$$\geq 5\sqrt{f_c'}\, \frac{A_{cp}}{f_{y\ell}} - \left(\frac{A_t}{s}\right) p_h \frac{f_{yt}}{f_{y\ell}} \qquad \text{[ACI Eq. (11-24)]}$$

where

$\dfrac{A_t}{s}$ = torsion stirrup area from Equation (4-11)

$$\geq 25b_w/f_{yt}$$
$$p_h = 2(x_1 + y_1)$$
$$A_{cp} = bh$$
$$b_w = \text{width of beam stem}$$
$$f_{yt} = \text{stirrup yield strength, psi}$$
$$f_{y\ell} = \text{longitudinal steel yield strength, psi}$$

The following should be noted with regard to the required torsional reinforcement:

- The additional longitudinal reinforcement must be distributed around the surface of the beam with a maximum spacing of 12 in., and there should be at least one longitudinal bar in each corner of the closed stirrups to help transfer the compression strut forces into the stirrups.
- The additional longitudinal reinforcement diameter should be at least 0.042 times the stirrup spacing (i.e., 0.042s), but not less than a $^3/_8$ in. diameter bar (ACI Code, Section 11.5.6.2) to prevent buckling of the longitudinal reinforcement due to the horizontal component of the diagonal compression strut force.
- The additional longitudinal rebar area should be added to the longitudinal rebar area required for bending.
- The closed torsional stirrups should be enclosed with 135° hooks (ACI Code, Section 11.5.4.1) and there should be at least one longitudinal bar enclosed by and at each corner of the stirrup. Note that 90° hooks are ineffective after the corners of the beam spall off due to torsion failure.

Torsion Design Procedure

The design procedure for torsion is as follows:

1. Determine the maximum factored concentrated or uniformly distributed torsional load and the corresponding factored gravity load that occurs simultaneously.
   Note that pattern or checkered live loading may need to be considered to maximize the torsional load and moment.
2. Determine the factored torsional moment, T_u, the factored shear, V_u, and the factored bending moment, M_u.
3. Determine the reinforcement required to resist the factored bending moment M_u.

4.   Calculate the concrete shear strength, ϕV_c.
5.   Determine the cracking torque, T_{cr} and if torsion can be neglected (check if $T_u \leq 0.25 T_{cr}$).
6.   Determine if the torsion in the member is caused by compatibility torsion or by equilibrium torsion:
   a.   For compatibility torsion, redistribution of internal forces is possible because the torsional moment is not required to maintain equilibrium; therefore, design the member for a reduced torque of ϕT_{cr}.
   b.   For equilibrium torsion, redistribution of internal forces is not possible because the torsional moment is required to maintain equilibrium; therefore, the member must be designed for the full torsional moment, T_u, calculated in step 1.
7.   Check the limits of the member cross section using ACI Code Equation (11-18) to prevent crushing of the diagonal concrete compression struts.
8.   Determine the required torsional stirrup area, A_t/s, the stirrup area required for shear, A_v/s, and the total stirrup area required for combined shear and torsion, A_{vt}/s. Check that maximum stirrup spacing is not exceeded, and check minimum stirrup area. Using the torsional moment diagram and shear force diagram, the required stirrup spacing can be laid out to match the variation in shear and torsional moment.
9.   Determine the additional longitudinal reinforcement required for torsion.
10.   Draw the detail of the torsional reinforcement.

Example 4-5 _____

Design of Beams for Torsion

The floor framing in the operating rooms in a hospital building consists of reinforced concrete beams $18'' \times 24''$ deep that support precast concrete planks as shown in Figure 4-22. The clear span of the beam is 27 ft between columns. The planks are 10 in. deep with 2-in. topping and supports stud wall partitions

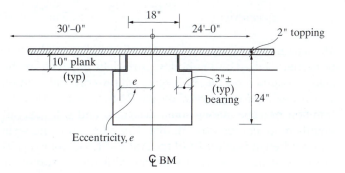

FIGURE 4-22   Beam section for Example 4-5.

that weigh 10 psf and mechanical/electrical equipment that weigh 5 psf. The weight of the precast planks is 70 psf. The centerline to centerline span of the planks is 30 ft on the left-hand side of the beam and 24 ft on the right-hand side of the beam. Design the beam for torsion and shear assuming normal weight concrete (i.e., $\lambda = 1.0$) and $f'_c = 4000$ psi. Assume the beam has already been designed for bending. The live load for hospital operating rooms is 60 psf.

Solution:

1. Determine the maximum factored concentrated or uniformly distributed torsional load and the corresponding factored gravity load that occurs simultaneously.

    Dead Load

$$10 \text{ in. plank} + 2 \text{ in. topping} = 95 \text{ psf}$$

$$\text{Mechanical and electrical} = 5 \text{ psf}$$

$$\text{Stud wall partitions} = 10 \text{ psf}$$

$$\text{Total dead load, } D = 110 \text{ psf}$$

$$\text{Floor live load (operating room), } L = 60 \text{ psf}$$

    Tributary Widths (TW) of Beam

$$\text{TW (due to the 30-ft-span hollow-core plank)} = \frac{30 \text{ ft}}{2} = 15 \text{ ft}$$

$$\text{TW (due to the 24-ft-span hollow-core plank)} = \frac{24 \text{ ft}}{2} = 12 \text{ ft}$$

    A review of Figure 4-22 shows that the torsion in this beam is equilibrium torsion caused by the eccentricity of the plank loads.

    Eccentricity of the Hollow-Core Plank Load

$$\text{Eccentricity, } e = \frac{18 \text{ in.}}{2} + \frac{3 \text{ in.}}{2} = 10.5 \text{ in.} = 0.88 \text{ ft}$$

    The maximum uniform torsional loading will occur due to checkerboard or partial loading on the hollow core slabs in which the full design live load is assumed on the 30-ft-span hollow-core slab and *one-half of the design live load is assumed* on the 24-ft-span hollow core slab. This is common practice among some designers and will generally result in a slightly more conservative design. This partial loading is similar to what is prescribed in Section 7.5 of the ASCE 7 Load Standard. The maximum torsion will be considered together with the corresponding maximum uniform vertical load that occurs at the same time.

The maximum factored uniform torsional load is

$$w_{tu} = \{[1.2(110 \text{ psf}) + 1.6(60 \text{ psf})](15 \text{ ft}) - [1.2(110 \text{ psf}) + 1.6(\tfrac{1}{2})(60 \text{ psf})](12 \text{ ft})\} \times 0.88 \text{ ft} = 1109 \text{ ft-lb/ft} = 1.11 \text{ ft-kips/ft}$$

The corresponding maximum factored uniform vertical load is

$$w_u = [1.2(110 \text{ psf}) + 1.6(60 \text{ psf})](15 \text{ ft}) + [1.2(110 \text{ psf}) + 1.6(\tfrac{1}{2})(60 \text{ psf})](12 \text{ ft})$$

$$= 5580 \text{ lb/ft} = 5.58 \text{ kips/ft}$$

2. Determine the factored torsional moment, T_u, the factored shear, V_u, and the factored bending moment, M_u:

   Assuming 2 layers of reinforcement, $d = 24 \text{ in.} - 3.5 \text{ in.} = 20.5 \text{ in.}$ For torsion and shear, use a reduced span commencing at d from the face of the beam supports. This reduced span is

   $$\ell = 27 \text{ ft} - \frac{2(20.5 \text{ in.})}{12} = 23.6 \text{ ft}$$

   Maximum design torsional moment,

   $$T_u = \frac{w_{tu}\ell}{2} = \frac{1.11 \text{ ft-kips/ft } (23.6 \text{ ft})}{2} = 13.1 \text{ ft-kips}$$

   Maximum design shear that occurs at the same time as the maximum torsion is

   $$V_u = \frac{w_u\ell}{2} = \frac{5.58 \text{ kips/ft } (23.6 \text{ ft})}{2} = 65.8 \text{ kips}$$

3. The reinforcement required to resist the bending moment is assumed to have previously been designed and is not calculated here.

4. The concrete shear strength is

   $$\phi V_c = 0.75(2)\sqrt{4000}(18 \text{ in.})(20.5 \text{ in.}) = 35,000 \text{ lb} = 35 \text{ kips}$$

5. Torsion can be neglected if the factored torsional moment is less than or equal to the concrete torsional strength, that is, if $T_u \le 0.25\phi T_{cr}$, where the concrete torsional strength is

   $$0.25\phi T_{cr} = \lambda\phi\sqrt{f_c'}\frac{(A_{cp})^2}{p_{cp}} = (1.0)(0.75)(\sqrt{4000})\frac{(18 \text{ in.} \times 24 \text{ in.})^2}{2(18 \text{ in.} + 24 \text{ in.})}$$

   $$= 105.4 \text{ in.-kips} = 8.8 \text{ ft-kips}$$

   Because $T_u = 13.1 \text{ ft-kips} > 8.8 \text{ ft-kips}$, this beam therefore, must be designed for torsion.

6. This is equilibrium torsion as redistribution of internal forces is not possible because the torsional moment is required to maintain equilibrium.

The member thus must be designed for the full torsional moment, T_u, calculated in step 1.

7. Check the limits of the member cross section using ACI Code Equation (11-18) to prevent crushing of the diagonal concrete compression struts:

$b = 18$ in.

$h = 24$ in.

Effective depth, $d = 24 - 3.5$ in. (assuming 2 layers of rebar) $= 20.5$ in.

$\theta = 45°$

$\phi = 0.75$

$x_1 = 18 \text{ in.} - (2 \text{ sides})\left(1.5\text{-in. cover} + \dfrac{0.5\text{-in. stirrup}}{2}\right) = 14.5 \text{ in.}$

$y_1 = 24 \text{ in.} - (2 \text{ sides})\left(1.5\text{-in. cover} + \dfrac{0.5\text{-in. stirrup}}{2}\right) = 20.5 \text{ in.}$

$f_{yt} = f_{yv} = 60{,}000$ psi

$A_{oh} = x_1 y_1 = (14.5 \text{ in.})(20.5 \text{ in.}) = 297.3 \text{ in.}^2$

$A_o = 0.85 A_{oh} = 0.85 x_1 y_1 = 0.85(14.5 \text{ in.})(20.5 \text{ in.}) = 252.7 \text{ in.}^2$

$p_h = 2(x_1 + y_1) = 2(14.5 + 20.5) = 70 \text{ in.}$

$A_{cp} = bh = (18 \text{ in.})(24 \text{ in.}) = 432 \text{ in.}^2$

The limits on the beam cross-sectional dimensions will now be checked using ACI Equation (11-18):

$$\sqrt{\left(\frac{V_u}{b_w d}\right)^2 + \left(\frac{T_u p_h}{1.7(A_{oh})^2}\right)^2} \le \phi\left(\frac{V_c}{b_w d} + 8\sqrt{f_c'}\right)$$

That is,

$$\sqrt{\left(\frac{65.8 \text{ kips}(1000)}{(18 \text{ in.})(20.5 \text{ in.})}\right)^2 + \left(\frac{(13.1 \text{ ft-kips})(12{,}000)(70 \text{ in.})}{1.7(297.3 \text{ in.}^2)^2}\right)^2}$$
$$\le \frac{35 \text{ kips}(1000)}{(18 \text{ in.})(24 \text{ in.})} + 8(0.75)\sqrt{4000}$$

$$192.8 \text{ psi} < 460.5 \text{ psi} \qquad\qquad \text{(O.K.)}$$

Thus, the diagonal concrete compression struts are not crushed and the size of the beam is adequate to resist the torsional moments.

8. Determine the required torsional stirrup area, A_t/s, the stirrup area required for shear, A_v/s, and the total stirrup area required for combined

shear and torsion, A_{vt}/s. Check that maximum stirrup spacing is not exceeded, and check minimum stirrup area.

From Equation (4-16), the torsional stirrup required is

$$\frac{A_t}{s} = \frac{T_u}{2\phi A_o f_{yt} \cot \theta} = \frac{13.1 \text{ ft-kips}(12,000)}{2(0.75)(252.7)(60,000)\cot 45°} = 0.0069$$

The stirrup area required to resist the maximum factored shear acting with the maximum torsion is

$$\frac{A_v}{s} = \frac{V_u - \phi V_c}{\phi f_{yv} d} = \frac{(65.8 \text{ kips} - 35 \text{ kips})(1000)}{(0.75)(60,000)(20.5 \text{ in.})} = 0.033$$

The total stirrup area required (2-leg stirrups) is calculated from

$$\frac{A_{vt}}{s} = \frac{A_v}{s} + \frac{2A_t}{s} = 0.033 + 2(0.0069) = 0.047$$

$$\geq \frac{50 b_w}{f_{yt}} = \frac{50(18 \text{ in.})}{60,000} = 0.015$$

$$\geq 0.75 \sqrt{f'_c} \frac{b_w}{f_{yt}} = 0.75 \sqrt{4000} \frac{18 \text{ in.}}{60,000} = 0.014 \qquad \text{(O.K.)}$$

Using No. 4 stirrups, A_{vt} (2 legs) = 2 (0.2 in.^{2}) = 0.4 in.^{2}, the spacing of the stirrups required to resist the maximum combined shear and torsion is calculated as

$$s = \frac{0.4 \text{ in.}^2}{0.047} = 8.5 \text{ in. (controls)}$$

$$\leq p_h/8 = 70 \text{ in.}/8 = 8.75 \text{ in.} \qquad \text{(O.K.)}$$

$$\leq 12 \text{ in.} \qquad \text{(O.K.)}$$

Therefore, use No. 4 closed stirrups at 8-in. on center.

The shear and torsion are at their maximum values at the face of the beam support and decrease linearly to zero at the midspan of the beam; the stirrup spacing thus can be varied accordingly, as done previously in the shear design examples.

9. Additional Longitudinal Reinforcement

b_w = width of beam stem = 18 in.

f_{yt} = stirrup yield strength = 60,000 psi

$f_{y\ell}$ = longitudinal steel yield strength = 60,000 psi

$$\frac{A_t}{s} = 0.0069 \text{ (as previously calculated)} < \frac{25 b_w}{f_{yt}} = \frac{25(18 \text{ in.})}{60,000} = 0.0075$$

Therefore use 0.0075.

From ACI Equations (11-22) and (11-24) the additional longitudinal reinforcement is calculated as

$$A_\ell = \left(\frac{A_t}{s}\right)p_h\frac{f_{yt}}{f_{y\ell}}\cot^2\theta = (0.0075 \text{ in.})(70 \text{ in.})\frac{60{,}000}{60{,}000}\cot^2 45° = 0.53 \text{ in.}^2$$

$$\geq 5\sqrt{f_c'}\frac{A_{cp}}{f_{y\ell}} - \left(\frac{A_t}{s}\right)p_h\frac{f_{yt}}{f_{y\ell}}$$

$$= 5\sqrt{4000}\frac{432 \text{ in.}^2}{60{,}000} - (0.0075 \text{ in.})(70 \text{ in.})\frac{60{,}000}{60{,}000} = 1.75 \text{ in.}^2$$

Therefore, the required additional longitudinal steel is $A_\ell = 1.75$ in.^{2}

This additional longitudinal reinforcement should be distributed at the corners of the beam but the spacing between these bars should be no greater than 12 in. Where the spacing exceeds 12 in., provide additional longitudinal bars at the midwidth or middepth of the beam as required. This longitudinal reinforcement is in addition to the reinforcement required to resist the bending moments on the beam.

If the additional reinforcement is concentrated on the top and bottom layers, therefore, the total areas of the top and bottom longitudinal reinforcement in the beam are calculated as

$$A_{s,\text{top}} = A_{s,\text{top}}(\text{due to bending}) + 0.5(1.75 \text{ in.}^2)$$
$$= A_{s,\text{top}}(\text{due to bending}) + 0.88 \text{ in.}^2$$

$$A_{s,\text{bottom}} = A_{s,\text{bottom}}(\text{due to bending}) + 0.5(1.75 \text{ in.}^2)$$
$$= A_{s,\text{bottom}}(\text{due to bending}) + 0.88 \text{ in.}^2$$

However, for the beam in this example, the spacing of the longitudinal reinforcement will exceed the maximum 12 in. because the center-to-center distance between the top and bottom rebars is approximately 18 in. The additional longitudinal reinforcement should thus be distributed as follows:

$$A_{s,\text{top}} = A_{s,\text{top}}(\text{due to bending}) + \left(\frac{1}{3}\right)(1.75 \text{ in.}^2)$$
$$= A_{s,\text{top}}(\text{due to bending}) + 0.58 \text{ in.}^2$$

$$A_{s,\text{midheight}} = \left(\frac{1}{3}\right)(1.75 \text{ in.}^2) = 0.58 \text{ in.}^2$$

$$A_{s,\text{bottom}} = A_{s,\text{bottom}}(\text{due to bending}) + \left(\frac{1}{3}\right)(1.75 \text{ in.}^2)$$
$$= A_{s,\text{bottom}}(\text{due to bending}) + 0.58 \text{ in.}^2$$

The *minimum* diameter of the longitudinal reinforcement is the largest of the following:

$$0.042\,s = 0.042(8 \text{ in.}) = 0.34 \text{ in.}$$

or ⅜ in. (controls)

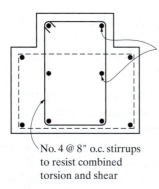

Provide additional
torsional longitudinal
rebar, in addition to bending
reinforcement

No. 4 @ 8" o.c. stirrups
to resist combined
torsion and shear

FIGURE 4-23 Beam torsional reinforcement detail.

10. Torsional reinforcement detail is shown in Figure 4-23.

PROBLEMS

4-1. A reinforced concrete beam of rectangular cross section is reinforced for moment only and subjected to a shear V_u of 9000 lb. Beam width b = 12 in., d = 7.25 in., f_c' = 3000 psi, and f_y = 60,000 psi. Is the beam satisfactory for shear?

4-2. An 8-in.-thick one-way slab is reinforced for positive moment with No. 6 bars at 6 in. on center. Cover is 1 in. Determine the maximum shear V_u permitted. Use f_c' = 4000 psi and f_y = 60,000 psi.

4-3. Assume that the beam of Problem 4-1 has an effective depth of 18 in. and is reinforced with No. 3 single-loop stirrups spaced at 10 in. on center. Determine the maximum shear V_u permissible.

4-4. The simply supported beam shown is on a clear span of 30 ft. The beam carries uniformly distributed service loads of 1.9 kips/ft live load and 0.7 kip/ft dead load (excluding the weight of the beam). Additionally, the beam carries two concentrated service loads of 8 kips dead load each, one load being placed 5 ft in from the face of each support. The beam is reinforced with No. 3 single-loop stirrups placed in the following pattern, starting at the face of the support (symmetry about midspan): one space at 2 in., five spaces at 8 in.,

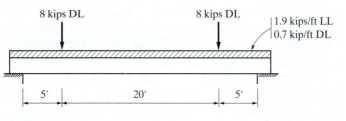

PROBLEM 4-4

four spaces at 10 in., and eight spaces at 12 in. Beam width $b = 15$ in., $h = 28$ in., and $d = 25.4$ in. Use $f_c' = 4000$ psi and $f_y = 60,000$ psi.

(a) Draw an elevation view of the beam showing the stirrup layout.

(b) Draw the V_u diagram.

(c) Calculate $\phi(V_c + V_s)$ for each spacing group (omit the 2-in. spacing).

(d) Superimpose the results from part (c) on the V_u diagram and comment on the results.

4-5. A uniformly loaded beam is subjected to a shear V_u of 60 kips at the face of the support. Clear span is 32 ft. Beam width $b = 12$ in., $d = 22$ in., $f_c' = 4000$ psi, and $f_y = 60,000$ psi. Determine the maximum spacing allowed for No. 3 single-loop stirrups at the critical section.

4-6. A simply supported, rectangular, reinforced concrete beam having $d = 24$ in. and $b = 15$ in. supports a uniformly distributed load w_u of 7.50 kips/ft as shown. The given load includes the beam weight. Assume No. 3 single-loop stirrups. The concrete strength is 4000 psi and the steel is grade 60.

(a) Select the stirrup spacing to use at the critical section (d distance from the face of support).

(b) Determine the maximum stirrup spacing allowed for this beam.

(c) Using the two spacings determined in parts (a) and (b), devise an appropriate stirrup spacing layout for this beam.

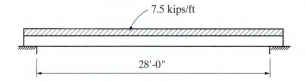

7.5 kips/ft

28'-0"

PROBLEM 4-6

4-7. A simply supported beam carries a total factored load w_u of 5.0 kips/ft. The span length is 28 ft center to center of supports, and the supports are 12-in. wide. Beam width $b = 14$ in., $d = 20$ in., $f_c' = 4000$ psi, and $f_y = 60,000$ psi. Determine spacings required for No. 3 stirrups and show the pattern with a sketch. (Recall that clear span is used for determining shears.)

4-8. A simply supported rectangular reinforced concrete beam, 13-in. wide and having an effective depth of 20 in., supports a total factored load (w_u) of 4.5 kips/ft on a 30-ft clear span. (The given load includes the weight of the beam.) Design the web reinforcement if $f_c' = 3000$ psi and $f_y = 40,000$ psi.

4-9. A rectangular reinforced concrete beam supports a total factored load (w_u) of 10 kips/ft on a simple clear span of 40 ft. (The given load includes the weight of the beam.) The effective depth $d = 40$ in., $b = 24$ in., $f_c' = 3000$ psi, and $f_y = 60,000$ psi. Design double-loop No. 3 stirrups.

4-10. Design stirrups for the beam shown. The supports are 12-in. wide, and the loads shown are service loads. The dead load includes the weight of the beam. Beam width $b = 16$ in., $d = 20$ in., $f'_c = 4000$ psi, and $f_y = 60,000$ psi. Sketch the stirrup pattern.

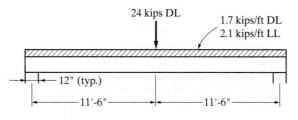

PROBLEM 4-10

4-11. For the beam shown, check moment and design single-loop stirrups. The loads shown are factored loads. Assume the supports to be 12-in. wide. Use $f'_c = 3000$ psi and $f_y = 60,000$ psi. The uniformly distributed load includes the beam weight.

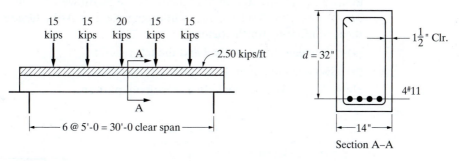

PROBLEM 4-11

4-12. Design stirrups for the beam shown. Service loads are 1.5 kips/ft dead load (includes beam weight) and 1.9 kips/ft live load. The supports are 12-in. wide. Beam width $b = 13$ in. and $d = 24$ in. for both top and bottom steel. Use $f'_c = 3000$ psi and $f_y = 60,000$ psi. Sketch the stirrup arrangement.

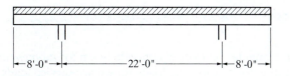

PROBLEM 4-12

4-13. Design stirrups for the beam shown. The supports are 12-in. wide, and the loads shown are factored design loads. The dead load includes the weight of the beam. Beam width $b = 14$ in., $d = 24$ in., $f'_c = 3000$ psi, and $f_y = 60,000$ psi. Sketch the stirrup pattern.

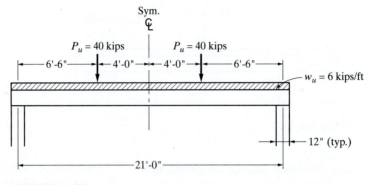

PROBLEM 4-13

4-14. The beam shown is supported on pedestals 20-in. wide. The loads are service loads, and the distributed load includes the beam weight. The clear span is $24'-4''$. Use $f'_c = 4000$ psi; the reinforcing steel is A615 Grade 60. Design shear reinforcing for the beam:

(a) Between the point load and the left reaction

(b) Between the point load and the right reaction

Show design sketches, including the stirrup pattern.

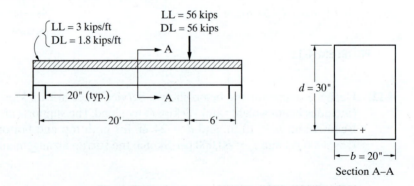

PROBLEM 4-14

4-15. Design the rectangular reinforced concrete beam for *moment and shear*. Use only tension steel for flexure. The loads shown are service loads. The uniform load is composed of 1 kip/ft dead load (does not include beam weight) and

1 kip/ft live load. The concentrated loads are dead load. Assume the supports to be 12-in. wide. Use $f_c' = 4000$ psi and $f_y = 60,000$ psi. Show design sketches, including the stirrup pattern.

4-16. Explain, with examples, the difference between equilibrium and compatibility torsion.

4-17. A rectangular reinforced concrete beam 14 in. × 20 in. deep is subject to a maximum factored torque of 24 ft-k. Calculate the cracking torque, T_{cr}, and determine if torsion can be neglected in the design of this beam. Assume normal-weight concrete (i.e., $\lambda = 1.0$) and $f_c' = 4000$ psi.

4-18. The floor framing in an office building consists of reinforced concrete beams 18 in. × 24 in. deep that support precast concrete planks on a 3-in. ledge similar to that shown in Figure 4-22. The clear span of the beam is 27 ft between columns. The planks are 10-in. deep with 2-in. topping and supports stud wall partitions that weigh 10 psf and mechanical/electrical equipment that weigh 5 psf. The weight of the precast planks is 70 psf. The centerline to centerline span of the planks is 30 ft on the left-hand side of the beam and 10 ft on the right-hand side of the beam. The live load on the 30-ft span is 50 psf while the live load on the 10-ft corridor span is 100 psf. Design the beam for torsion and shear assuming normal weight concrete (i.e., $\lambda = 1.0$), $f_c' = 4000$ psi and $f_y = 60,000$ psi. Assume the beam has already been designed for bending.

Chapter 5

Development, Splices, and Simple-Span Bar Cutoffs

5-1 DEVELOPMENT LENGTH: INTRODUCTION

One of the fundamental assumptions of reinforced concrete design is that at the interface of the concrete and the steel bars, perfect bonding exists and no slippage occurs. Based on this assumption, it follows that some form of bond stress exists at the contact surface between the concrete and the steel bars. In beams, this bond stress is caused by the change in bending moment along the length of the beam and the accompanying change in the tensile stress in the bars and has historically been termed *flexural bond*. The actual distribution of bond stresses along the reinforcing steel is highly complex, due primarily to the presence of concrete cracks. Research has indicated that large local variations in bond stress are caused by flexural and diagonal cracks, and very high bond stresses have been measured adjacent to these cracks. These high bond stresses may result in small local slips adjacent to the cracks, with resultant widening of cracks as well as increased deflections. Generally, this will be

PHOTO 5-1   90° bends for development of No. 18 bars. Seabrook station, New Hampshire.

harmless as long as failure does not propagate all along the bar with resultant complete loss of bond. It is possible, if *end anchorage* is reliable, that the bond can be severed along the entire length of bar, excluding the anchorage, without endangering the carrying capacity of the beam. The resulting behavior is similar to that of a tied arch.

End anchorage may be considered reliable if the bar is embedded into concrete a prescribed distance known as the *development length*, ℓ_d, of the bar. If in the beam the actual extended length of a bar is equal to or greater than this required development length, no premature bond failure will occur—that is, the predicted strength of the beam will not be controlled by bond but rather by some other factor.

Hence the main requirement for safety against bond failure is that the length of the bar from any point of given steel stress f_s (or, as a maximum, f_y) to its nearby free end must be at least equal to its development length. If this requirement is satisfied, the magnitude of the flexural bond stress along the beam is of only secondary importance, because the integrity of the member is assured even in the face of possible minor local failures. If the actual available length is inadequate for full development, however, special anchorages, such as hooks, must be provided to ensure adequate strength.

Current design methods based on the ACI Code (318-08) disregard high localized bond stress even though it may result in localized slip between steel and concrete adjacent to the cracks. Instead, attention is directed toward providing adequate length of embedment, past the location at which the bar is fully stressed, which will ensure development of the full strength of the bar.

5-2   DEVELOPMENT LENGTH: TENSION BARS

The ACI Code, Section 12.2.1, specifies that the development length ℓ_d for deformed bars and deformed wires in tension shall be determined using either the tabular criteria of Section 12.2.2 or the general equation of Section 12.2.3, but in either case, ℓ_d shall not be less than 12 in.

The general equation of Section 12.2.3 (ACI Equation [12-1]) offers a simple approach that allows the user to see the effect of all variables controlling the development length. The tabular criteria of Section 12.2.2 also offer a simple, conservative approach that recognizes commonly used practical construction techniques. Based on a sampling of numerous cases, the authors have found that significantly shorter development lengths are computed using ACI Equation (12-1) of Section 12.2.3. Therefore the development length ℓ_d computation used in this text will be based on ACI Equation (12-1):

$$\ell_d = \frac{3}{40}\left(\frac{f_y}{\lambda\sqrt{f_c'}}\right)\left[\frac{\psi_t\psi_e\psi_s}{\left(\dfrac{c_b + K_{tr}}{d_b}\right)}\right]d_b$$

in which the term $(c_b + K_{tr})/d_b$ shall not be taken greater than 2.5 and where

ℓ_d = development length (in.)

f_y = specified yield strength of non-prestressed reinforcement (psi)

f_c' = specified compressive strength of concrete (psi); the value of $\sqrt{f_c'}$ shall not exceed 100 psi (ACI Code, Section 12.1.2)

d_b = nominal diameter of bar or wire (in.)

The other factors used in Equation (12-1) are defined as follows (ψ is lowercase Greek psi):

1. ψ_t is a reinforcement location factor that accounts for the position of the reinforcement in freshly placed concrete.

Where horizontal reinforcement is so placed that more than 12 in. of fresh concrete is cast in the member below the development length or splice, use ψ_t = 1.3 (ACI Code, Section 12.2.4). This is because this condition lends itself to the formation of entrapped air and moisture on the underside of the bars, resulting in partial loss of bond between the concrete and steel.

For other reinforcement, use ψ_t = 1.0.

2. ψ_e is a coating factor reflecting the effects of epoxy coating. Studies of the anchorage of epoxy-coated bars show that bond strength is reduced because the coating prevents adhesion and friction between the bar and the concrete.

For epoxy-coated reinforcement having cover less than $3d_b$ or clear spacing between bars less than $6d_b$, use ψ_e = 1.5.

For all other conditions, use ψ_e = 1.2.

For uncoated and galvanized reinforcement, use ψ_e = 1.0.

The product of ψ_t and ψ_e need not be taken greater than 1.7 (ACI Code, Section 12.2.4).

3. ψ_s is a reinforcement size factor.

Where No. 6 and smaller bars are used, use ψ_s = 0.8.

Where No. 7 and larger bars are used, use ψ_s = 1.0.

4.   Lambda (λ) is the lightweight-aggregate concrete factor and has been discussed in Section 1-6 of Chapter 1. For purposes of development length calculation where lightweight concrete is used, λ shall not exceed 0.75 unless average f_{ct} is specified, in which case it can be calculated (see Section 1-6).

5.   The factor c_b represents a spacing or cover dimension (in.).

The value of c_b will be the smaller of either the distance from the center of the bar to the nearest concrete surface (cover) or one-half the center-to-center spacing of the bars being developed (spacing).

The bar spacing will be the actual center-to-center spacing between the bars if adjacent bars are all being developed at the same location. If, however, an adjacent bar has been developed at another location, the spacing to be used will be greater than the actual spacing to the adjacent bar. Note in Figure 5-1 that the spacing for bars Y may be taken the same as for bars X, because bars

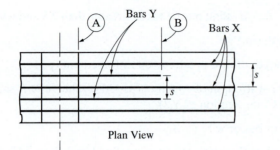

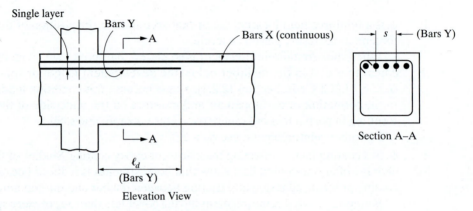

FIGURE 5-1   Spacing criteria for bars being developed.

Y are developed in length AB, whereas bars X are developed at a location other than AB.

6.   The transverse reinforcement index K_{tr} is computed from

$$\frac{40A_{tr}}{sn}$$

where

A_{tr} = total cross-sectional area of all transverse reinforcement that is within the spacing s and that crosses the potential plane of splitting through the reinforcement being developed (in.^{2})

s = maximum center-to-center spacing of transverse reinforcement within ℓ_d (in.)

n = number of bars or wires being developed along the plane of splitting

To simplify the design, the ACI Code permits the use of $K_{tr} = 0$. This is conservative and may be used even if transverse reinforcement is present. To further simplify

Table 5-1   Coefficient K_D for ACI Code Equation (12-1)

$$K_D = \frac{3}{40}\frac{f_y}{\sqrt{f_c'}}$$

f_c' (psi)	$f_y = 40{,}000$ psi	$f_y = 50{,}000$ psi	$f_y = 60{,}000$ psi	$f_y = 75{,}000$ psi
3000	54.8	68.5	82.2	102.7
4000	47.4	59.3	71.2	88.9
5000	42.4	53.0	63.6	79.5
6000	38.7	48.4	58.1	72.6

development length computations, we designate a portion of ACI Equation (12-1) as K_D, where

$$K_D = \frac{3}{40}\frac{f_y}{\sqrt{f_c'}}$$

Values of K_D as a function of various combinations of f_y and f_c' are tabulated in Table 5-1.

A reduction in the development length ℓ_d is permitted where reinforcement is in excess of that required by analysis (except where anchorage or development for f_y is specifically required or where the design includes provisions for seismic considerations). We designate this reduction factor as K_{ER}. Excess reinforcement factor K_{ER} does not apply for development of positive moment reinforcement at supports (ACI Code, Section 12.11.2) or for development of shrinkage and temperature reinforcement (ACI Code, Section 7.12.2.3). Although K_{ER} is not reflected in ACI Equation (12-1), it may be calculated from

$$K_{ER} = \frac{A_s \text{ required}}{A_s \text{ provided}}$$

and subsequently applied to the ℓ_d computed from ACI Equation (12-1).

When bundled bars are used, the ACI Code, Section 12.4, stipulates that calculated development lengths are to be made for individual bars within a bundle (either in tension or compression) and then increased by 20% for three-bar bundles and by 33% for four-bar bundles. Bundled bars consist of a group of not more than four parallel reinforcing bars in contact with each other and assumed to act as a unit. For determining the appropriate factors as discussed previously, a unit of bundled bars shall be treated as a single bar of a diameter derived from the equivalent total area. Additional criteria for bundled bars are furnished in the ACI Code, Section 7.6.6.

Summary of Procedures for Calculation of ℓ_d [Using ACI Equation (12-1)]

1. Determine K_D from Table 5-1.
2. Determine applicable factors (use 1.0 unless otherwise determined).
    a. Use $\psi_t = 1.3$ for top reinforcement, when applicable.

b. Coating factor ψ_e applies to epoxy-coated bars. Use $\psi_e = 1.5$ if cover $< 3d_b$ or clear space $< 6d_b$. Use $\psi = 1.2$ otherwise.

c. Use $\psi_s = 0.8$ for No. 6 bars and smaller.

d. Use $\lambda = 0.75$ for lightweight concrete with f_{ct} not specified and $\lambda = 1.0$ for normal-weight concrete. Use

$$\lambda = f_{ct}/(6.7\sqrt{f_c'}) \leq 1.0$$

for lightweight concrete with f_{ct} specified.

3. Check $\psi_t \psi_e \leq 1.7$.

4. Determine c_b, the smaller of cover or half-spacing (both referenced to the center of the bar).

5. Calculate $K_{tr} = 40A_{tr}/(sn)$, or use $K_{tr} = 0$ (conservative).

6. Check $(c_b + K_{tr})/d_b \leq 2.5$.

7. Calculate K_{ER} if applicable:

$$K_{ER} = \frac{A_s \text{ required}}{A_s \text{ provided}}$$

8. Calculate ℓ_d:

$$\ell_d = \frac{K_D}{\lambda}\left[\frac{\psi_t\psi_e\psi_s}{\left(\dfrac{c_b + K_{tr}}{d_b}\right)}\right]K_{ER}\,d_b \geq 12 \text{ in.}$$

Example 5-1 _____

Calculate the tensile development length ℓ_d required for No. 8 top bars (more than 12 in. of fresh concrete to be cast below the bars) in a lightweight-aggregate concrete beam as shown in Figure 5-2. The clear cover is 2 in., and the clear space between bars is 3 in. Use $f_y = 60,000$ psi and $f_c' = 4000$ psi. Stirrups are No. 4 bars. All bars are uncoated. f_{ct} is not specified.

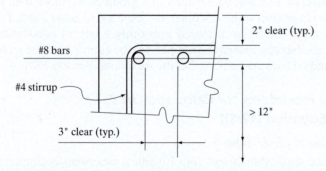

FIGURE 5-2   Partial cross section.

Solution:

Use ACI Equation (12-1) and follow the procedural outline that precedes this example.

1. From Table 5-1, $K_D = 71.2$.
2. Establish values for the factors ψ_t, ψ_e, ψ_s, and λ.
    a. $\psi_t = 1.3$ (the bars are top bars).
    b. The bars are uncoated; $\psi_e = 1.0$.
    c. The bars are No. 8; $\psi_s = 1.0$.
    d. Lightweight-aggregate concrete is used; $\lambda = 0.75$.
3. The product $\psi_t \times \psi_e = 1.3 < 1.7$. (O.K.)
4. Determine c_b. Based on cover (center of bar to nearest concrete surface), consider the clear cover, the No. 4 stirrup diameter, and one-half the diameter of the No. 8 bar:

$$c_b = 2 + 0.5 + 0.5 = 3.0 \text{ in.}$$

Based on bar spacing (one-half the center-to-center distance),

$$c_b = 0.5[3 + 2(0.5)] = 2.0 \text{ in.}$$

Therefore use $c_b = 2.0$ in.

5. In the absence of data needed for a calculation, K_{tr} may be conservatively taken as zero.
6. Check $(c_b + K_{tr})/d_b \le 2.5$:

$$\frac{c_b + K_{tr}}{d_b} = \frac{2.0 + 0}{1.0} = 2.0 < 2.5 \qquad \text{(O.K.)}$$

7. The excess reinforcement factor is assumed not applicable and is omitted.
8. Calculate ℓ_d:

$$\ell_d = \frac{K_D}{\lambda} \left[\frac{\psi_t \psi_e \psi_s}{\left(\dfrac{c_b + K_{tr}}{d_b} \right)} \right] d_b$$

$$= \frac{71.2}{0.75} \left[\frac{1.3(1.0)(1.0)}{2.0} \right] (1.0) = 61.7 \text{ in.} > 12 \text{ in.} \qquad \text{(O.K.)}$$

Example 5-2 _____

Calculate the development length required for the No. 9 bars in the top of a 15-in.-thick reinforced concrete slab (see Figure 5-3). Note that these bars are the tension reinforcement for negative moment in the slab at the supporting beam. As this is a slab, no stirrups are used. Use $f_y = 60{,}000$ psi and $f_c' = 4000$ psi (normal-weight concrete). The bars are epoxy coated.

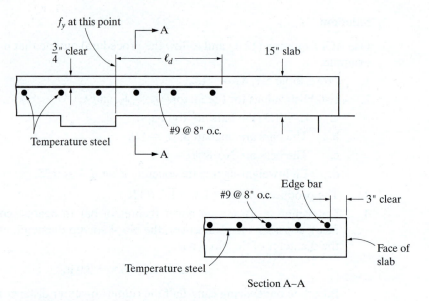

FIGURE 5-3   Sketch for Example 5-2.

Solution:

1.   From Table 5-1, $K_D = 71.2$.
2.   Establish values for the factors ψ_t, ψ_e, ψ_s, and λ.
   a. $\psi_t = 1.3$ (the bars are top bars).
   b.   The bars are epoxy coated. Compare cover ($\frac{3}{4}$ in.) with $3d_b$. If necessary, calculate clear space and compare with $6d_b$.

$$3d_b = 3(1.13) = 3.39 \text{ in.}$$

$$0.75 \text{ in.} < 3.39 \text{ in.}$$

   Therefore use $\psi_e = 1.5$.
   c.   The bars are No. 9. Use $\psi_s = 1.0$.
   d.   Normal-weight concrete is used. $\lambda = 1.0$.
3.   Check the product of ψ_t and ψ_e:

$$\psi_t \times \psi_e = 1.3(1.5) = 1.95 > 1.7$$

   Therefore use $\psi_t \times \psi_e = 1.7$.
4.   Determine c_b. Based on cover (center of bar to nearest concrete surface), consider the clear cover and one-half the diameter of the No. 9 bar:

$$c_b = 0.75 + \frac{1.128}{2} = 1.314 \text{ in.}$$

Based on bar spacing (one-half the center-to-center distance),

$$c_b = 0.5(8.0) = 4.0 \text{ in.}$$

Therefore use $c_b = 1.314$ in.

5. K_{tr} is taken as zero. (There is no transverse reinforcement crossing the plane of splitting.)

6.   Check $(c_b + K_{tr})/d_b \le 2.5$:

$$\frac{c_b + K_{tr}}{d_b} = \frac{1.314 + 0}{1.128} = 1.165 < 2.5 \qquad \text{(O.K.)}$$

7.   The excess reinforcement factor is assumed not applicable and is omitted.

8.   Calculate ℓ_d (recall that $\psi_t \times \psi_e$ will be taken as 1.7 from step 3):

$$\ell_d = \frac{K_D}{\lambda}\left[\frac{\psi_t\psi_e\psi_s}{\left(\dfrac{c + K_{tr}}{d_b}\right)}\right]d_b$$

$$= \frac{71.2}{1.0}\left[\frac{1.7(1.0)}{1.165}\right](1.128) = 117.2 \text{ in.} > 12 \text{ in.} \qquad \text{(O.K.)}$$

Example 5-3 _____

Calculate the development length required for the interior two No. 7 bars in the beam shown in Figure 5-4. The two No. 7 outside bars are continuous for the full length of the beam. Use $f_y = 60,000$ psi and $f_c' = 4000$ psi (normal-weight concrete.) The bars are uncoated. Assume that, from the design of this member, the required tension steel area was 2.28 in.[2].

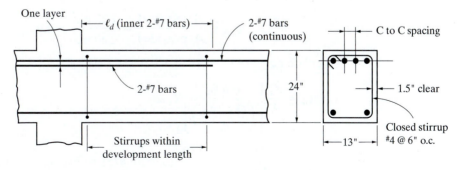

FIGURE 5-4   Sketch for Example 5-3.

Solution:

1. From Table 5-1, $K_D = 71.2$.
2. Establish values for the factors ψ_t, ψ_e, ψ_s, and λ.
   a. $\psi_t = 1.3$ (the bars are top bars).
   b. The bars are uncoated; $\psi_e = 1.0$.
   c. The bars are No. 7; $\psi_s = 1.0$.
   d. Normal-weight concrete is used; $\lambda = 1.0$.
3. The product $\psi_t \times \psi_e = 1.3 < 1.7$. (O.K.)
4. Determine c_b. Based on cover (center of bar to nearest concrete surface), consider the clear cover, the No. 4 stirrup diameter, and one-half the diameter of the No. 7 bar:

$$c_b = 1.5 + 0.5 + \frac{0.875}{2} = 2.44 \text{ in.}$$

Based on bar spacing (one-half the center-to-center distance),

$$c_b = \frac{13 - 2(1.5) - 2(0.5) - 2\left(\dfrac{0.875}{2}\right)}{3(2)} = 1.354 \text{ in.}$$

Therefore use $c_b = 1.354$ in.

5. Using data on stirrups from Figure 5-4:

$$K_{tr} = \frac{40A_{tr}}{(sn)} = \frac{40(0.40)}{(6)(2)} = 1.333$$

6. Check $(c_b + K_{tr})/d_b \leq 2.5$:

$$\frac{c_b + K_{tr}}{d_b} = \frac{1.354 + 1.333}{0.875} = 3.07 > 2.5$$

Therefore, use 2.5.

7. Calculate the excess reinforcement factor:

$$K_{ER} = \frac{A_s \text{ required}}{A_s \text{ provided}} = \frac{2.28 \text{ in.}^2}{2.40 \text{ in.}^2} = 0.95$$

8. Calculate ℓ_d:

$$\ell_d = \frac{K_D}{\lambda}\left[\frac{\psi_t\psi_e\psi_s}{\left(\dfrac{c_b + K_{tr}}{d_b}\right)}\right]K_{ER}d_b$$

$$= \frac{71.2}{1.0}\left[\frac{1.3(1.0)(1.0)}{2.5}\right](0.95)(0.875) = 30.8 \text{ in.} > 12 \text{ in.} \qquad \text{(O.K.)}$$

5-3 DEVELOPMENT LENGTH: COMPRESSION BARS

Whereas tension bars produce flexural tension cracking in the concrete, this effect is not present with compression bars, and thus shorter development lengths are allowed. For deformed bars in compression, the development length ℓ_{dc} is calculated from

$$\ell_{dc} = \left(\frac{0.02f_y}{\lambda\sqrt{f'_c}}\right)d_b \geq 0.0003f_yd_b$$

in inches, where all quantities are as described previously for tension development length. This value of ℓ_{dc} is tabulated in Table A-12 for $\lambda = 1.0$, grade 60 bars, and various values of f'_c. The required compression development length may be further reduced by multiplying ℓ_{dc} by the following modification factors:

1. Reinforcement in excess of that required:

$$\frac{A_s \text{ required}}{A_s \text{ provided}}$$

2. Bars enclosed within a spiral that is not less than $\frac{1}{4}$ in. in diameter and not more than 4 in. in pitch or within No. 4 ties in conformance with ACI, Section 7.10.5, and spaced at not more than 4 in. on center: Use 0.75.

The value of ℓ_{dc} shall not be less than 8 in.

5-4 DEVELOPMENT LENGTH: STANDARD HOOKS IN TENSION

In the event that the desired development length in tension cannot be furnished, it is necessary to provide mechanical anchorage at the end of the bars. Although the ACI Code (Section 12.6) allows any mechanical device to serve as anchorage if its adequacy is verified by testing, anchorage for reinforcement is usually accomplished by means of a 90° or 180° hook. The dimensions and bend radii for these hooks have been standardized by the ACI Code. Additionally, 90° and 135° hooks have been standardized for stirrups and tie reinforcement. Hooks in compression bars are ineffective and cannot be used as anchorage. Standard reinforcement hooks are shown in Figure 5-5. The bend diameters are measured on the inside of the bar.

The ACI Code, Section 12.5, specifies that the development length ℓ_{dh} (see Figure 5-5) for deformed bars in tension, which terminate in a standard hook, be computed from the following expression:

$$\ell_{dh} = \left[\frac{0.02\,\psi_e f_y}{\lambda\sqrt{f'_c}}\right]d_b$$

where $\psi_e = 1.2$ for epoxy-coated reinforcing (1.0 otherwise)

$\lambda = 0.75$ for lightweight concrete (1.0 otherwise)

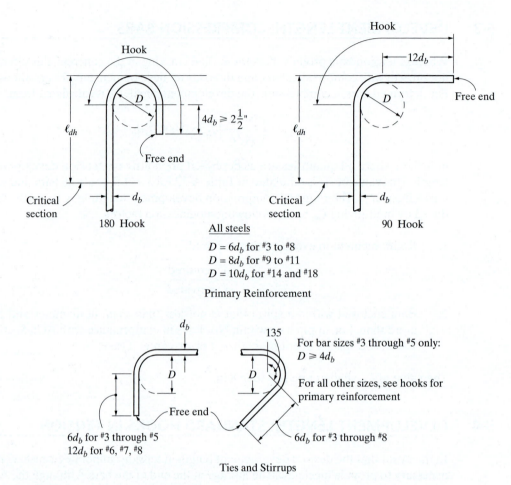

FIGURE 5-5 ACI standard hooks.

Table A-13 gives values of ℓ_{dh} for various values of f'_c, $f_y = 60{,}000$ psi, $\psi_e = 1.0$, and $\lambda = 1.0$. The length ℓ_{dh} may be further reduced by multiplication by the following modification factors. The final ℓ_{dh} shall not be less than $8d_b$ nor less than 6 in.

1. Cover. For bars No. 11 and smaller with side cover (normal to the plane of the hook) not less than $2\frac{1}{2}$ in., and for a 90° hook with cover on the extension beyond the hook not less than 2 in.: 0.7.

2. Enclosure for 90° hooks for No. 11 and smaller bars within ties or stirrups perpendicular to the bar being developed, spaced not more than $3d_b$ along ℓ_{dh}; or enclosed within ties or stirrups parallel to the bar being developed, spaced not more than $3d_b$ along the tail extension plus the bend: 0.8.

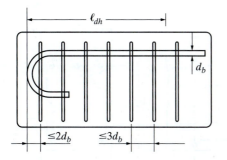

FIGURE 5-6   Enclosure for 180° hook; modification factor = 0.8°.

3.   Enclosure for 180° hooks for No. 11 and smaller bars within ties or stirrups perpendicular to the bar being developed, with spacing not greater than $3d_b$ along ℓ_{dh} of the hook: 0.8.

4.   Where anchorage or development of f_y is not specifically required, reinforcement in excess of that required by analysis:

$$\frac{A_{s(\text{required})}}{A_{s(\text{provided})}}$$

In the preceding items 2 and 3, d_b is the diameter of the hooked bar, and the first tie or stirrup shall enclose the bent portion of the hook, within $2d_b$ of the outside of the bend. This is illustrated in Figure 5-6.

In addition, the ACI Code, Section 12.5.4, establishes criteria for hooked bars that terminate at the discontinuous ends of members such as simply supported beams, free ends of cantilevers, and ends of members that frame into a joint where the member does not extend beyond the joint. If the full strength (f_y) of the hooked bar must be developed and if both the side cover and the top (or bottom) cover over the hook are less than $2\frac{1}{2}$ in., closed ties or stirrups, perpendicular to the bar being developed, spaced at $3d_b$ maximum are required over the development length ℓ_{dh}. The first tie or stirrup shall enclose the bent portion of the hook, within $2d_b$ of the bend. This does not apply to the discontinuous ends of slabs with concrete confinement provided by the slab continuous on both sides perpendicular to the plane of the hook. Also, the 0.8 MF of the preceding items 2 and 3 does not apply.

Example 5-4 _____

Determine the anchorage or development length required for the tension (top) bars for the conditions shown in Figure 5-7. Use $f_c' = 3000$ psi (normal-weight concrete) and $f_y = 60,000$ psi. The No. 8 bars may be categorized as top bars. Assume a side cover on the main bars of $2\frac{1}{2}$ in. minimum. Bars are uncoated.

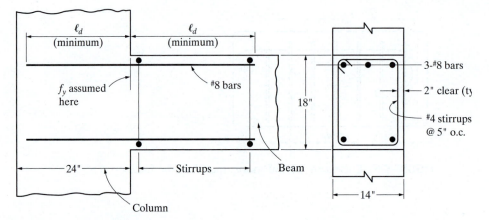

FIGURE 5-7   Sketch for Example 5-4.

Solution:

a.   Anchorage of top bars into the exterior column

1.   From Table 5-1, $K_D = 82.2$.

2.   Establish values for the factors ψ_t, ψ_e, ψ_s, and λ.

   a. $\psi_t = 1.3$ (the bars are top bars).

   b.   The bars are uncoated; $\psi_e = 1.0$.

   c.   The bars are No. 8; $\psi_s = 1.0$.

   d.   Normal-weight concrete is used; $\lambda = 1.0$.

3.   The product $\psi_t \times \psi_e = 1.3 < 1.7$ (O.K.).

4.   Determine c_b. Based on cover (center of bar to nearest concrete surface), consider the clear cover, the No. 4 stirrup diameter, and one-half the diameter of the No. 8 bar:

$$c_b = 2.0 + 0.5 + \frac{1.0}{2} = 3.0 \text{ in.}$$

Based on bar spacing (one-half the center-to-center distance),

$$c_b = \frac{14 - 2(2.0) - 2(0.5) - 2(0.5)}{2(2)} = 2.0 \text{ in.}$$

Therefore use $c_b = 2.0$ in.

5.   Figure 5-7 shows stirrups in the beam. However, there are no stirrups in the column and K_{tr} is taken as 0.

6.   Check $(c_b + K_{tr})/d_b \leq 2.5$:

$$\frac{c_b + K_{tr}}{d_b} = \frac{2.0 + 0}{1.0} = 2.0 < 2.5 \qquad \text{(O.K.)}$$

7. The excess reinforcement factor is assumed not applicable and is omitted.

8. Calculate ℓ_d:

$$\ell_d = \frac{K_D}{\lambda}\left[\frac{\psi_t\psi_e\psi_s}{\left(\dfrac{c_b + K_{tr}}{d_b}\right)}\right]d_b$$

$$= \frac{82.2}{1.0}\left[\frac{1.3(1.0)(1.0)}{2.0}\right](1.0) = 53.4 \text{ in.} > 12 \text{ in.} \qquad \text{(O.K.)}$$

Because 53.4 in. > 24 in. column width, use a standard hook, either a 90° hook or a 180° hook.

b. Anchorage using a standard 180° hook

1. The development length ℓ_{dh} for the hook shown in Figure 5-8 is calculated from

$$\ell_{dh} = \frac{0.02\psi_e f_y}{\lambda\sqrt{f_c'}}d_b$$

The bars are uncoated and the concrete is normal weight. Therefore, both ψ_e and λ are 1.0:

$$\ell_{dh} = \frac{0.02(60,000)}{\sqrt{3000}}(1.00) = 21.9 \text{ in.}$$

(Check this with Table A-13.)

2. The only applicable modification factor is based on side cover (normal to the plane of the hook) of 2½ in. Use a modification factor of 0.7 (ACI Code, Section 12.5.3a).

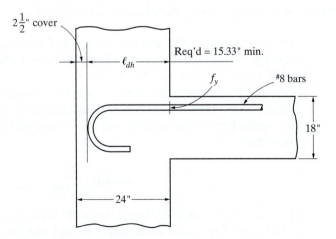

FIGURE 5-8   Sketch for Example 5-4.

3.   The required development length is then calculated from

$$\ell_{dh} = 21.9(0.7) = 15.33 \text{ in.}$$

Check minimum:

$$\text{minimum } \ell_{dh} = 8d_b \geq 6 \text{ in.}$$

$$8d_b = 8 \text{ in.} < 15.33 \text{ in.} \qquad \text{(O.K.)}$$

The minimum width of column required is

$$15.33 + 2.5 = 17.83 \text{ in.} < 24 \text{ in.} \qquad \text{(O.K.)}$$

The hook therefore will fit into the column and the detail is satisfactory.

Anchorage into beam: The development length required if bars are straight is (conservatively) 53.4 in., as determined previously. Therefore, the bars must extend at least this distance into the span. The ACI Code has additional requirements for the extension of tension bars in areas of negative moments. These requirements are covered in the discussion on continuous construction in Chapter 6.

5-5   DEVELOPMENT OF WEB REINFORCEMENT

Anchorage of web reinforcement must be furnished in accordance with the ACI Code, Section 12.13. Stirrups must be carried as close to the compression and tension surfaces as possible. Close proximity to the compression face is necessary as flexural tension cracks penetrate deeply as ultimate load is approached.

The ACI Code stipulates that ends of single-leg, simple U-, or multiple U-stirrups shall be anchored by one of the following means (see Figure 5-9):

1.   For a No. 5 bar or smaller, and for Nos. 6, 7, and 8 bars of $f_y = 40,000$ psi or less, anchorage is provided by a standard stirrup hook, bent around a longitudinal bar (ACI Code, Section 12.13.2.1).

2.   For Nos. 6, 7, and 8 stirrups with $f_y > 40,000$ psi, anchorage is provided by a standard stirrup hook bent around a longitudinal bar *plus* an embedment between midheight of the member and the outside end of the hook equal to or greater than

$$0.014 d_b \frac{f_y}{\left(\lambda \sqrt{f_c'}\right)}$$

A 135° or a 180° hook is preferred, but a 90° hook is acceptable provided that the free end of the hook is extended the full $12d_b$ as required in ACI Code, Section 7.1.3 (ACI Code, Section 12.13.2.2).

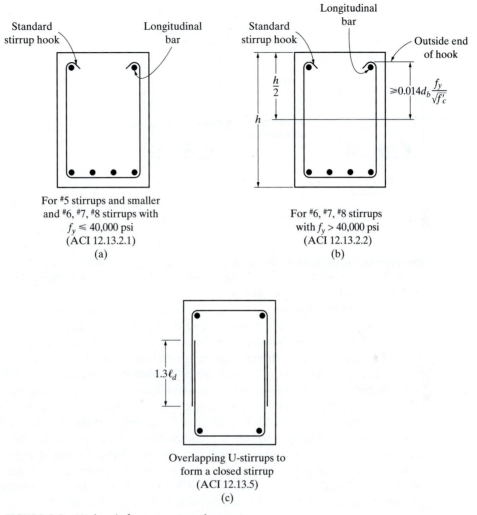

FIGURE 5-9   Web reinforcement anchorage.

It should be noted that the ACI standard hooks for ties and stirrups, as shown in Figure 5-9, include 90° and 135° hooks only. This does not imply that the 180° hook is not acceptable. The 135° hook is easier to fabricate. Many stirrup bending machines now in use are not designed to fabricate a 180° hook. The anchorage strength of either a 135° or 180° hook is approximately the same.

In addition, the ACI Code, Section 12.13.5, establishes criteria with respect to lapping of double U-stirrups or ties (without hooks) to form a closed stirrup. Legs shall be considered properly spliced when lengths of lap are 1.3 ℓ_d, as depicted in Figure 5-9c. Each bend of each simple U-stirrup must enclose a longitudinal bar. If the lap of 1.3 ℓ_d cannot fit within the depth of a shallow member, provided that the depth of

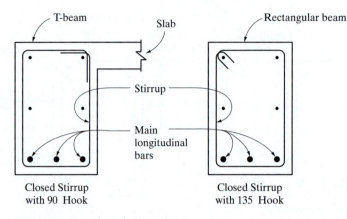

FIGURE 5-10   Closed stirrups.

the member is at least 18 in., double U-stirrups may be used if each U-stirrup extends the full available depth of the member and the force in each leg does not exceed 9000 lb (that is, $A_b f_y \leq 9000$ lb).

Where torsional reinforcing is required or desired, a commonly used alternative stirrup is the one-piece closed stirrup. Use of the one-piece closed stirrup is disadvantageous, however, in that the entire beam reinforcing (longitudinal steel and stirrups) may have to be prefabricated as a cage and then placed as a unit. This may not be practical if the longitudinal bars have to be passed between column bars. Alternatively, and at a greater cost, the longitudinal bars could be threaded through the closed stirrups and column bars. Two commonly used types of one-piece closed stirrups are shown in Figure 5-10. If spalling of the member at the transverse torsional reinforcement anchorage (hooks) is restrained by a flange or similar member, 90° standard hooks around a longitudinal bar are allowed, as shown in Figure 5-10a. Otherwise, 135° standard hooks around a longitudinal bar are required (Figure 5-10b). Numbers 6, 7, and 8 stirrups with $f_y > 40,000$ psi require additional anchorage, as previously described. See ACI Code, Section 11.5.4.2, for full details.

5-6   SPLICES

The need to splice reinforcing steel is a reality due to the limited lengths of steel available. All bars are readily available in lengths up to 60 ft; No. 3 and No. 4 bars will tend to bend in handling when longer than 40 ft, however. Typical stock straight lengths are as follows:

No. 3 bar: 20, 40, 60 ft
No. 4 bar: 30, 40, 60 ft
Nos. 5 to 18 bars: 60 ft

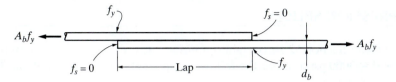

FIGURE 5-11   Stress transfer in tension lap splice.

Splicing may be accomplished by welding, by mechanical means, or, most commonly for No. 11 bars and smaller, by lapping bars, as shown in Figure 5-11. Lap splices may not be used for bars larger than No. 11 except for compression splices at footings, as provided in the ACI Code, Section 15.8.2.3, and for compression splices of bars of different sizes, as provided in the ACI Code, Section 12.16.2. The splice composed of lapped bars is usually more economical than the other types. The lapped bars are commonly tied in contact with each other. They may, however, be spaced apart up to one-fifth of the lap length, with an upper limit of 6 in. Splices in regions of maximum moment preferably should be avoided.

The ACI Code (Section 1.2.1) requires that the design drawings show the location and length of lap splices and the type and location of mechanical and welded splices of reinforcement.

5-7   TENSION SPLICES

The required length of lap is based on the class in which the splice is categorized. The required lap length increases with increased stress and increased amount of steel spliced in close proximity. Lap length is expressed in terms of the tensile development length ℓ_d for the particular bar, as shown in Section 5-2. The 12-in. minimum for the ℓ_d calculation is not considered, nor is the excess reinforcement factor considered for tension splices, as the splice classification already reflects any excess reinforcement at the splice location.

The ACI Code, Section 12.15, directs that lap splices be class B (lap length = $1.3\ell_d$) except that class A (lap length = $1.0\ell_d$) splices are allowed if (1) the area of reinforcement provided is twice that required for the entire length of the splice *and* (2) not more than 50% of the total reinforcement is spliced within the required lap length.

The minimum length of lap for tension lap splices is 12 in. (ACI Code, Section 12.15.1). Requirements for mechanical and welded splices for tensile reinforcement are contained in the ACI Code, Sections 12.14.3 and 12.15. In addition, splices in "tension tie members" must be made with a full welded splice or full mechanical splice, and their locations must be staggered a distance of at least 30 in., in accordance with the ACI Code, Section 12.15.6. Staggering of all tension splices in all types of members is encouraged.

5-8   COMPRESSION SPLICES

The ACI Code, Section 12.16, contains requirements for lap splices for compression bars. For $f'_c = 3000$ psi or more, the following lap lengths, in multiples of bar diameters d_b, are required:

$$f_y = 40,000 \text{ psi}: 20d_b$$

$$f_y = 60,000 \text{ psi}: 30d_b$$

$$f_y = 75,000 \text{ psi}: 44d_b$$

but not less than 12 in. For $f'_c < 3000$ psi, the length of lap should be increased by one-third. Within ties of specific makeup or spirals (ACI Code, Sections 12.17.2.4 and 12.17.2.5), these laps may be reduced to 0.83 or 0.75, respectively, of the foregoing values, but must not be less than 12 in.

Compression splices may also be of the end-bearing type, where bars are cut square, then butted together and held in concentric contact by a suitable device. End-bearing splices must not be used except in members containing closed ties, closed stirrups, or spirals. Welded splices and mechanical connections are also acceptable and are subject to the requirements of the ACI Code, Section 12.16. Special splice requirements for columns are furnished by the ACI Code, Section 12.17.

5-9   SIMPLE-SPAN BAR CUTOFFS AND BENDS

The maximum required A_s for a beam is needed only where the moment is maximum. This maximum steel may be reduced at points along a bending member where the bending moment is smaller. This is usually accomplished by either stopping or bending the bars in a manner consistent with the theoretical requirements for the strength of the member, as well as the requirements of the ACI Code.

Bars can theoretically be stopped or bent in flexural members whenever they are no longer needed to resist moment. The ACI Code, Section 12.10.3, however, requires that each bar be extended beyond the point at which it is no longer required for flexure for a distance equal to the effective depth of the member or $12d_b$, whichever is greater, except at supports of simple spans and at free ends of cantilevers. In effect, this prohibits the cutting off of a bar at the theoretical cutoff point but can be interpreted as permitting bars to be bent at the theoretical cutoff point. If bars are to be bent, a general practice that has evolved is to commence the bend at a distance equal to one-half the effective depth beyond the theoretical cutoff point. The bent bar should be anchored or made continuous with reinforcement on the opposite face of the member. It should be pointed out, however, that the bending of reinforcing bars in slabs and beams to create *truss bars* (see types 3 through 7, Figure 13-4) has fallen into disfavor over the years because of placement problems and the labor involved. It is more common to use straight bars and place them in accordance with the strength requirements.

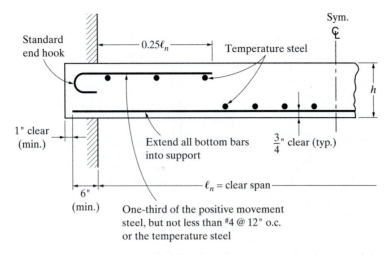

FIGURE 5-12   Recommended bar details: one-way simple-span slabs; tensile-reinforced simple-span beams similar.

With simple-span flexural members of constant dimensions, we can assume that the required A_s varies directly with the bending moment and that the shape of a *required A_s curve* is identical with that of the moment diagram. The moment diagram (or curve) may then be used as the required A_s curve by merely changing the vertical scale. Steel areas may be used as ordinates, but it is more convenient to use the required number of bars (assuming that all bars are of the same diameter). The necessary correlation is established by using the maximum ordinate of the curve as the maximum required A_s in terms of the required number of bars. Because of possible variations in the shape of the moment diagram, either a graphical or a mathematical approach may be more appropriate. A graphical approach requires that the moment diagram be plotted to scale. Example 5-5 lends itself to a mathematical solution.

In determining bar cutoffs, it should be remembered that the stopping of bars should be accomplished by using a symmetrical pattern so that the remaining bars will also be in a symmetrical pattern. In addition, the ACI Code, Section 12.11.1, requires that for simple spans, at least one-third of the positive moment steel extend into the support a distance of 6 in. In practice, this requirement is generally exceeded. Normally, recommended bar details for single-span solid concrete slabs indicate that all bottom bars should extend into the support. See Figure 5-12 for recommended bar details.

Economy sometimes dictates that reinforcing steel should be cut off in a simple span. Example 5-5 shows one approach to this problem.

Example 5-5 _____

A simple-span, uniformly loaded beam, shown in Figure 5-13a and b, requires six No. 7 bars for tensile reinforcement. If the effective depth d is 18 in., determine where the bars may be stopped. Use $f'_c = 4000$ psi and $f_y = 60,000$ psi.

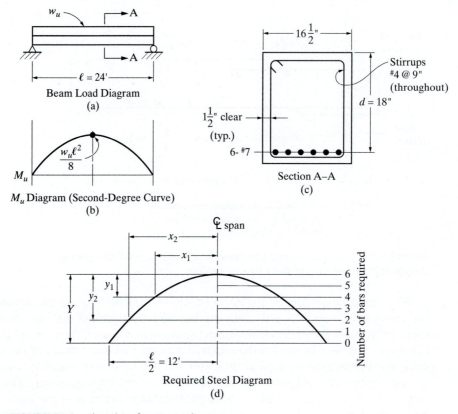

FIGURE 5-13   Sketches for Example 5-5.

Assume that there is no excess steel (required A_s = furnished A_s = 3.60 in.^{2}). Assume normal-weight concrete and that the stirrups extend to the end of the beam at the support. The bars are uncoated.

Solution:

We will try to establish a bar cutoff scheme whereby the two center bars are cut first, followed by the cutting of the other two inside bars. The two corner bars are to run the full length of the beam. First, we check the minimum steel requirement. From Table A-5:

$$A_{s,min} = 0.0033\,bd = 0.0033(16.5)(18) = 0.98 \text{ in.}^2$$

The two corner bars provide a steel area of 1.20 in.^{2} Therefore the minimum steel area requirement is met.

1.   Determine where the first two bars may be cut off. The first two bars may be stopped where only four are required. This distance from the center-line of the span is designated x_1 in Figure 5-13d.

As the moment diagram is a second-degree curve (a parabola), offsets to the line tangent to the curve at the point of maximum moment vary as the squares of the distances from the centerline of the span. The solution for the distance x_1 may be formulated as follows:

$$\frac{(x_1)^2}{\left(\dfrac{\ell}{2}\right)^2} = \frac{y_1}{Y}$$

$$\frac{(x_1)^2}{12^2} = \frac{2 \text{ bars}}{6 \text{ bars}}$$

from which $x_1 = 6.93$ ft. This locates the *theoretical* point where two bars may be terminated. Additionally, bars must be extended past this point a distance d or $12d_b$, whichever is larger. Thus

$$12 \text{ bar diameters} = 12(0.875)$$

$$= 10.5 \text{ in.}$$

Because $d = 18$ in., the bars should be extended 18 in. (1.50 ft). Then the minimum distance from the centerline of the span to the cutoff of the first two bars is

$$6.93 + 1.50 = 8.43 \text{ ft}$$

2. Determine where the next two bars may be cut off. The two remaining bars, which are the corner bars, will continue into the support. With reference to Figure 5-13d:

$$\frac{(x_2)^2}{12^2} = \frac{4}{6}$$

$$x_2 = 9.80 \text{ ft}$$

Therefore the minimum distance from the centerline of the span to the cutoff of the next two bars is

$$9.80 + 1.50 = 11.30 \text{ ft}$$

3. A check must be made of the required development length ℓ_d for the first two bars. Because the bar cutoff occurs 8.43 ft from the centerline of the span, ℓ_d must be less than 8.43 ft for the stress f_y to be developed at the centerline.

a. From Table 5-1, $K_D = 71.2$.

b. Establish values for the factors ψ_t, ψ_e, ψ_s, and λ.

　　1. $\psi_t = 1.0$ (the bars are not top bars).

　　2. The bars are uncoated; $\psi_e = 1.0$.

　　3. The bars are No. 7; $\psi_s = 1.0$.

　　4. Normal-weight concrete is used; $\lambda = 1.0$.

c. The product $\psi_t \times \psi_e = 1.0 < 1.7$. (O.K.)

d.　Determine c_b. Based on cover (center of bar to nearest concrete surface), consider the clear cover, the No. 4 stirrup, and one-half the diameter of the No. 7 bar:

$$c_b = 1.5 + 0.5 + \frac{0.875}{2} = 2.44 \text{ in.}$$

Based on bar spacing (one-half the center-to-center distance),

$$c_b = \frac{16.5 - 2(1.5) - 2(0.5) - 2\left(\dfrac{0.875}{2}\right)}{5(2)} = 1.163 \text{ in.}$$

Therefore use $c_b = 1.163$ in.

e.　Using data on stirrups from Figure 5-13:

$$K_{tr} = \frac{40A_{tr}}{sn} = \frac{40(0.40)}{(9)(2)} = 0.889$$

f.　Check $(c_b + K_{tr})/d_b \leq 2.5$:

$$\frac{c_b + K_{tr}}{d_b} = \frac{1.163 + 0.889}{0.875} = 2.35 < 2.5 \qquad \text{(O.K.)}$$

g.　The excess reinforcement factor is neglected (conservative).

h.　Calculate ℓ_d:

$$\ell_d = \frac{K_D}{\lambda}\left[\frac{\psi_t\psi_e\psi_s}{\left(\dfrac{c_b + K_{tr}}{d_b}\right)}\right]d_b$$

$$= \frac{71.2}{1.0}\left[\frac{1.0(1.0)(1.0)}{2.35}\right](0.875) = 26.5 \text{ in.} > 12 \text{ in.} \qquad \text{(O.K.)}$$

26.5 in. = 2.21 ft

As the required tensile development length of 2.21 ft is much less than the distance from the centerline to the actual cutoff point (8.43 ft), the cutoff for the first two bars is satisfactory. Similarly, if we consider the second two bars, for which ℓ_d can conservatively be taken as 2.21 ft (conservative, as the c_b distance, based on bar spacing, is larger for these two bars than for the first two cut bars), the point at which they are stressed to maximum (the theoretical cutoff point for the first two bars), measured from the centerline, is 6.93 ft. The second two bars must extend at least ℓ_d past this point. Again measured from the centerline, the bars must extend at least

$$6.93 + 2.21 = 9.14 \text{ ft}$$

Because this is less than the actual distance to the cutoff point (11.30 ft), the cutoff for the second pair is satisfactory (see Figure 5-14).

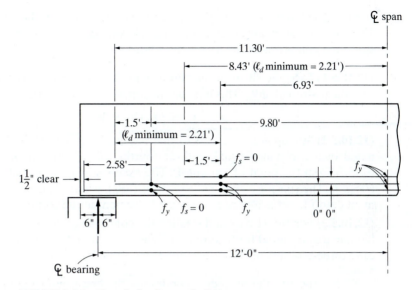

FIGURE 5-14   Sketch for Example 5-5.

4.   The remaining pair of corner bars, representing one-third of the positive moment steel, must continue into the support. As the corner bars are stressed to f_y at a distance of 9.80 ft from the centerline of the span (where the second pair of bars may be theoretically cut), however, the development length of these bars must be satisfied from the end of the bar back to the point of maximum stress f_y. The development length required is again conservatively taken as 2.21 ft. The straight development length furnished (or available) is 2.58 ft (see Figure 5-14). Because 2.58 ft > 2.21 ft, there is adequate development length available. If the development length available were insufficient, either of two solutions could be used: (a) Provide a standard 180° hook at the end of each of the two corner bars or (b) extend all four of the outside bars to the end of the beam (thereby cutting only the two center bars).

Example 5-5 could also be solved by plotting the M_u diagram to scale and superimposing on the M_u diagram the values of ϕM_n for four bars and two bars. The theoretical cut points are established where the ϕM_n lines intersect the M_u curve. For instance, the theoretical cut point for the two center bars is established where the ϕM_n line for four bars intersects the M_u curve. This method has several advantages: (1) Any excess flexural steel in the design will result in the two center bars being cut closer to the midpoint of the beam, thus saving steel; (2) the calculation of ϕM_n better reflects the available moment strength than does the division of the M_u diagram ordinate into six equal parts (in Example 5-5); and (3) it can be used for any shape M_u diagram.

The bars in Example 5-5 are terminating in a tension zone. When this occurs, the member undergoes a reduction in shear capacity. Therefore special shear and stirrup calculations must be performed in accordance with the ACI Code, Section 12.10.5, which stipulates that no flexural reinforcement shall be terminated in a tension zone unless *one* of the following conditions is satisfied:

(12.10.5.1) Factored shear at the cutoff point does not exceed two-thirds of the design shear strength ϕV_n. This may be written as

$$V_u \leq \frac{2}{3}\phi(V_c + V_s)$$

(12.10.5.2) Stirrup area in excess of that required is provided along each bar terminated over a distance equal to three-fourths the effective depth of the member (0.75*d*) from the point of bar cutoff. The excess stirrup area shall be not be less than $60b_w s/f_y$. The spacing *s* shall not exceed $d/(8\beta_b)$, where β_b is the ratio of reinforcement cutoff to total area of tension reinforcement at the section.

(12.10.5.3) For No. 11 bars and smaller, the continuing reinforcement provides double the area required for flexure at the cutoff point and factored shear does not exceed three-quarters of the design shear strength ϕV_n.

    With respect to the preceding conditions, the first is interpreted as a check only. Although the V_s term could be varied, the code is not specific concerning the length of beam over which this V_s must exist. The third condition involves moving the cutoff point to another location. The second condition is a design method whereby additional stirrups may be introduced if the shear strength is inadequate as determined by the ACI Code, Section 12.10.5.1. This condition will be developed further.

    It is convenient to determine the *number* of additional stirrups to be added in the length 0.75*d* along the end of the bar from the cutoff point. The required excess stirrup area is

$$A_v = \frac{60b_w s}{f_{yt}}$$

Assuming that the size of the stirrup (and therefore A_v) is known, the maximum spacing is

$$s = \frac{A_v f_{yt}}{60b_w}$$

Therefore, the number of stirrups N_s to be added in the length 0.75*d* is

$$N_s = \frac{0.75d}{s} + 1 = \frac{0.75d}{\left(\dfrac{A_v f_{yt}}{60b_w}\right)} + 1 = \left(\frac{45b_w d}{A_v f_{yt}}\right) + 1$$

Also, because maximum $s = d/(8\beta_b)$,

$$N_{s(\min)} = \frac{0.75d}{\left(\dfrac{d}{8\beta_b}\right)} + 1 = 6\beta_b + 1$$

The larger resulting N_s controls.

Example 5-6

In the beam shown in Figure 5-15, the location at which the top layer of two No. 9 bars is to be terminated is in a tension zone. At the cutoff point, $V_u = $ 52 kips and the No. 3 stirrups are spaced at 11 in. on center. Check shear in accordance with the ACI Code, Section 12.10.5, and redesign the stirrup spacing if necessary. Assume that $f'_c = 3000$ psi and $f_y = 60{,}000$ psi.

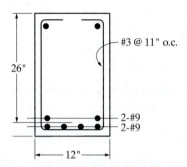

#3 @ 11" o.c.

26"

2-#9
2-#9

12"

FIGURE 5-15   Sketch for Example 5-6.

Solution:

In accordance with the ACI Code, Section 12.10.5.1, check $V_u \le \frac{2}{3}\phi(V_c + V_s)$:

$$\frac{2}{3}\phi(V_c + V_s) = \frac{2}{3}(0.75)\left(2\sqrt{f'_c}bd + \frac{A_v f_{yt} d}{s}\right)$$

$$= \frac{2}{3}(0.75)\left[2\sqrt{3000}\,(12)(26) + \frac{(0.22)(60{,}000)(26)}{11}\right]$$

$$= 32{,}700 \text{ lb}$$

$$= 32.7 \text{ kips} < 52 \text{ kips} \qquad\qquad\qquad \text{(N.G.)}$$

Therefore, add excess stirrups over a length of $0.75d$ along the terminated bars from the cut end in accordance with ACI 12.10.5.2:

$$N_s = \left(\frac{45 b_w d}{A_v f_{yt}} + 1\right) \text{ or } (6\beta_b + 1)$$

$$\frac{45(12)(26)}{0.22(60{,}000)} + 1 = 2.06$$

$$6(\tfrac{1}{3}) + 1 = 3$$

Add three stirrups over a length of $0.75d = 19.5$ in. Find the new spacing required in the 19.5-in. length:

$$\frac{19.5}{11} = 1.77 \text{ stirrups at 11-in. spacing}$$

$$\underline{+\ 3.00} \text{ additional stirrups}$$

$$4.77 \text{ stirrups in 19.5-in. length}$$

The new stirrup spacing is

$$\frac{19.5}{4.77} = 4.09 \text{ in.}$$

Use 4 in.

The stirrup pattern originally designed should now be altered to include the 4-in. spacing along the last 19.5 in. of the cut bars.

The problems associated with terminating bars in a tension zone may be avoided by extending the bars in accordance with the ACI Code, Section 12.10.5.3, or by extending them into the support.

In summary, a general representation of the bar cutoff requirements for positive moment steel in a simple span may be observed in Figure 5-16. If bars A are to be cut off, they must (1) project ℓ_d past the point of maximum positive moment and (2) project beyond their theoretical cutoff point a distance equal to the effective depth of the member or 12 bar diameters, whichever is greater. The remaining positive moment bars B must extend ℓ_d past the theoretical cutoff point of bars A and extend at least 6 in. into the support.

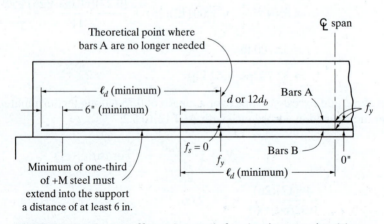

FIGURE 5-16 Bar cutoff requirements for simple spans (positive moment steel).

5-10   CODE REQUIREMENTS FOR DEVELOPMENT OF POSITIVE MOMENT STEEL AT SIMPLE SUPPORTS

The ACI Code, Section 12.11.3, contains requirements concerning the development of straight, positive moment bars *at simple supports* and at points of inflection. The intent of this code section is the same as the check on the development of the two bars extending into the support in Example 5-5. The method of Example 5-5 may be used if we are working with the actual moment diagram. The code approach does not require the use of a moment diagram.

The ACI Code requirement places a restriction on the size of the bar that may be used such that

$$\ell_d \leq \frac{M_n}{V_u} + \ell_a \qquad\qquad \text{[ACI Eq. (12-5)]}$$

where

M_n = nominal moment strength $\left[A_s f_y \left(d - \dfrac{a}{2} \right) \right]$ assuming all reinforcement at the section to be stressed to f_y

V_u = total applied design shear force at the section

ℓ_a = (at a support) the embedment length beyond the center of the support

ℓ_a = (at a point of inflection) the effective depth of the member or $12d_b$, whichever is greater

This requirement need not be satisfied for reinforcement terminating beyond the centerline of simple supports by a standard hook or a mechanical anchorage at least equivalent to a standard hook.

The effect of this code restriction is to require that bars be small enough so that they can become fully developed before the applied moment has increased to the magnitude where they *must* be capable of carrying f_y. The M_n/V_u term approximates the distance from the section in question to the location where applied moment M_u exists that is equal to ϕM_n (and where f_y must exist in the bars).

Therefore the distance from the end of the bar to the point where the bar must be fully developed is $(M_n/V_u) + \ell_a$, and bars must be chosen so that their ℓ_d is less than this distance. The code allows M_n/V_u to be increased by 30% when the ends of the reinforcement are confined by a compressive reaction such as is found in a simply supported beam (a beam supported by a wall).

Example 5-7

At the support of a simply supported beam, a cross section exists as shown in Figure 5-17. Check the bar diameters in accordance with the ACI Code, Section 12.11.3. Assume a support width of 12 in., 1½-in. cover, f'_c = 4000 psi, and f_y = 60,000 psi. Assume that V_u at the support is 80 kips. Normal-weight concrete is used. The stirrups begin at 3 in. from the face of the support. The bars are uncoated.

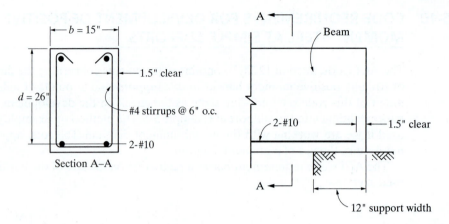

FIGURE 5-17   Sketch for Example 5-7.

Solution:

Because this beam has its ends confined by a compressive reaction, we will check ACI Equation (12-3):

$$\ell_d \leq 1.3\left(\frac{M_n}{V_u}\right) + \ell_a$$

We next calculate the required tensile development length ℓ_d:

1.   From Table 5-1, $K_D = 71.2$.
2.   Establish values for the factors ψ_t, ψ_e, ψ_s, and λ.
     a. $\psi_t = 1.0$ (the bars are not top bars).
     b.   The bars are uncoated; $\psi_e = 1.0$.
     c.   The bars are No. 10; $\psi_s = 1.0$.
     d.   Normal-weight concrete is used; $\lambda = 1.0$.
3.   The product $\psi_t \times \psi_e = 1.0 < 1.7$ (O.K.).
4.   Determine c. Based on cover (center of bar to nearest concrete surface), consider the clear cover, the No. 4 stirrup, and one-half the diameter of the No. 10 bar:

$$c_b = 1.5 + 0.5 + \frac{1.27}{2} = 2.64 \text{ in.}$$

Based on bar spacing (one-half the center-to-center distance),

$$c_b = \frac{15 - 2(1.5) - 2(0.5) - 2\left(\dfrac{1.27}{2}\right)}{2} = 4.87 \text{ in.}$$

Therefore use $c_b = 2.64$ in.

5. K_{tr} is taken as zero, as stirrups do not extend to the ends of the bars.

6. Check $(c_b + K_{tr})/d_b \leq 2.5$:

$$\frac{c_b + K_{tr}}{d_b} - \frac{2.64 + 0}{1.27} = 2.08 < 2.5 \qquad \text{(O.K.)}$$

7. The excess reinforcement factor is not applicable and is omitted.

8. Calculate ℓ_d:

$$\ell_d = \frac{K_D}{\lambda}\left[\frac{\psi_t\psi_e\psi_s}{\left(\dfrac{c_b + K_{tr}}{d_b}\right)}\right]d_b$$

$$= \frac{71.2}{1.0}\left[\frac{1.0(1.0)(1.0)}{2.08}\right](1.27) = 43.5 \text{ in.} > 12 \text{ in.} \qquad \text{(O.K.)}$$

Now check ACI Equation (12-5):

$$\rho = \frac{A_s}{bd} = \frac{2.54}{15(26)} = 0.0065$$

$$\bar{k} = 0.3676 \text{ ksi}$$

$$M_n = bd^2\bar{k} = 15(26)^2(0.3676) = 3728 \text{ in.-kips}$$

The maximum permissible required ℓ_d is

$$1.3\frac{M_n}{V_u} + \ell_a = 1.3\left(\frac{3728}{80}\right) + 4.5 = 65.1 \text{ in.}$$

$$65.1 \text{ in.} > 43.5 \text{ in.} \qquad \text{(O.K.)}$$

Therefore, the bar diameter is adequately small and the bar can be developed as required. (If the required ℓ_d were in excess of 65.1 in., the use of a standard hook beyond the centerline of support would satisfy the development requirement and this code section would not apply. Also, the use of smaller bars would result in a smaller required ℓ_d.)

PROBLEMS

For the following problems, unless otherwise noted, concrete is normal weight and steel is uncoated grade 60 ($f_y = 60,000$ psi).

5-1. Determine the tension development length required for the No. 8 bars in the T-beam shown in Figure 3-11 of Example 3-4. Use $f'_c = 3000$ psi. Assume that the No. 3 stirrups are spaced at 8 in. throughout. Concrete is normal weight. Neglect the compression steel.

5-2. A 12-in.-thick concrete wall is supported on a continuous footing as shown. Use $f'_c = 3000$ psi. Determine if the development length is adequate if the steel is No. 6 bars at 8 in. o.c. Assume that the critical section for moment (f_y is developed) is at the face of the wall.

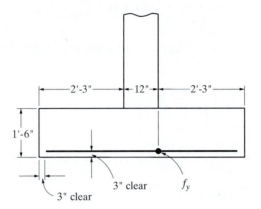

PROBLEM 5-2

5-3. The exterior balcony/canopy shown is to be constructed of lightweight concrete (f_{ct} not specified). The bars are epoxy coated. The design-required tension steel A_s was 0.67 in.^{2}. Determine the required development length ℓ_d from the point of maximum stress in the bars. Specify the minimum required side cover. Shrinkage and temperature steel is not shown. $f'_c = 4000$ psi, $f_y = 60,000$ psi.

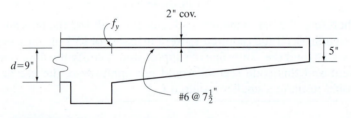

PROBLEM 5-3

5-4. Calculate the required development length (ℓ_d) into the beam for the negative moment steel shown so as to develop the tensile strength of the steel at the face of the column. Required $A_s = 2.75$ in.^{2}, and $f'_c = 4000$ psi.

5-5. Considering the anchorage of the beam bars into the column, determine the largest bar that can be used without a hook. Use $f'_c = 4000$ psi. Clear space between bars is 3 in. (minimum). Side cover is 2 in.

5-6. Seven No. 11 vertical compression bars extend from a column into the supporting footing. Use $f'_c = 5000$ psi. The A_s required was 10.2 in.^{2} There is no lateral reinforcing enclosing the vertical bars in the footing. Determine the required compression development length.

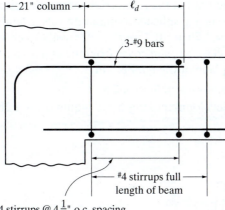

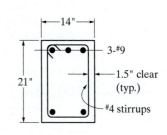

#4 stirrups @ $4\frac{1}{2}$" o.c. spacing
for development length (ℓ_d)

PROBLEM 5-4

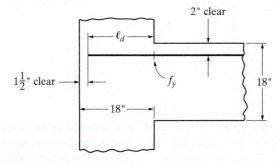

PROBLEM 5-5

5-7. Determine the development length required in the column for the bars shown. If the available development length is not sufficient to develop the tensile strength of the steel (f_y), design an anchorage using a 180° hook and check its adequacy. Use $f_c' = 4000$ psi. The clear space between the No. 10 bars is 3 in. with a side cover of 2½ in.

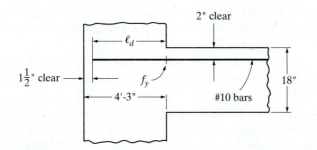

PROBLEM 5-7

5-8. The tension bars in the stem of the reinforced concrete retaining wall are No. 9 at 7 in. o.c. and are to be lap spliced to similar dowels extending up from the footing. Required $A_s = 1.50$ in.^{2}/ft. Use $f_c' = 3000$ psi. Cover on the bars from the rear face of the wall is 2 in., and cover from the end of the wall to the edge bar, measured in the plane of the bars, is 3 in.

(a) Calculate the required length of splice.

(b) Find the hook development length ℓ_{dh} required for a 180° hook in the footing.

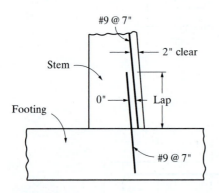

PROBLEM 5-8

5-9. Design the tension splices at points A and B in the beam shown, assuming, for purposes of this problem, that the splices must be located as shown. Positive M_u at A is 120 ft-kips; negative M_u at B is 340 ft-kips. Assume that 50% of the steel is spliced at the designated locations. Use $f_c' = 4000$ psi. Assume $\phi = 0.90$ for moment.

5-10. A wall is reinforced with No. 7 bars at 12 in. o.c. in each face as shown. The bars are in compression and are to be spliced to dowels of the same size and spacing in the footing. Use $f_c' = 3000$ psi. Determine the required length of splice.

5-11. Number 9 compression bars in a column are to be lap spliced. There is no excess steel, and the bars are enclosed by a spiral of $\frac{3}{8}$-in.-diameter wire having a pitch of 3 in. Use $f_c' = 4000$ psi. Determine the required length of splice.

5-12. For the overhanging beam shown, find the theoretical cutoff point for the center No. 9 bar. Assume the section of maximum moment to be at the face of support. Use $f_c' = 5000$ psi.

5-13. A simply supported, uniformly loaded beam carries a total factored load of 4.8 kips/ft (this includes the beam weight) on a clear span of 34 ft. The cross section has been designed and is shown. The required tension steel area for the design was 5.48 in.^{2} The supports are 12-in. wide. Use $f_c' = 3000$ psi.

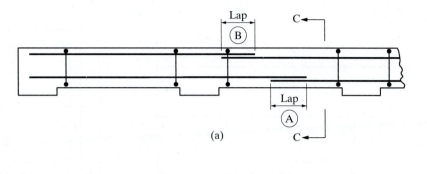

(a)

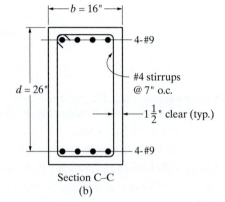

Section C–C
(b)

PROBLEM 5-9

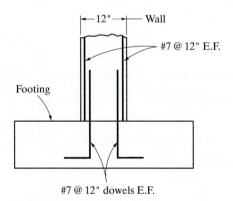

PROBLEM 5-10

(a) Determine the cutoff location for the two center bars (indicated in the sketch).

(b) Design the shear reinforcement and check whether code requirements are met for bars terminated in a tension zone. Redesign the stirrups if necessary.

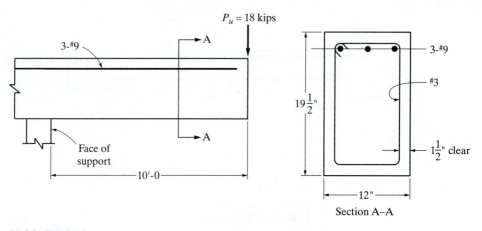

PROBLEM 5-12

5-14. The beam shown is to carry a total factored uniform load of 10 kips/ft (includes beam weight) and factored concentrated loads of 20 kips. Use $f_c' = 4000$ psi. The supports are 12-in. wide.

(a) Design a rectangular beam (tension steel only).

(b) Determine bar cutoffs (use the graphical approach).

(c) Design the shear reinforcement and check bar cutoffs in tension zones if necessary.

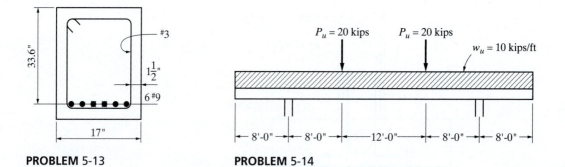

PROBLEM 5-13   **PROBLEM** 5-14

Chapter 6

Continuous Construction Design Considerations

6-1   INTRODUCTION

6-2   CONTINUOUS-SPAN BAR CUTOFFS

6-3   DESIGN OF CONTINUOUS FLOOR SYSTEMS

6-1   INTRODUCTION

A common form of concrete cast-in-place building construction consists of a continuous one-way slab cast monolithically with supporting continuous beams and girders. In this type of system, all members contribute in carrying the floor load to the supporting columns (see Figure 3-1). The slab steel runs through the beams, the beam steel runs through the girders, and the steel from both the beams and girders runs through the columns. The result is that the whole floor system is tied together, forming a highly indeterminate and complex type of rigid structure. The behavior of the members is affected by their rigid connections. Not only will loads applied directly *on* a member produce moment, shear, and a definite deflected shape, but loads applied to *adjacent* members will produce similar effects because of the rigidity of the connections. The shears and moments transmitted through a joint will depend on the relative stiffnesses of all the members framing into that joint. With this type of condition, a precise evaluation of moments and shears resulting from a floor loading is excessively time-consuming and is outside the scope of this text. Several commercial computer programs are available to facilitate these analysis computations.

In an effort to simplify and expedite the design phase, the ACI Code, Section 8.3.3, permits the use of standard moment and shear equations whenever the span and loading conditions satisfy stipulated requirements. This approach applies to continuous non-prestressed one-way slabs and beams. It is an approximate method and may be used for buildings of the usual type of construction, spans, and story heights. The ACI moment equations result from the product of a coefficient and $w_u \ell_n^2$. Similarly, the ACI shear equations result from the product of a coefficient and $w_u \ell_n$. In these equations w_u is the factored design uniform load and ℓ_n is the *clear*

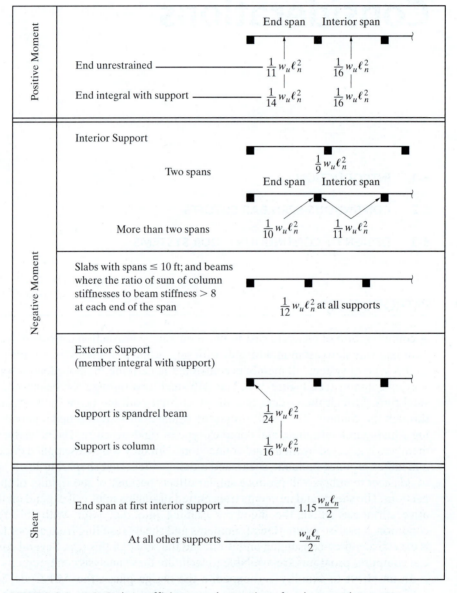

FIGURE 6-1   ACI Code coefficients and equations for shear and moment for continuous beams and one-way slabs.

span for positive moment (and shear) and the average of two adjacent clear spans for negative moment. The application of the equations is limited to the following:

1.   The equations can be used for two or more approximately equal spans (with the larger of two adjacent spans not exceeding the shorter by more than 20%).
2.   Loads must be uniformly distributed.
3.   The maximum allowable ratio of live load to dead load is 3:1 (based on service loads).
4.   Members must be prismatic.

These shear and moment equations generally give reasonably conservative values for the stated conditions. If more precision is required, or desired, for economy, or because the stipulated conditions are not satisfied, a more theoretical and precise analysis must be made. The moment and shear equations are depicted in Figure 6-1. Their use will be demonstrated later in this chapter.

6-2   CONTINUOUS-SPAN BAR CUTOFFS

Using a design approach similar to that for simple spans, the area of main reinforcing steel required at any given point is a function of the design moment. As the moment varies along the span, the steel may be modified or reduced in accordance with the theoretical requirements of the member's strength and the requirements of the ACI Code.

Bars can theoretically be stopped or bent in flexural members whenever they are no longer needed to resist moment. A general representation of the bar cutoff requirements for continuous spans (both positive and negative moments) is shown in Figure 6-2.

In continuous members the ACI Code, Section 12.11.1, requires that a minimum of one-fourth of the positive moment steel be extended into the support a distance of at least 6 in. The ACI Code, Section 12.12.3, also requires that at least one-third of the negative moment steel be extended beyond the extreme position of the point of inflection a distance not less than one-sixteenth of the clear span, the effective depth of the member d, or 12 bar diameters, whichever is greater. If negative moment bars C (Figure 6-2a) are to be cut off, they must extend at least a full development length ℓ_d beyond the face of the support. In addition, they must extend a distance equal to the effective depth of the member or 12 bar diameters, whichever is larger, beyond the theoretical cutoff point defined by the moment diagram. The remaining negative moment bars D (minimum of one-third of total negative steel) must extend at least ℓ_d beyond the theoretical point of cutoff of bars C and, in addition, must extend a distance equal to the effective depth of the member, 12 bar diameters, or one-sixteenth of the clear span, whichever is the greater, past the point of inflection. Where negative moment bars are cut off before reaching the point of inflection, the situation is analogous to the simple beam cutoffs where the reinforcing bars are being terminated in a tension zone. The reader is referred to Example 5-5.

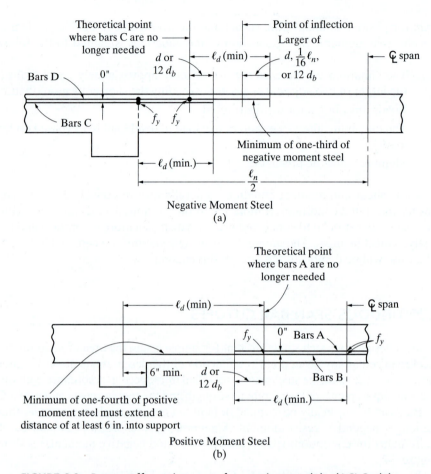

FIGURE 6-2   Bar cutoff requirements for continuous slabs (ACI Code).

In Figure 6-2b, if positive moment bars A are to be cut off, they must project ℓ_d past the point of maximum positive moment as well as a distance equal to the effective depth of the member or 12 bar diameters, whichever is larger, beyond their theoretical cutoff point. Recall that the location of the theoretical cutoff point depends on the amount of steel to be cut and the shape of the applied moment diagram. The remaining positive moment bars B must extend ℓ_d past the theoretical cutoff point of bars A and extend at least 6 in. into the support. Comments on terminating bars in a tension zone again apply. Additionally, the size of the positive moment bars at the point of inflection must meet the requirements of the ACI Code, Section 12.11.3.

Because the determination of cutoff and bend points constitutes a relatively time-consuming chore, it has become customary to use defined cutoff points that experience has indicated are safe. These defined points may be used where the ACI moment coefficients have application but must be applied with judgment where parameters vary. The recommended bar details and cutoffs for continuous spans are shown in Figure 6-3.

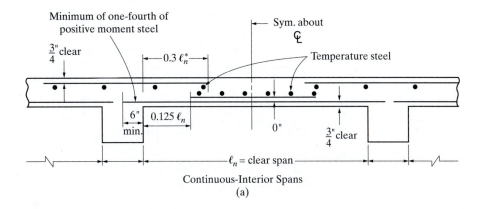

Continuous-Interior Spans
(a)

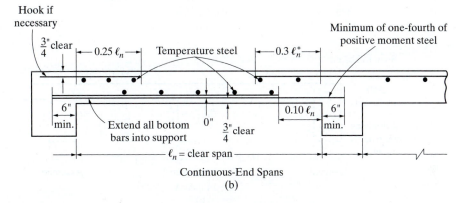

Continuous-End Spans
(b)

*If adjacent spans have different span lengths, use the larger of the two.

FIGURE 6-3   Recommended bar details and cutoffs, one-way slabs; tensile reinforced beams similar.

6-3   DESIGN OF CONTINUOUS FLOOR SYSTEMS

One common type of floor system consists of a continuous, cast-in-place, one-way reinforced concrete slab supported by monolithic, continuous reinforced concrete beams. Assuming that the floor system parameters and loading conditions satisfy the criteria for application of the ACI Code coefficients, the design of the system may be based on these coefficients. Example 6-1 furnishes a complete design of a typical one-way slab and beam floor system.

Example 6-1

The floor system shown in Figure 6-4 consists of a continuous one-way slab supported by continuous beams. The service loads on the floor are 25 psf dead

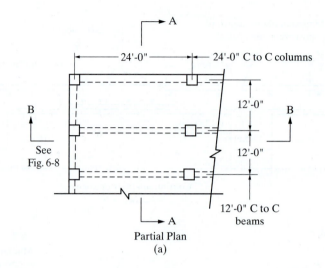

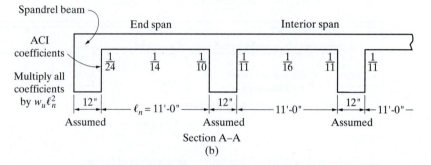

FIGURE 6-4 Sketches for Example 6-1.

load (does not include weight of slab) and 250 psf live load. Use $f_c' = 3000$ psi (normal-weight concrete) and $f_y = 60,000$ psi. The bars are uncoated.

a.   Design the continuous one-way floor slab.
b.   Design the continuous supporting beam.

Solution:

The primary difference in this design from previous flexural designs is that, because of continuity, the ACI coefficients and equations will be used to determine design shears and moments.

a.   Continuous one-way floor slab
    1.   Determine the slab thickness. The slab will be designed to satisfy the ACI minimum thickness requirements from Table 9.5(a) of the Code and this thickness will be used to estimate slab weight.

With both ends continuous,

$$\text{minimum } h = \frac{1}{28}\ell_n = \frac{1}{28}(11)(12) = 4.71 \text{ in.}$$

With one end continuous,

$$\text{minimum } h = \frac{1}{24}\ell_n = \frac{1}{24}(11)(12) = 5.5 \text{ in.}$$

Try a 5½-in.-thick slab. Design a 12-in.-wide segment ($b = 12$ in.).

2. Determine the load:

$$\text{slab dead load} = \frac{5.5}{12}(150) = 68.8 \text{ psf}$$

$$\text{total dead load} = 25.0 + 68.8 = 93.8 \text{ psf}$$

$$w_u = 1.2w_{DL} + 1.6w_{LL}$$

$$= 1.2(93.8) + 1.6(250)$$

$$= 112.6 + 400$$

$$= 512.6 \text{ psf} \quad \text{(design load)}$$

Because we are designing a slab segment that is 12-in. wide, the foregoing loading is the same as 512.6 lb/ft or 0.513 kip/ft.

3. Determine the moments and shears. Moments are determined using the ACI moment equations. Refer to Figures 6-1 and 6-4. Thus

$$+M_u = \frac{1}{14}w_u\ell_n^2 = \frac{1}{14}(0.513)(11)^2 = 4.43 \text{ ft-kips}$$

$$+M_u = \frac{1}{16}w_u\ell_n^2 = \frac{1}{16}(0.513)(11)^2 = 3.88 \text{ ft-kips}$$

$$-M_u = \frac{1}{10}w_u\ell_n^2 = \frac{1}{10}(0.513)(11)^2 = 6.21 \text{ ft-kips}$$

$$-M_u = \frac{1}{11}w_u\ell_n^2 = \frac{1}{11}(0.513)(11)^2 = 5.64 \text{ ft-kips}$$

$$-M_u = \frac{1}{24}w_u\ell_n^2 = \frac{1}{24}(0.513)(11)^2 = 2.59 \text{ ft-kips}$$

Similarly, the shears are determined using the ACI shear equations. In the end span at the face of the first interior support,

$$V_u = 1.15\frac{w_u\ell_n}{2} = 1.15(0.513)\left(\frac{11}{2}\right) = 3.24 \text{ kips}$$

whereas at all other supports,

$$V_u = \frac{w_u \ell_n}{2} = 0.513\left(\frac{11}{2}\right) = 2.82 \text{ kips}$$

4.   Design the slab. Using the assumed slab thickness of 5½ in., find the approximate d. Assume No. 5 bars for main steel and ¾-in. cover for the bars in the slab. Thus

$$d = 5.5 - 0.75 - 0.31 = 4.44 \text{ in.}$$

5.   Design the steel reinforcing. Assume a tension-controlled section ($\epsilon_t \geq 0.005$) and $\phi = 0.90$. Select the point of maximum moment. This is a negative moment and occurs in the end span at the first interior support, and

$$M_u = \frac{w_u \ell_n^2}{10} = 6.21 \text{ ft-kips}$$

$$\phi M_n = \phi b d^2 \overline{k}$$

Because for design purposes $M_u = \phi M_n$ as a limit, then

$$\text{required } \overline{k} = \frac{M_u}{\phi b d^2} = \frac{6.21(12)}{0.90(12)(4.44)^2}$$

$$= 0.3500 \text{ ksi}$$

From Table A-8,

$$\rho = 0.0063 < \rho_{\max} = 0.01355 \quad\quad\quad\quad \text{(O.K.)}$$

$$\text{required } A_s = \rho b d = 0.0063(12)(4.44)$$

$$A_s = 0.34 \text{ in.}^2$$

As the steel area required at all other points will be less, the preceding process will be repeated for the other points. The expression

$$\text{required } \overline{k} = \frac{M_u}{\phi b d^2}$$

can be simplified because all values are constant except M_u:

$$\text{required } \overline{k} = \frac{M_u(12)}{0.9(12)(4.44)^2} = \frac{M_u}{17.74}$$

where M_u must be in ft-kips. In the usual manner, the required steel ratio ρ and the required steel area A_s may then be determined. The results of these calculations are listed in Table 6-1.

TABLE 6-1   Slab Steel Area Requirements

Location	Moment equation	$\bar{k}$ (ksi)	Required ρ	A_s (in.^{2}/ft)
End span				
At spandrel	$-\frac{1}{24}w_u\ell_n^2$	0.1460	0.0025	0.13
Midspan	$+\frac{1}{14}w_u\ell_n^2$	0.2497	0.0044	0.24
Interior spans				
Interior support	$-\frac{1}{11}w_u\ell_n^2$	0.3179	0.0057	0.30
Midspan	$+\frac{1}{16}w_u\ell_n^2$	0.2187	0.0038	0.20

Minimum reinforcement for slabs of constant thickness is that required for shrinkage and temperature reinforcement:

$$\text{minimum required } A_s = 0.0018bh$$

$$= 0.0018(12)(5.5) = 0.12 \text{ in.}^2$$

Also, maximum $\rho = 0.01355$, corresponding to a net tensile strain ϵ_t of 0.005 (see Table A-8). Therefore the slab steel requirements for flexure as shown in Table 6-1 are within acceptable limits. The shrinkage and temperature steel may be selected based on the preceding calculation:

$$\text{use No. 3 bars at 11 in. o.c. } (A_s = 0.12 \text{ in.}^2)$$

Recall that the maximum spacing allowed is the smaller of $5h$ or 18 in. Because $5h = 5(5.5) = 27.5$ in., the 18 in. would control and the spacing is acceptable.

6. Check the shear strength. From step 3, maximum $V_u = 3.24$ kips at the face of the support. A check of shear at the face of the support, rather than at the critical section that is at a distance equal to the effective depth of the member from the face of the support, is conservative. Slabs are not normally reinforced for shear; therefore

$$\phi V_n = \phi V_c = \phi 2\sqrt{f_c'}b_w d$$

$$= \frac{0.75(2\sqrt{3000})(12)(4.44)}{1000} = 4.38 \text{ kips}$$

$$\phi V_n > V_u$$

Therefore the thickness is O.K.

7. Select the main steel. Using Table A-4, establish a pattern in which the number of bar sizes and the number of different spacings are kept to a minimum. The maximum spacing for the main steel is 16.5 in. (the smaller of $3h$ or 18 in.). A work sketch (see Figure 6-5) is recommended to establish steel pattern and cutoff points. With regard to the steel selection, note that in the positive moment areas

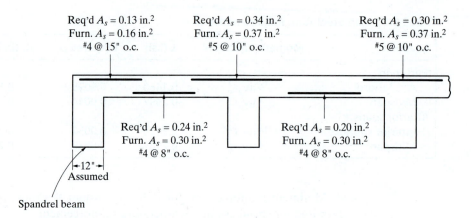

Note: Bar cutoffs and temperature steel may be observed in the design sketch, Fig. 6-7.

FIGURE 6-5   Work sketch for Example 6-1.

of both the end span and interior span, No. 4 bars at 9 in. could be used. If alternate bars were terminated, however, the spacing of the bars remaining would exceed the maximum spacing of 16.5 in. The use of No. 4 bars at 8 in. avoids this problem. The steel selected is conservative.

8.   Check anchorage into the spandrel beam. The steel is No. 4 bars at 15 in. o.c. Refer to the procedure for development length calculation in Section 5-2.

   a.   From Table 5-1, $K_D = 82.2$.

   b.   Establish values for the factors ψ_t, ψ_e, ψ_s, and λ.

   1. $\psi_t = 1.3$ (the bars are top bars).

   2.   The bars are uncoated; $\psi_e = 1.0$.

   3.   The bars are No. 4; $\psi_s = 0.8$.

   4.   Normal-weight concrete is used; $\lambda = 1.0$.

   c.   The product $\psi_t \times \psi_e = 1.3 < 1.7$. (O.K.)

   d.   Determine c_b. Based on cover (center of bar to nearest concrete surface),

$$c_b = \frac{3}{4} + \frac{0.5}{2} = 1 \text{ in.}$$

   Based on bar spacing (one-half the center-to-center distance),

$$c_b = \tfrac{1}{2}(15) = 7.5 \text{ in.}$$

   Therefore use $c_b = 1.0$ in.

   e. K_{tr} is taken as zero. There is no transverse steel that crosses the potential plane of splitting.

f. Check $(c_b + K_{tr})/d_b \leq 2.5$:

$$\frac{c_b + K_{tr}}{d_b} = \frac{1.0 + 0}{0.5} = 2.0 < 2. \qquad \text{(O.K.)}$$

g. Calculate the excess reinforcement factor:

$$K_{ER} = \frac{A_s \text{ required}}{A_s \text{ provided}} = \frac{0.130}{0.160} = 0.813$$

h. Calculate ℓ_d:

$$\ell_d = \frac{K_D}{\lambda} \left[\frac{\psi_t \psi_e \psi_s}{\left(\dfrac{c_b + K_{tr}}{d_b} \right)} \right] K_{ER} d_b$$

$$= \frac{82.2}{1.0} \left[\frac{1.3(1.0)(0.8)}{2.0} \right] (0.813)(0.5)$$

$$= 17.4 \text{ in.} > 12 \text{ in.} \qquad \text{(O.K.)}$$

Use $\ell_d = 18$ in. (minimum).

9. Because the 18-in. length cannot be furnished, a hook will be provided. Determine if a 180° standard hook will be adequate.

a. Calculate ℓ_{dh}.

$$\ell_{dh} = \left(\frac{0.02 \psi_e f_y}{\lambda \sqrt{f_c'}} \right) d$$

b. The bars are uncoated and the concrete is normal weight. Therefore ψ_e and λ are both 1.0.

c. Modification factors are as follows:

1. Assume the concrete side cover is 2½ in. normal to the plane of the hook; use 0.7.

2. For excess steel, use

$$\frac{A_s \text{ required}}{A_s \text{ provided}} = \frac{0.13}{0.16} = 0.813$$

d. Therefore, the required development length is

$$\ell_{dh} = \left(\frac{0.02(1.0)(60,000)}{(1.0)\sqrt{3000}} \right) (0.50)(0.7)(0.813) = 6.23 \text{ in.}$$

Minimum ℓ_{dh} is 6 in. or $8d_b$, whichever is greater:

$$8d_b = 8(\tfrac{1}{2}) = 4 \text{ in.}$$

Therefore the minimum is 6 in.:

$$6.23 \text{ in.} > 6 \text{ in.} \qquad \text{(O.K.)}$$

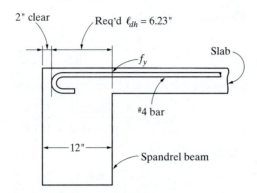

FIGURE 6-6   Hook detail for Example 6-1.

      e.   Check the total width of beam required at the discontinuous end (see Figure 6-6):

$$6.23 + 2 = 8.23 \text{ in.} < 12 \text{ in.} \qquad \text{(O.K.)}$$

   10.   Determine the bar cutoff points. For the normal type of construction for which the typical bar cutoff points shown in Figure 6-3 are used, the cutoff points are located so that all bars terminate in compression zones. Thus, the requirements of the ACI Code, Section 12.10.5, need not be checked, and the recommended bar cutoff points, as shown in Figure 6-3, are used.

   11.   Prepare the design sketches. The final design sketch for the slab is shown in Figure 6-7. For clarity, the interior and end spans are shown separately.

  b.   Continuous supporting beam: The second part of Example 6-1 involves the design of the continuous supporting beam. From Figure 6-4a, it is seen that these beams span between columns. The ACI coefficients to be used for moment determination are shown in Figure 6-8.

    1.   Determine the loading:

$$\text{service live load} = 250 \text{ psf} \times 12 = 3000 \text{ lb/ft}$$

$$\text{service dead load} = 25 \text{ psf} \times 12 = 300 \text{ lb/ft}$$

$$\text{weight of slab} = \left(\frac{5.5}{12}\right)(150)(12) = 825 \text{ lb/ft}$$

Assuming a beam width of 12 in. and an overall depth of 30 in. for purposes of member weight estimate (see Figure 6-9),

$$\text{weight of beam} = \frac{12(30 - 5.5)}{144}(150) = 306.3 \text{ lb/ft}$$

$$\text{total service live load} = 3000 \text{ lb/ft}$$

$$\text{total service dead load} = 1431.3 \text{ lb/ft} \quad \text{say, } 1431 \text{ lb/ft}$$

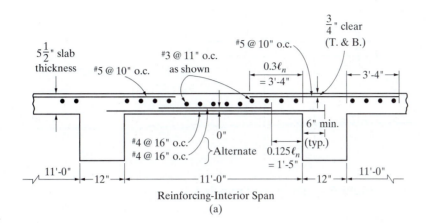

Reinforcing-Interior Span
(a)

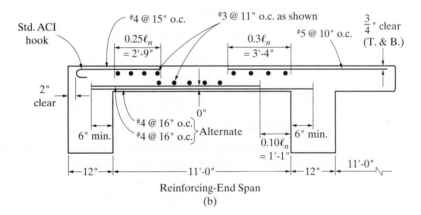

Reinforcing-End Span
(b)

FIGURE 6-7   Design sketches for Example 6-1.

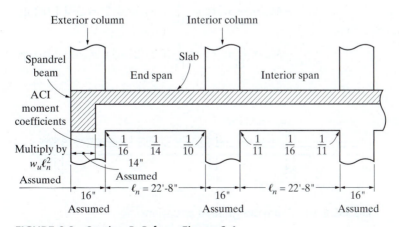

FIGURE 6-8   Section B–B from Figure 6-4a.

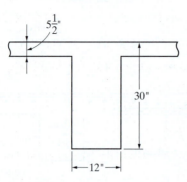

FIGURE 6-9   Sketch for Example 6-1.

2.   Calculate the design load:

$$w_u = 1.2w_{DL} + 1.6w_{LL}$$

$$= 1.2(1431) + 1.6(3000)$$

$$= 1717 + 4800$$

$$= 6517 \text{ lb/ft} \quad \text{say, } 6.5 \text{ kips/ft}$$

The loaded beam is depicted in Figure 6-10.

3.   Calculate the design moments and shears. The design moments and shears are calculated by using the ACI equations:

$$+M_u = \frac{1}{14}w_u\ell_n^2 = \frac{1}{14}(6.5)(22.67)^2 = 238.6 \text{ ft-kips}$$

$$+M_u = \frac{1}{16}w_u\ell_n^2 = \frac{1}{16}(6.5)(22.67)^2 = 208.8 \text{ ft-kips}$$

$$-M_u = \frac{1}{10}w_u\ell_n^2 = \frac{1}{10}(6.5)(22.67)^2 = 334.1 \text{ ft-kips}$$

$$-M_u = \frac{1}{11}w_u\ell_n^2 = \frac{1}{11}(6.5)(22.67)^2 = 303.7 \text{ ft-kips}$$

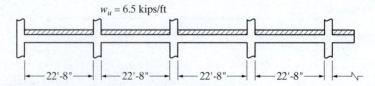

FIGURE 6-10   Beam design load and spans.

$$V_u = \frac{w_u \ell_n}{2} = \frac{6.5(22.67)}{2} = 73.7 \text{ kips}$$

$$V_u = 1.15 \frac{w_u \ell_n}{2} = 1.15(6.5)\left(\frac{22.67}{2}\right) = 84.7 \text{ kips}$$

4. Design the beam. Establish concrete dimensions based on the maximum bending moment. This occurs in the end span at the first interior support where the negative moment $M_u = w_u \ell_n^2/10$. Since the top of the beam is in tension, the design will be that of a rectangular beam.

   a. Maximum moment (negative) = 334.1 ft-kips.

   b. From Table A-5, assume that $\rho = 0.0090$ (which is less than ρ_{max} of 0.01355 from Table A-8). A check of minimum steel required will be made shortly.

   c. From Table A-8, $\bar{k} = 0.4828$ ksi.

   d. Assume that $b = 12$ in.:

$$\text{required } d = \sqrt{\frac{M_u}{\phi b \bar{k}}} = \sqrt{\frac{334.1(12)}{0.9(12)(0.4828)}}$$

$$= 28.0 \text{ in.}$$

$$d/b \text{ ratio} = \frac{28.0}{12} = 2.34, \quad \text{which is within the acceptable range}$$

   e. Check the estimated beam weight assuming one layer of No. 11 bars and No. 3 stirrups:

$$\text{required } h = 28.0 + \frac{1.41}{2} + 0.38 + 1.5 = 30.6 \text{ in.}$$

Use $h = 31$ in. with an assumed d of 28 in. Also, check the minimum h from the ACI Code, Table 9.5(a):

$$\text{minimum } h = \frac{1}{18.5}(22.67)(12) = 14.7 \text{ in.} < 31 \text{ in.} \quad \text{(O.K.)}$$

Note that the estimated beam weight based on $b = 12$ in. and $h = 30$ in. is slightly on the low side but may be considered acceptable.

   f. Design the steel reinforcing for points of negative moment as follows:

     1. At the first interior support, based on an assumed ρ of 0.0090 (see the preceding step 4),

$$\text{required } A_s = \rho b d$$

$$= 0.0090(12)(28) = 3.0 \text{ in.}^2$$

From Table A-5,

$$A_{s,\min} = 0.0033(12)(28) = 1.11 \text{ in.}^2$$

$$3.0 \text{ in.}^2 > 1.11 \text{ in.}^2 \qquad\qquad (\text{O.K.})$$

2.   At the other interior supports, $-M_u = 303.7$ ft-kips:

$$\text{required } \overline{k} = \frac{M_u}{\phi bd^2} = \frac{303.7(12)}{0.9(12)(28)^2} = 0.4304 \text{ ksi}$$

From Table A-8, $\rho = 0.0079$. Check ρ with $\rho_{\max}$:

$$0.0079 < 0.01355 \qquad\qquad (\text{O.K.})$$

Therefore

$$\text{required } A_s = \rho bd$$

$$= 0.0079(12)(28) = 2.65 \text{ in.}^2$$

Check the required A_s with $A_{s,\min}$. From Table A-5,

$$A_{s,\min} = 0.0033(12)(28) = 1.11 \text{ in.}^2$$

$$2.65 \text{ in.}^2 > 1.11 \text{ in.}^2 \qquad\qquad (\text{O.K.})$$

3.   At the exterior support (exterior column), $-M_u = 208.8$ ft-kips:

$$\text{required } \overline{k} = \frac{M_u}{\phi bd^2} = \frac{208.8(12)}{0.9(12)(28)^2} = 0.2959 \text{ ksi}$$

From Table A-8, $\rho = 0.0053$. Check ρ with $\rho_{\max}$:

$$0.0053 < 0.01355 \qquad\qquad (\text{O.K.})$$

Therefore

$$\text{required } A_s = \rho bd$$

$$= 0.0053(12)(28) = 1.78 \text{ in.}^2$$

Check the required A_s with $A_{s,\min}$. From Table A-5,

$$A_{s,\min} = 0.0033(12)(28) = 1.11 \text{ in.}^2$$

$$1.78 \text{ in.}^2 > 1.11 \text{ in.}^2 \qquad\qquad (\text{O.K.})$$

g.   Design steel reinforcing for points of positive moment as follows: At points of positive moment, the top of the beam is in compression; therefore, the design will be that of a T-beam.

   1.   End-span positive moment:

   a.   Design moment = 238.6 ft-kips = M_u.

   b.   Effective depth $d = 28$ in. (see negative moment design).

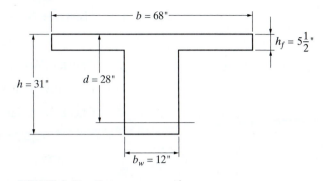

FIGURE 6-11 Beam cross section.

c. Effective flange width:

$$\tfrac{1}{4}\text{ span length} = 0.25(22.67)(12) = 68 \text{ in.}$$

$$b_w + 16h_f = 12 + 16(5.5) = 100 \text{ in.}$$

$$\text{beam spacing} = 144 \text{ in.}$$

Use an effective flange width $b = 68$ in. (see Figure 6-11).

d. Assuming total flange in compression, $\epsilon_t = 0.005$, and $\phi = 0.90$:

$$\phi M_{nf} = \phi(0.85f'_c)bh_f\left(d - \frac{h_f}{2}\right)$$

$$= 0.9(0.85)(3)(68)(5.5)\left(\frac{28 - 5.5/2}{12}\right) = 1806 \text{ ft-kips}$$

e. Because $1806 > 238.6$, the member behaves as a wide rectangular T-beam with $b = 68$ in. and $d = 28$ in.

f.
$$\text{required } \bar{k} = \frac{M_u}{\phi bd^2}$$

$$= \frac{238.6(12)}{0.9(68)(28)^2} = 0.0597 \text{ ksi}$$

g. From Table A-8,

$$\text{required } \rho = 0.0010$$

h. The required steel area is

$$\text{required } A_s = \rho bd$$

$$= 0.0010(68)(28) = 1.90 \text{ in.}^2$$

i.   Use three No. 8 bars ($A_s = 2.37$ in.^{2}):

required $b = 9.0$ in.   (O.K.)

j.   Check d. With a No. 3 stirrup and 1½-in. cover,

$$d = 31 - 1.5 - 0.38 - 1.00/2 = 28.6 \text{ in.} > 28 \text{ in.} \quad \text{(O.K.)}$$

k.   Check the required A_s with $A_{s,\min}$. From Table A-5,

$$A_{s,\min} = 0.0033(12)(28) = 1.11 \text{ in.}^2$$

$$2.37 \text{ in.}^2 > 1.11 \text{ in.}^2 \qquad \text{(O.K.)}$$

l.   Check ϵ_t and ϕ. From Table A-8, with $\rho = 0.001$, $\epsilon_t \geq$ 0.005. Therefore, the assumed ϕ of 0.90 is O.K.

2.   Interior span positive moment:

a.   Design moment $= 208.8$ ft-kips $= M_u$.

b.   through (e) See the end-span positive moment computations. Use an effective flange width $b = 68$ in. and an effective depth $d = 28$ in. Also, for total flange in compression, $\phi M_n = 1806$ ft-kips $> M_u$. Therefore, this member also behaves as a rectangular T-beam.

f. $$\text{required } \overline{k} = \frac{M_u}{\phi b d^2}$$

$$= \frac{208.8(12)}{0.9(68)(28)^2} = 0.0522 \text{ ksi}$$

g.   From Table A-8,

required $\rho = 0.0010$

h.   The required steel area is

required $A_s = \rho b d$

$$= 0.0010(68)(28) = 1.90 \text{ in.}^2$$

i.   Use three No. 8 bars ($A_s = 2.37$ in.^{2}):

required $b = 9.0$ in.   (O.K.)

j.   through 1. are identical to those for the end-span positive moment.

5.   Check the distribution of negative moment steel. The ACI Code, Section 10.6.6, requires that where flanges are in tension, a part of the main tension reinforcement be distributed over the effective flange width or a width equal to one-tenth of the span, whichever is smaller. The use of smaller bars spread out into part of the flange

will also be advantageous where a beam is supported by a spandrel girder or exterior column and the embedment length for the negative moment steel is limited. Thus

$$\frac{\text{span}}{10} = \frac{22.67(12)}{10} = 27 \text{ in.}$$

$$\text{effective flange width} = b = 68 \text{ in.}$$

Therefore distribute the negative moment bars over a width of 27 in. Figure 6-12 shows suitable bars and patterns to use to satisfy the foregoing code requirement and furnish the cross-sectional area of steel required for flexure.

The ACI Code also stipulates that if the effective flange width exceeds one-tenth of the span, some longitudinal reinforcement shall be provided in the outer portions of the flange. In this design no additional steel will be furnished. In the authors' opinion, this requirement is satisfied by the slab temperature and shrinkage steel (see Figures 6-7 and 6-18).

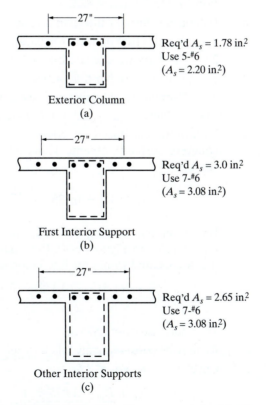

Req'd A_s = 1.78 in.²
Use 5-#6
(A_s = 2.20 in.²)

Exterior Column
(a)

Req'd A_s = 3.0 in.²
Use 7-#6
(A_s = 3.08 in.²)

First Interior Support
(b)

Req'd A_s = 2.65 in.²
Use 7-#6
(A_s = 3.08 in.²)

Other Interior Supports
(c)

FIGURE 6-12   Negative moment steel for beam of Example 6-1.

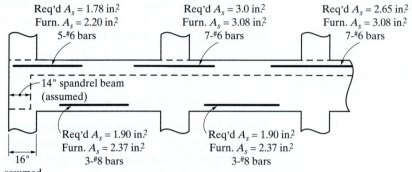

FIGURE 6-13   Work sketch for beam design of Example 6-1.

6. Prepare the work sketch. A work sketch is developed in Figure 6-13, which includes the bars previously chosen.

7. Check the anchorage into the exterior column.

   a. From Table 5-1, $K_D = 82.2$.

   b. Establish values for the factors ψ_t, ψ_e, ψ_s, and λ.

      1. $\psi_t = 1.3$ (the bars are top bars).

      2. The bars are uncoated; $\psi_e = 1.0$.

      3. The bars are No. 6; $\psi_s = 0.8$.

      4. Normal-weight concrete is used; $\lambda = 1.0$.

   c. The product $\psi_t \times \psi_e = 1.3 < 1.7$. (O.K.)

   d. Determine c_b. Based on cover (center of bar to nearest concrete surface), assume $1\frac{1}{2}$-in. clear cover and a No. 3 stirrup:

$$c_b = 1.5 + 0.375 + \frac{0.75}{2} = 2.25 \text{ in.}$$

   Based on bar spacing (one-half the center-to-center distance), refer to Figures 6-12. Consider the three No. 6 bars that are located within the No. 3 loop stirrup. Here

$$c_b = \left[\frac{12 - 2(1.5) - 2(0.375) - 0.75}{2} \right]\left(\frac{1}{2}\right) = 1.875 \text{ in.}$$

   Therefore use $c_b = 1.875$ in.

   e. K_{tr} may be conservatively taken as zero.

   f. Check $(c_b + K_{tr})/d_b \leq 2.5$:

$$\frac{c_b + K_{tr}}{d_b} = \frac{1.875 + 0}{0.75} = 2.5 (\leq 2.5) \qquad \text{(O.K.)}$$

g. Determine the excess reinforcement factor:

$$K_{ER} = \frac{A_s \text{ required}}{A_s \text{ provided}} = \frac{1.78}{2.20} = 0.809$$

h. Calculate ℓ_d:

$$\ell_d = \frac{K_D}{\lambda} \left[\frac{\psi_t \psi_e \psi_s}{\left(\dfrac{c_b + K_{tr}}{d_b} \right)} \right] d_b K_{ER}$$

$$= \frac{82.2}{1.0} \left[\frac{1.3(1.0)(0.8)}{2.5} \right] (0.75)(0.809)$$

$$= 20.7 \text{ in.} > 12 \text{ in.} \qquad\qquad (\text{O.K.})$$

Use $\ell_d = 21$ in. (minimum).

With 2.0 in. clear at the end of the bar, the embedment length available in the column is $16.0 - 2.0 = 14.0$ in.

8. Because 21 in. > 14.0 in., a hook is required. Determine if a 90° standard hook will be adequate.

a. From Table A-13, $\ell_{dh} = 16.4$ in.

b. Modification factors (MF) to be used are

1. Assume concrete cover ≥ 2½ in. and cover on the bar extension beyond the hook = 2 in.; use 0.7.

2. For excess steel, use

$$\frac{\text{required } A_s}{\text{provided } A_s} = \frac{1.78}{2.20} = 0.809$$

c. The required development length for the hook is

$$\ell_{dh} = 16.4(0.7)(0.809) = 9.29 \text{ in.}$$

Minimum ℓ_{dh} is 6 in. or $8d_b$, whichever is greater:

$$8d_b = 8(\tfrac{3}{4}) = 6 \text{ in.}$$

$$9.29 \text{ in.} > 6 \text{ in.} \qquad\qquad (\text{O.K.})$$

d. Check the total width of column required (see Figure 6-14):

$$9.29 + 2 = 11.3 \text{ in.} < 16 \text{ in.} \qquad\qquad (\text{O.K.})$$

For other points along the continuous beam, use bar cutoff points recommended in Figure 6-3 and as shown in Figure 6-18.

9. Prepare the stirrup design. Established values are $b_w = 12$ in., effective depth $d = 28$ in., $f'_c = 3000$ psi, and $f_y = 60,000$ psi.

a. The shear force V_u diagram may be observed in Figure 6-15. Because the diagram is unsymmetrical with respect to the

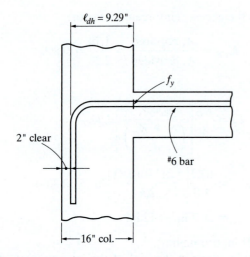

FIGURE 6-14   Anchorage at column.

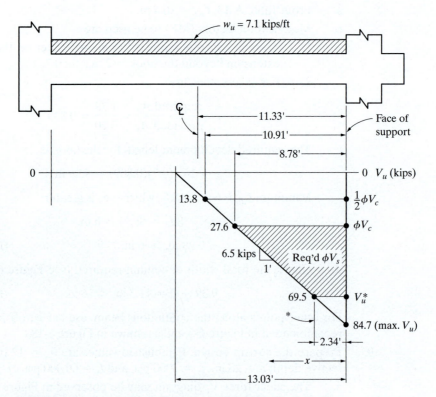

FIGURE 6-15 V_u diagram for stirrup design (Example 6-1).

centerline of the span, the stirrup design will be based on the shear in the interior portion of the end span where the maximum values occur. The resulting stirrup pattern will be used throughout the continuous beam. Only the applicable portion of the V_u diagram is shown.

b.   Determine if stirrups are required:

$$\phi V_c = \phi 2\sqrt{f_c'} b_w d = \frac{0.75(2\sqrt{3000})(12)(28)}{1000} = 27.6 \text{ kips}$$

$$\tfrac{1}{2}\phi V_c = \tfrac{1}{2}(27.6) = 13.8 \text{ kips}$$

At the critical section d distance (28 in.) from the face of the support,

$$V_u^* = 84.7 - \frac{28}{12}(6.5) = 69.5 \text{ kips}$$

(Quantities at the critical section are designated with an asterisk.) Stirrups are required because

$$V_u^* > \tfrac{1}{2}\phi V_c (69.4 \text{ kips} > 13.8 \text{ kips})$$

c.   Find the length of span over which stirrups are required. Stirrups are required to the point where

$$V_u = \tfrac{1}{2}\phi V_c = 13.8 \text{ kips}$$

From Figure 6-15 and referencing from the face of the support, $V_u = 13.8$ kips at

$$\frac{84.7 - 13.8}{6.5} = 10.91 \text{ ft}$$

The distance from the face of the support to where $V_u = \phi V_c = 27.6$ kips is

$$\frac{84.7 - 27.6}{6.5} = 8.78 \text{ ft}$$

d.   On the V_u diagram, designate the area between the ϕV_c line, the V_u^* line, and the sloping V_u line as "Req'd ϕV_s." At locations between 2.34 ft and 8.78 ft from the face of the support, the required ϕV_s varies. Designating the slope of the V_u diagram as m (kips/ft) and taking x (ft) from the face of the support $(2.34 \leq x \leq 8.78)$ yields

$$\text{required } \phi V_s = \text{ maximum } V_u - \phi V_c - mx$$

$$= 84.7 - 27.6 - 6.5x$$

$$= 57.1 - 6.5x$$

e.   Assume a No. 3 vertical stirrup ($A_v = 0.22$ in.^{2}):

$$\text{required } s^* = \frac{A_v f_{yt} d}{V_s}$$

$$= \frac{\phi A_v f_{yt} d}{\text{required } \phi V_s^*}$$

$$= \frac{\phi A_v f_{yt} d}{V_u^* - \phi V_c}$$

$$= \frac{0.75(0.22)(60)(28)}{69.5 - 27.6} = 7.57 \text{ in.}$$

Use 6½-in. spacing between the critical section and the face of the support.

f.   Establish ACI Code maximum spacing requirements:

$$4\sqrt{f_c'}b_w d = \frac{4\sqrt{3000}(12)(28)}{1000} = 73.6 \text{ kips}$$

Calculating V_s^* at the critical section yields

$$\phi V_s^* = V_u^* - \phi V_c$$

$$= 69.5 - 27.6 = 41.9 \text{ kips}$$

$$V_s^* = \frac{\phi V_s^*}{\phi} = \frac{41.9}{0.75} = 55.9 \text{ kips}$$

Because 55.9 kips < 73.6 kips, the maximum spacing should be the smaller of $d/2$ or 24 in.:

$$\frac{d}{2} = \frac{28}{2} = 14 \text{ in.}$$

Also check:

$$s_{max} = \frac{A_v f_{yt}}{0.75\sqrt{f_c'}\, b_w} \leq \frac{A_v f_{yt}}{50 b_w}$$

$$\frac{A_v f_{yt}}{0.75\sqrt{f_c'}\, b_w} = \frac{0.22(60,000)}{0.75\sqrt{3000}(12)} = 26.7 \text{ in.}$$

and

$$\frac{A_v f_{yt}}{50 b_w} = \frac{0.22(60,000)}{50(12)} = 22.0 \text{ in.}$$

Therefore, use a maximum spacing of 14 in.

g.   Determine the spacing requirements based on shear strength to be furnished. The denominator of the following formula

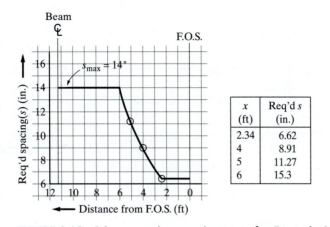

FIGURE 6-16   Stirrup spacing requirements for Example 6-1.

for required spacing uses the expression for required ϕV_s from step d:

$$\text{required } s = \frac{\phi A_v f_{yt} d}{\text{required } \phi V_s} = \frac{0.75(0.22)(60)(28)}{57.1 - 6.5x} = \frac{277.2}{57.1 - 6.5x}$$

The results for several arbitrary values of x are shown tabulated and plotted in Figure 6-16.

h.   Using Figure 6-16, the stirrup pattern shown in Figure 6-17 is developed. Despite the lack of symmetry in the shear diagram,

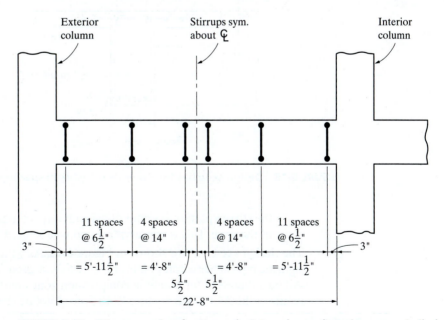

FIGURE 6-17   Stirrup spacing for Example 6-1 end span (interior spans similar).

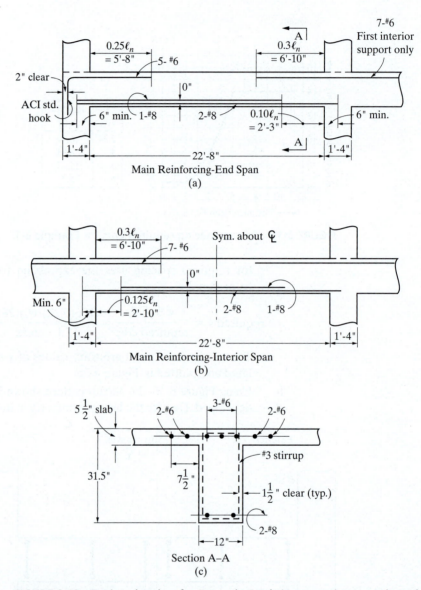

Main Reinforcing-End Span
(a)

Main Reinforcing-Interior Span
(b)

Section A–A
(c)

FIGURE 6-18   Design sketches for Example 6-1 (stirrup spacings not shown).

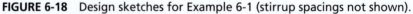

the stirrup pattern is symmetrical with respect to the centerline of the span. This is conservative and will be used for all spans.

The design sketches are shown in Figure 6-18. As with the slab design, the typical bar cutoff points of Figure 6-3 are used for this beam. All bars therefore terminate in compression zones, and the requirements of the ACI Code, Section 12.10.5, need not be checked.

PROBLEMS

For the following problems, all concrete is normal weight and all steel is grade 60 (f_y = 60,000 psi).

6-1. For the one-way slab shown, determine all moments and shears for service loads of 100 psf dead load (includes the slab weight) and 300 psf live load.

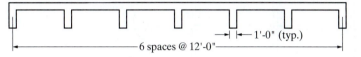

PROBLEM 6-1

6-2. For the continuous beam shown, determine all moments and shears for service loads of 2.00 kips/ft dead load (includes weight of beam and slab) and 3.00 kips/ft live load.

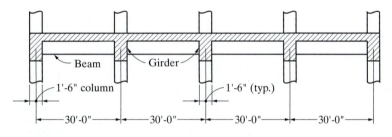

PROBLEM 6-2

6-3. Using the ACI coefficients, design a continuous reinforced concrete beam that will span four supports. The end spans are 24-ft long; the center span is 28-ft long (spans are measured center to center of supports). The exterior support (a spandrel beam) and the interior supporting girders have widths b of 18 in. The service loads are 0.90 kip/ft dead load (not including the weight of the beam) and 1.10 kip/ft live load. Use f'_c = 3000 psi. Use only tension reinforcing. Design for moment only. Use the recommended bar cutoffs shown in Figure 6-3.

6-4. A floor system is to consist of beams, girders, and a slab; a partial floor plan is shown. Service loads are to be 45 psf dead load (does *not* include the weight of the floor system) and 160 psf live load. Use f'_c = 4000 psi.

    (a) Design the continuous one-way slab.

    (b) Design the continuous beam along column line 2.

Be sure to include complete design sketches.

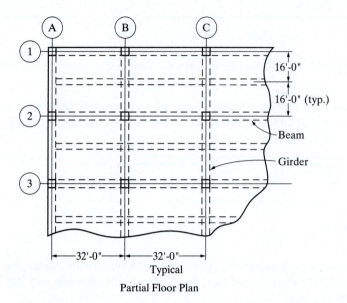

Typical

Partial Floor Plan

PROBLEM 6-4

Serviceability

7-1   INTRODUCTION

The ACI Code, Section 9.1, requires that bending members have structural strength adequate to support the anticipated factored design loads and that they have adequate performance at service load levels. Adequate performance, or *serviceability*, relates to deflections and cracking in reinforced concrete beams and slabs. It is important to realize that serviceability is to be assured at *service load levels*, not at ultimate strength. At service loads, deflections should be held to specified limits because of many considerations, among which are aesthetics, effects on nonstructural elements such as windows and partitions, undesirable vibrations, and proper functioning of roof drainage systems. Any cracking should be limited to hairline cracks for reasons of appearance and to ensure protection of reinforcement against corrosion.

7-2   DEFLECTIONS

Guidelines for the control of deflections are found in the ACI Code, Section 9.5. In addition, Table 9.5(b) of the Code indicates the maximum permissible deflections.

227

For the purpose of following the Code guidelines, either of two methods may be used: (1) using the minimum thickness (or depth of member) criteria as established in Table 9.5(a) of the Code, which will result in sections that are sufficiently deep and stiff so that deflections will not be excessive; and (2) calculating expected deflections using standard deflection formulas in combination with the Code provisions for moment of inertia and the effects of the load/time history of the member.

Minimum thickness (depth) guidelines are simple and direct and should be used whenever possible. Note that the tabulated minimum thicknesses apply to non-prestressed, one-way members that do not support and are not attached to partitions or other construction likely to be damaged by large deflections. For members not within these guidelines, deflections *must* be calculated.

For the second method, in which deflections are calculated, the ACI Code stipulates that the members should have their deflections checked at *service load levels*. Therefore the *properties* at service load levels must be used. Under service loads, concrete flexural members still exhibit generally elastic-type behavior (see Figure 1-1) but will have been subjected to cracking in tension zones at any point where the applied moment is large enough to produce tensile stress in excess of concrete tensile strength. The cross section for moment-of-inertia determination then has the shape shown in Figure 7-1.

The moment of inertia of the cracked section of Figure 7-1 is designated I_{cr}. It is determined based on the assumption that the concrete is cracked to the neutral axis. In other words, the concrete is assumed to have no tensile strength, and the small tension zone below the neutral axis and above the upper limit of cracking is neglected.

The moment of inertia of the cracked section described represents one end of a range of values that may be used for deflection calculations. At the other end of the range, as a result of a small bending moment and a maximum flexural stress less than the modulus of rupture, the full uncracked section may be considered in determining the moment of inertia to resist deflection. This is termed the *moment of inertia of the gross cross section* and is designated I_g.

In reality, both I_{cr} and I_g occur in a bending member in which the maximum moment is in excess of the cracking moment. The I_{cr} occurs at or near the cracks, whereas the I_g occurs between the cracks. Research has indicated that due to the presence of the two extreme conditions, a more realistic value of a moment of inertia

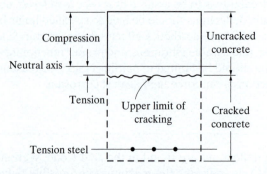

FIGURE 7-1 Typical cracked cross section.

lies somewhere between these values. The ACI Code recommends that deflections be calculated using an *effective moment of inertia, I_e,* where

$$I_g > I_e > I_{cr}$$

Once the effective moment of inertia is determined, the member deflection may be calculated by using standard deflection expressions. The effective moment of inertia will depend on the values of the moment of inertia of the gross section and the moment of inertia of the cracked section. The gross (or uncracked) moment of inertia for a rectangular shape may be easily calculated from

$$I_g = \frac{bh^3}{12}$$

This expression neglects the presence of any reinforcing steel.

7-3 CALCULATION OF I_{cr}

The moment of inertia of the cracked cross section can also be calculated in the normal way once the problem of the differing materials (steel and concrete) is overcome. To accomplish this, the steel area will be replaced by an equivalent area of concrete A_{eq}. This is a fictitious concrete that *can* resist tension. The determination of the magnitude of A_{eq} is based on the theory (from strength of materials) that when two differing elastic materials are subjected to equal strains, the stresses in the materials will be in proportion to their moduli of elasticity.

In Figure 7-2 ϵ_1 is the compressive strain in the concrete at the top of the beam and ϵ_2 is a tensile strain at the level of the steel. Using the notation

$$f_s = \text{tensile steel stress}$$

$$f_{c(\text{tens})} = \text{theoretical tensile concrete stress at the level of the steel}$$

$$E_s = \text{steel modulus of elasticity (29,000,000 psi)}$$

$$E_c = \text{concrete modulus of elasticity}$$

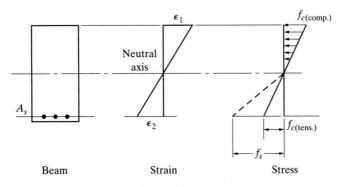

FIGURE 7-2 Beam bending: elastic theory.

the following relationships can be established:

$$\epsilon_2 = \frac{f_s}{E_s} \quad \text{and} \quad \epsilon_2 = \frac{f_{c(\text{tens})}}{E_c}$$

Equating these two expressions and solving for f_s yields

$$\frac{f_s}{E_s} = \frac{f_{c(\text{tens})}}{E_c}$$

$$f_s = \frac{E_s}{E_c}[f_{c(\text{tens})}]$$

The ratio E_s/E_c is normally called the *modular ratio* and is denoted n. Therefore

$$f_s = n f_{c(\text{tens})}$$

Values of n may be taken as the nearest whole number (but not less than 6). Values of n for normal-weight concrete are tabulated in Table A-6.

As we are replacing the steel (theoretically) with an equivalent concrete area, the equivalent concrete area A_{eq} must provide the same tensile resistance as that provided by the steel.

Therefore

$$A_{eq} f_{c(\text{tens})} = f_s A_s$$

Substituting, we obtain

$$A_{eq} f_{c(\text{tens})} = n f_{c(\text{tens})} A_s$$

from which

$$A_{eq} = n A_s$$

This defines the equivalent area of concrete with which we are replacing the steel. Another way of visualizing this is to consider the steel to be transformed into an equivalent concrete area of nA_s. The resulting *transformed concrete cross section* is composed of a single (although hypothetical) material and may be dealt with in the normal fashion for neutral-axis and moment-of-inertia determinations. Figure 7-3

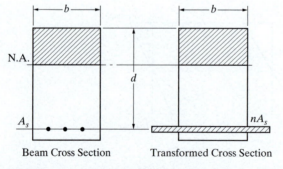

Beam Cross Section   Transformed Cross Section

FIGURE 7-3   Method of transformed section.

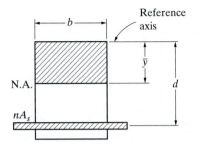

FIGURE 7-4   Effective area.

depicts the transformed section. Because the steel is normally assumed to be concentrated at its centroid, which is a distance d from the compression face, the replacing equivalent concrete area must also be assumed to act at the same location.
Therefore the representation of this area is shown as a thin rectangle extending out
past the beam sides. Recalling that the cross section is assumed cracked up to the
neutral axis, the resulting effective area is as shown in Figure 7-4 and the neutral axis
will be located a distance $\bar{y}$ down from a reference axis at the top of the section.

The neutral axis may be determined by taking a summation of moments of the effective areas about the reference axis:

$$\bar{y} = \frac{\Sigma(Ay)}{\Sigma A} = \frac{(b\bar{y})\dfrac{\bar{y}}{2} + nA_s d}{b\bar{y} + nA_s}$$

$$b(\bar{y}^2) + nA_s\bar{y} = \frac{b(\bar{y}^2)}{2} + nA_s d$$

$$\frac{b(\bar{y}^2)}{2} + nA_s\bar{y} - nA_s d = 0$$

This is a quadratic equation of the form $ax^2 + bx + c = 0$, and it may be solved either
by completion of the square, as was done in Example 3-7, or by using the formula for
roots of a quadratic equation, which will result in the following useful expression:

$$\bar{y} = \frac{nA_s\left[\sqrt{1 + 2\dfrac{bd}{nA_s}} - 1\right]}{b}$$

Once the neutral axis is located, the moment of inertia (I_{cr}) may be found using the
familiar transfer formula from engineering mechanics.

Example 7-1 _____

Find the cracked moment of inertia for the cross section shown in Figure 7-5a.
Use $A_s = 2.00$ in.^{2}, $n = 9$ (Table A-6), and $f'_c = 3000$ psi.

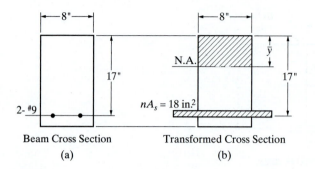

FIGURE 7-5   Sketches for Example 7-1.

Solution:

With reference to the transformed section, shown in Figure 7-5b, the neutral axis is located as follows:

$$\bar{y} = \frac{nA_s\left[\sqrt{1 + 2\dfrac{bd}{nA_s}} - 1\right]}{b}$$

$$= \frac{9(2.0)\left[\sqrt{1 + 2\dfrac{(8)(17)}{9(2.0)}} - 1\right]}{8}$$

$$= 6.78 \text{ in.}$$

The moment of inertia of the cracked section may now be found (all units are inches):

$$I_{cr} = \frac{b\bar{y}^3}{3} + nA_s(d - \bar{y})^2$$

$$= \frac{8(6.78)^3}{3} + 18(17 - 6.78)^2$$

$$= 831.1 + 1880$$

$$= 2711 \text{ in.}^4$$

If the beam cross section contains compression steel, this steel may also be transformed and the neutral-axis location and cracked moment-of-inertia calculations can be carried out as before. Because the compression steel displaces concrete that is in compression, it should theoretically be transformed using $(n - 1)A_s'$ rather than nA_s'. For deflection calculations, which are only approximate, however, the use of nA_s' will not detract from the accuracy expected. The resulting transformed section will appear as shown in Figure 7-6.

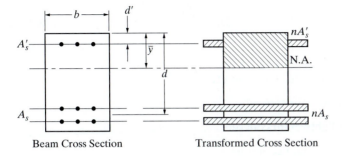

Beam Cross Section   Transformed Cross Section

FIGURE 7-6   Doubly reinforced beam.

The neutral axis location $\bar{y}$ may be determined from the solution of

$$\frac{b}{2}\bar{y}^2 + nA'_s\bar{y} - nA'_sd' - nA_sd + nA_s\bar{y} = 0$$

and the moment of inertia with respect to the neutral axis from

$$I_{cr} = \frac{b\bar{y}^3}{3} + nA_s(d - \bar{y})^2 + nA'_s(\bar{y} - d')^2$$

7-4   IMMEDIATE DEFLECTION

Immediate deflection is the deflection that occurs as soon as load is applied on the member. For all practical purposes, the member is elastic. The ACI Code, Section 9.5.2.3, states that this deflection may be calculated using a concrete modulus of elasticity E_c as specified in Section 8.5.1 and an effective moment of inertia I_e computed as follows:

$$I_e = \left\{ \left(\frac{M_{cr}}{M_a} \right)^3 I_g + \left[1 - \left(\frac{M_{cr}}{M_a} \right)^3 \right] I_{cr} \right\} \le I_g \qquad \text{[ACI Eq. (9-8)]}$$

where

> I_e = effective moment of inertia

> I_{cr} = moment of inertia of the cracked section transformed to concrete

> I_g = moment of inertia of the gross (uncracked) concrete cross section about the centroidal axis, neglecting all steel reinforcement

> M_a = maximum moment in the member at the stage for which the deflection is being computed

> M_{cr} = moment that would initially crack the cross section computed from

$$M_{cr} = \frac{f_r I_g}{y_t}$$

[ACI Eq. (9-9)]

where

f_r = modulus of rupture for the concrete = $7.5 \lambda \sqrt{f_c'}$

λ = 1.0 for normal-weight concrete

    = 0.85 for sand-lightweight concrete

    = 0.75 for all-lightweight concrete

Values for the modulus of rupture for normal-weight concrete are tabulated in Table A-6.

y_t = distance from the neutral axis of the uncracked cross section (neglecting steel) to the extreme tension fiber

Inspection of the formula for the effective moment of inertia will show that if the maximum moment is low with respect to the cracking moment M_{cr}, the moment of inertia of the gross section I_g will be the dominant factor. If the maximum moment is large with respect to the cracking moment, however, the moment of inertia of the cracked section I_{cr} will be dominant. In any case, I_e will lie somewhere between I_{cr} and I_g. For continuous beams, the use of the average value of the effective moments of inertia existing at sections of critical positive and negative moments is recommended. The use of midspan sectional properties for simple and continuous spans, and at the support for cantilevers, will also give satisfactory results.

The actual calculation of deflections will be made using the standard deflection methods for elastic members. Deflection formulas of the type found in standard handbooks may be suitable, or we may use more rigorous techniques when necessary.

7-5   LONG-TERM DEFLECTION

In addition to deflections that occur immediately, reinforced concrete members are subject to added deflections that occur gradually over long periods. These additional deflections are due mainly to creep and shrinkage and may eventually become excessive. The additional (or long-term) deflections are computed based on two items: (1) the amount of sustained dead and live load and (2) the amount of compression reinforcement in the beam. The additional long-term deflections may be estimated as follows:

$$\Delta_{LT} = \lambda_\Delta \Delta_i = \left(\frac{\xi}{1 + 50\rho'} \right) \Delta_i$$

where

Δ_{LT} = additional long-term deflection

Δ_i = immediate deflection due to sustained loads

λ_Δ = a multiplier for additional long-term deflections [ACI Eq. (9-11)]

ρ' = non-prestressed compression reinforcement ratio (A_s'/bd)

ξ = time-dependent factor for sustained loads: 5 years or more   2.0
                                                        12 months   1.4
                                                        6 months   1.2
                                                        3 months   1.0

Some judgment will be required in determining just what portion of the live loads should be considered *sustained*. In a residential application, 20% sustained live load might be a logical estimate, whereas in storage facilities, 100% sustained live load would be reasonable.

The calculated deflections must not exceed the maximum permissible deflections that are found in the ACI Code, Table 9.5(b). This table sets permissible deflections in terms of fractions of span length. These limitations guard against damage to the various parts of the system (both structural and nonstructural parts) as a result of excessive deflection. In the case of attached nonstructural elements, only the deflection that takes place after such attachment needs to be considered.

Example 7-2 _____

Determine which of the ACI Code deflection criteria will be satisfied by the non-prestressed reinforced concrete beam having a cross section shown in Figure 7-7 and subjected to maximum moments of M_{DL} = 20 ft-kips and M_{LL} = 15 ft-kips. Assume a 50% sustained live load and a sustained load time period of more than 5 years. Assume normal-weight concrete. Use f_c' = 3000 psi and f_y = 60,000 psi. The beam is on a simple span of 30 ft.

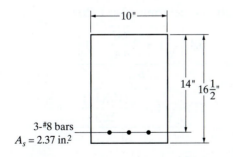

FIGURE 7-7   Sketch for Example 7-2.

Solution:

1.   The maximum *service* moments at midspan are

$$M_{DL} = 20 \text{ ft-kips} \quad \text{and} \quad M_{LL} = 15 \text{ ft-kips}$$

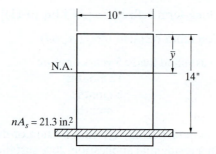

FIGURE 7-8   Transformed section for Example 7-2.

2.   Check the beam depth based on the ACI Code Table 9.5(a):

$$\text{minimum } h = \frac{\ell}{16} = \frac{30(12)}{16} = 22.5 \text{ in.}$$

Because 22.5 in. > 16.5 in., deflection must be calculated.

3.   The effective moment of inertia will be calculated using

$$I_e = \left(\frac{M_{cr}}{M_a}\right)^3 I_g + \left[1 - \left(\frac{M_{cr}}{M_a}\right)^3\right] I_{cr}$$

Therefore we first must compute the various terms within the expression. The moment of inertia of the cracked transformed section will be determined with reference to Figure 7-8.

The steel area of 2.37 in.^{2} and a modular ratio n of 9 (see Table A-6) will result in a transformed area of

$$nA_s = 9(2.37) = 21.3 \text{ in.}^2$$

The neutral-axis location is determined as follows:

$$\bar{y} = \frac{nA_s\left[\sqrt{1 + 2\dfrac{bd}{nA_s}} - 1\right]}{b}$$

$$= \frac{21.3\left[\sqrt{1 + 2\dfrac{(10)(14)}{21.3}} - 1\right]}{10}$$

$$= 5.88 \text{ in.}$$

The moment of inertia of the cracked transformed section is then determined:

$$I_{cr} = \frac{10(5.88)^3}{3} + 21.3(14 - 5.88)^2$$

$$= 2082 \text{ in.}^4$$

Determination of the moment of inertia of the gross section results in

$$I_g = \frac{bh^3}{12} = \frac{1}{12}(10)(16.5)^3 = 3743 \text{ in.}^4$$

The moment that would initially crack the cross section may be determined next:

$$M_{cr} = \frac{f_r I_g}{y_t}$$

where

$$y_t = \frac{16.5}{2} = 8.25 \text{ in.}$$

$$f_r = 7.5\lambda \sqrt{f'_c} = 0.411 \text{ ksi} \quad \text{(from Table A-6 for } \lambda = 1\text{)}$$

Therefore

$$M_{cr} = \frac{0.411(3743)}{8.25(12)} = 15.5 \text{ ft-kips}$$

We will assume that $M_a = 35$ ft-kips on the basis that it is the maximum moment that the beam must carry that will establish the crack pattern. This is not in strict accordance with the ACI Code, which indicates that M_a should be the maximum moment occurring at the stage the deflection is computed. It is logical that, as the cracking pattern is irreversible, the use of the effective moment of inertia based on the full maximum moment is more realistic. This approach is conservative and will furnish a lower I_e, which will subsequently result in a larger computed deflection.

4. Determine the effective moment of inertia:

$$I_e = \left(\frac{M_{cr}}{M_a}\right)^3 I_g + \left[1 - \left(\frac{M_{cr}}{M_a}\right)^3\right] I_{cr}$$

$$= \left(\frac{15.5}{35}\right)^3 3743 + \left[1 - \left(\frac{15.5}{35}\right)^3\right](2082) = 2226 \text{ in.}^4$$

5. Compute the immediate dead load deflection ($M_{DL} = 20$ ft-kips):

$$\Delta = \frac{5w\ell^4}{384 E_c I_e} = \frac{5M\ell^2}{48 E_c I_e}$$

where M is the moment due to a uniform load and E_c may be found in Table A-6. Thus

$$\Delta = \frac{5(20 \text{ ft-kips})(30 \text{ ft})^2(1728 \text{ in.}^3/\text{ft}^3)}{48(3120 \text{ kips/in.}^2)(2226 \text{ in.}^4)} = 0.467 \text{ in.}$$

6.   Compute the immediate live load deflection (M_{LL} = 15 ft-kips). By proportion,

$$\Delta = \frac{15}{20}(0.467) = 0.350 \text{ in.}$$

7.   The total immediate DL + LL deflection is

$$0.467 + 0.350 = 0.817 \text{ in.}$$

8.   The long-term (LT) deflection (DL + sustained LL) multiplier is

$$\lambda_\Delta = \frac{\xi}{1 + 50\rho'} = \frac{2.0}{1 + 0} = 2.0$$

Because M_{DL} = 20 ft-kips and 50% M_{LL} = 7.5 ft-kips, the sustained moment for long-term deflection = 27.5 ft-kips. Then

$$\Delta_{LT} = \frac{27.5}{20}(0.467)(2.0) = 1.28 \text{ in.}$$

9.   A comparison of actual deflections to maximum permissible deflections may now be made. In the comparison, made in Table 7-1, the maximum permissible deflections are from the ACI Code, Table 9.5(b). Case 1 in Table 7-1 applies to flat roofs not supporting or attached to nonstructural elements likely to be damaged by large deflections. Case 2 applies to floors not supporting or attached to nonstructural elements likely to be damaged by large deflections. As the permissible deflection is not exceeded in case 1 or case 2, the beam of Example 7-2 is limited to usage as defined by those two cases.

TABLE 7-1   Permissible Versus Actual Deflection (Example 7-2)

Case	Maximum permissible deflection*	Actual computed deflection
1	$\dfrac{\ell}{180} = \dfrac{30(12)}{180} = 2$ in.	Δ_{LL} = 0.35 in. (immediate LL)
2	$\dfrac{\ell}{360} = \dfrac{30(12)}{360} = 1$ in.	Δ_{LL} = 0.35 in. (immediate LL)
3	$\dfrac{\ell}{480} = \dfrac{30(12)}{480} = 0.75$ in.	$\Delta_{LL} + \Delta_{LT}$ = 0.35 + 1.28 = 1.63 in.
4	$\dfrac{\ell}{240} = \dfrac{30(12)}{240} = 1.5$ in.	$\Delta_{LL} + \Delta_{LT}$ = 0.35 + 1.28 = 1.63 in.

*From ACI 318-08, Table 9.5(b).

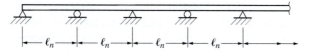

FIGURE 7-9   Continuous beam.

7-6   DEFLECTIONS FOR CONTINUOUS SPANS

To compute deflections for continuous spans subject to uniformly distributed loads such as the beam in Figure 7-9, the following approximate approach may be used:

$$\Delta = \frac{5w\ell_n^4}{384E_cI_e} - \frac{M\ell_n^2}{8E_cI_e}$$

where

M = negative moment at suppports (based on service loads) for span being investigated; if values are different, use average moment

ℓ_n = clear span

E_c = modulus of elasticity for concrete (see Table A-6 for normal-weight concrete)

w = uniformly distributed service load

I_e = effective moment of inertia; use the average value of I_e at positive moment area and I_e at negative moment area

In a similar manner, for long-term deflections the long-term deflection multiplier should be averaged for the different locations.

7-7   CRACK CONTROL

With the advent of higher-strength reinforcing steels, where more strain is required to produce the higher stresses, cracking of reinforced concrete flexural members has become more troublesome.

It seems logical that cracking would have an effect on corrosion of the reinforcing steel. However, there is no clear correlation between corrosion and surface crack widths in the usual range found in structures with reinforcement stresses at service load levels. Further, there is no clear experimental evidence available regarding the crack width beyond which a corrosion danger exists. Exposure tests indicate that concrete quality, adequate consolidation, and ample concrete cover may be more important in corrosion considerations than is crack width.

Rather than a small number of large cracks, it is more desirable to have only hairline cracks and to accept more numerous cracks, if necessary. To achieve this, the current ACI Code (Section 10.6) directs that the flexural tension reinforcement be well distributed in the maximum tension zones of a member. Section 10.6.4 contains a provision for maximum spacing s that is intended to control surface cracks to a width that is generally acceptable in practice. The maximum spacing is limited to

$$s = 15\left(\frac{40,000}{f_s}\right) - 2.5c_c \le 12\left(\frac{40,000}{f_s}\right) \qquad \text{[ACI Eq. (10-4)]}$$

where

s = center-to-center spacing of flexural tension reinforcement nearest to the tension face, in.

f_s = calculated stress, psi. This may be taken as ⅔ of the specified yield strength.

c_c = clear cover from the nearest surface in tension to the surface of the flexural tension reinforcement, in.

ACI 318-08, Section 10.6.5, cautions that if a structure is designed to be watertight or if it is to be subjected to very aggressive exposure, the provisions of Section 10.6.4 are not sufficient and special investigations and precautions are required.

Example 7-3 _____

Check the steel distribution for the beam shown in Figure 7-10 to establish whether reasonable control of flexural cracking is accomplished in accordance with the ACI Code, Section 10.6. Use $f_y = 60,000$ psi. Assume $d = 30$ in.

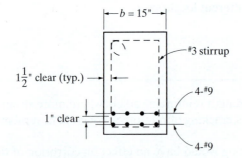

FIGURE 7-10   Sketch for Example 7-3.

Solution:

1.   Calculate the center-to-center spacing between the No. 9 bars:

$$s = \frac{15 - 2(1.5) - 2(0.375) - 2\left(\dfrac{1.128}{2}\right)}{3} = 3.37 \text{ in.}$$

2.   Assume positive moment and calculate the concrete clear cover from the bottom (tension) face of the beam to the surface of the nearest tension reinforcement:

$$c_c = 1.5 + 0.375 = 1.875 \text{ in.}$$

3.   Calculate f_s using ⅔ of f_y:

$$f_s = \frac{2}{3}f_y = \frac{2}{3}(60{,}000) = 40{,}000 \text{ psi}$$

4.   Calculate maximum spacing allowed using ACI Equation (10-4):

$$s = 15\left(\frac{40{,}000}{f_s}\right) - 2.5c_c = 15\left(\frac{40{,}000}{40{,}000}\right) - 2.5(1.875) = 10.31 \text{ in.}$$

Check the upper limit for ACI Equation (10-4):

$$12\left(\frac{40{,}000}{f_s}\right) = 12\left(\frac{40{,}000}{40{,}000}\right) = 12 \text{ in.} > 10.31 \text{ in.} \qquad \text{(O.K.)}$$

And lastly:

$$3.37 \text{ in.} < 10.31 \text{ in.} \qquad \text{(O.K.)}$$

When beams are relatively deep, there exists the possibility for surface cracking in the tension zone areas away from the main reinforcing. ACI 318-08, Section 10.6.7, requires, for beams having depths h in excess of 36 in., the placing of longitudinal skin reinforcing along both side faces for a distance $h/2$ from the tension face of the beam. The spacing s between these longitudinal bars or wires shall not exceed the spacing as provided in ACI 318-08, Section 10.6.4 (ACI Equation [10-4]). Bar sizes ranging from No. 3 to No. 5 (or welded wire reinforcement with a minimum area of 0.1 in.² per foot of depth) are typically used.

Example 7-4

Select skin reinforcement for the cross section shown in Figure 7-11a. Flexural tension reinforcement is 5 No. 9 bars and $f_y = 60{,}000$ psi.

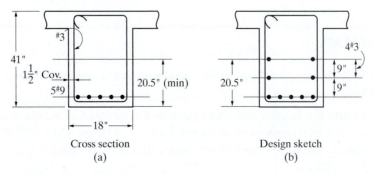

FIGURE 7-11   Sketch for Example 7-4.

Solution:

Because $h > 36$ in., skin reinforcement is required. The skin reinforcing must cover the tension surface for a minimum of $h/2$, or 20.5 in., up from the tension face of the beam. This is shown in Figure 7-11a. Assume No. 3 bars ($A_b = 0.11$ in.^{2}) for the skin reinforcing and calculate the maximum spacing s as follows:

$$c_c = 1.5 + 0.375 = 1.875 \text{ in.}$$

$$f_s = \frac{2}{3}f_y = \frac{2}{3}(60{,}000) = 40{,}000 \text{ psi}$$

$$s = 15\left(\frac{40{,}000}{f_s}\right) - 2.5c_c = 15\left(\frac{40{,}000}{40{,}000}\right) - 2.5(1.875) = 10.31 \text{ in.}$$

Check upper limit:

$$12\left(\frac{40{,}000}{f_s}\right) = 12\left(\frac{40{,}000}{40{,}000}\right) = 12 \text{ in.} > 10.31 \text{ in.} \qquad \text{(O.K.)}$$

The distance from the tension face of the beam to the centroid of the tension reinforcement is

$$1.5 + 0.375 + \frac{1.128}{2} = 2.44 \text{ in.}$$

Therefore, the required number of spaces N is

$$N = \frac{20.5 - 2.44}{10.31} = 1.75 \text{ spaces} \quad \text{(Use 2 spaces)}$$

The actual spacing provided is

$$\frac{20.5 - 2.44}{2 \text{ spaces}} \approx 9 \text{ in.} < 10.31 \text{ in.} \qquad \text{(O.K.)}$$

The design is shown in Figure 7-11b.

PROBLEMS

For the following problems, unless otherwise noted, assume normal-weight concrete.

7-1. Locate the neutral axis and calculate the moment of inertia for the cracked transformed cross sections shown. Use $f_c' = 4000$ psi and $f_y = 60{,}000$ psi.

7-2. Find I_g and I_{cr} for the T-beam shown. The effective flange width is 60 in., and $f_c' = 4000$ psi.

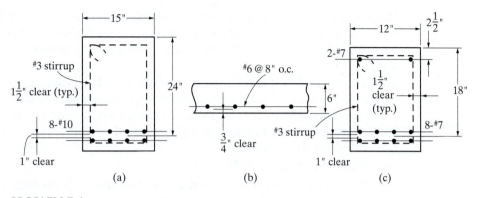

PROBLEM 7-1

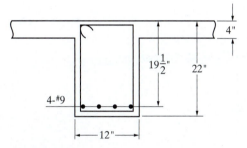

PROBLEM 7-2

7-3. The beam of cross section shown is on a simple span of 20 ft and carries service loads of 1.5 kips/ft dead load (includes beam weight) and 1.0 kip/ft live load. Use $f'_c = 3000$ psi and $f_y = 60,000$ psi.

   (a) Compute the immediate deflection due to dead load and live load.

   (b) Compute the long-term deflection due to the dead load. Assume the time period for sustained loads to be in excess of 5 years.

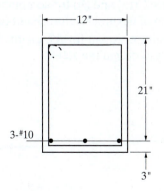

PROBLEM 7-3

7-4. Rework Problem 7-3 using a beam on a simple span of 26 ft, service load of 0.8 kip/ft dead load (includes beam weight) over the full span, and a point live load of 12 kips at midspan. Assume tensile reinforcing to be three No. 9 bars and $f'_c = 4000$ psi.

7-5. The floor beam shown is on a simple span of 16 ft. The beam supports non-structural elements likely to be damaged by large deflections. The service loads are 0.6 kip/ft dead load (does not include the beam weight) and 1.40 kips/ft live load. Assume that the live load is 60% sustained for 6-month periods. Use $f'_c = 3000$ psi and $f_y = 60,000$ psi.

(a) Check the beam for deflections.

(b) If the beam is unsatisfactory, redesign it so that it meets both flexural strength and deflections requirements.

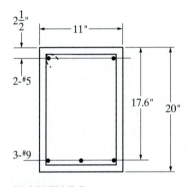

PROBLEM 7-5

7-6. Calculate the expected long-term deflection due to dead load and sustained live load for (a) the slab and (b) the beam of the floor system designed for Problem 6-4. Assume 10% sustained live load and a time period in excess of 5 years.

7-7. Check the cross sections of Problems 7-1(a) and (b) for acceptability under the ACI Code provisions for distribution of flexural reinforcement (crack control).

7-8. Check the distribution of flexural reinforcement for the members designed in Problems 2-21 and 2-25. If necessary, redesign the steel.

Walls

8-1   INTRODUCTION

Walls are generally used to provide lateral support for an earth fill, embankment, or some other material and to support vertical loads. Some of the more common types of walls are shown in Figure 8-1. One primary purpose for these walls is to maintain a difference in the elevation of the ground surface on each side of the wall. The earth whose ground surface is at the higher elevation is commonly called the *backfill*, and the wall is said to retain this backfill.

All the walls shown in Figure 8-1 have applications in either building or bridge projects. They do not necessarily behave in an identical manner under load, but still serve the same basic function of providing lateral support for a mass of earth or other material that is at a higher elevation behind the wall than the earth or other material in front of the wall. Hence they all may be broadly termed *retaining structures* or *retaining walls*. Some retaining walls may support vertical loads in addition to the lateral loads from the retained materials.

The gravity wall (Figure 8-1a) depends mostly on its own weight for stability. It is usually made of plain concrete and is used for walls up to approximately 10 ft in height. The semigravity wall is a modification of the gravity wall in which small amounts of reinforcing steel are introduced. This, in effect, reduces the massiveness of the wall.

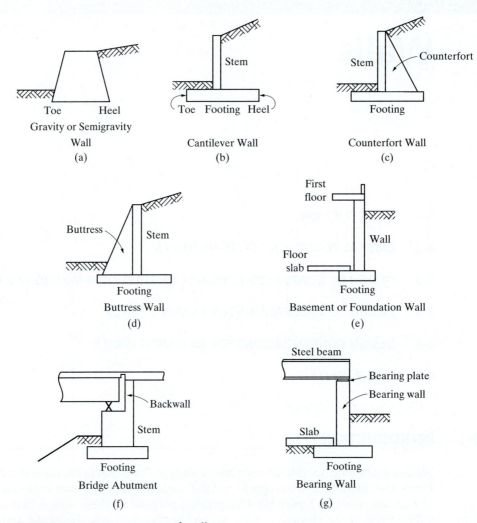

FIGURE 8-1 Common types of walls.

The cantilever wall (Figure 8-1b and Photo 8-1) is the most common type of re-taining structure and generally is used for walls in the range from 10 to 25 ft in height. It is so named because its individual parts (toe, heel, and stem) behave as, and are designed as, cantilever beams. Aside from its stability, the capacity of the wall is a function of the strength of its individual parts.

The counterfort wall (Figure 8-1c) may be economical when the wall height is in excess of 25 ft. The counterforts are spaced at intervals and act as tension members to support the stem. The stem is then designed as a continuous member spanning horizontally between the counterforts.

The buttress wall (Figure 8-1d) is similar to the counterfort wall except that the buttresses are located on the side of the stem opposite to the retained material and act as compression members to support the stem. The counterfort wall is more

PHOTO 8-1   Cantilever retaining wall (h_w = 12′-0″).

commonly used because it has a clean, uncluttered exposed face and allows for more efficient use of space in front of the wall.

The basement or foundation wall (Figure 8-1e) may act as a cantilever retaining wall. The first floor may provide an additional horizontal reaction similar to the basement floor slab, however, thereby making the wall act as a vertical beam. This wall would then be designed as a simply supported member spanning between the first floor and the basement floor slab.

The bridge abutment (Figure 8-1f) is similar in some respects to the basement wall. The bridge superstructure induces horizontal as well as vertical loads, thus altering the normal cantilever behavior.

The bearing wall (Figure 8-1g) may exist with or without lateral loads. A bearing wall may be defined as a wall that supports any vertical load in addition to its own weight. Depending on the magnitudes of the vertical and lateral loads, the wall may have to be designed for combined bending and axial compression. Bearing walls and basement walls are further discussed later in this chapter.

8-2   LATERAL FORCES ON RETAINING WALLS

The design of a retaining wall must account for all the applied loads. The load that presents the greatest problem and is of primary concern is the lateral earth pressure induced by the retained soil. The comprehensive earth pressure theories evolving from the original Coulomb and Rankine theories can be found in almost any text-book on soil mechanics.

The magnitude and direction of the pressures as well as the pressure distribution exerted by a soil backfill upon a wall are affected by many variables. These variables include, but are not limited to, the type of backfill used, the drainage of the backfill material, the level of the water table, the slope of the backfill material, added loads applied on the backfill, the degree of soil compaction, and movement of the wall caused by the action of backfill.

An important consideration is that water must be prevented from accumulating in the backfill material. Walls are rarely designed to retain saturated material, which means that proper drainage must be provided. It is generally agreed that the best backfill material behind a retaining structure is a well-drained, cohensionless material. Hence it is the condition that is usually specified and designed for. Materials that contain combinations of types of soil will act like the predominant material.

The lateral earth pressure can exist and develop in three different categories: active state, at rest, and passive state. If a wall is absolutely rigid, earth pressure at rest will develop. If the wall should deflect or move a very small amount away from the backfill, active earth pressure will develop and in effect reduce the lateral earth pressure occurring in the at-rest state. Should the wall be forced to move toward the backfill for some reason, passive earth pressure will develop and increase the lateral earth pressure appreciably above that occurring in the at-rest state. As indicated, the magnitude of earth pressure at rest lies somewhere between active and passive earth pressures.

Under normal conditions, earth pressure at rest is of such a magnitude that the wall deflects slightly, thus relieving itself of the at-rest pressure. The active pressure results. For this reason, retaining walls are generally designed for active earth pressure due to the retained soil.

Because of the involved nature of a rigorous analysis of an earth backfill and the variability of the material and conditions, assumptions and approximations are made with respect to the nature of lateral pressures on a retaining structure. It is common practice to assume linear active and passive earth pressure distributions. The pressure intensity is assumed to increase with depth as a function of the weight of the soil in a manner similar to that which would occur in a fluid. Hence this horizontal pressure of the earth against the wall is frequently called an *equivalent fluid pressure*. Experience has indicated that walls designed on the basis of these assumptions and those of the following discussion are safe and rela-tively economical.

Level Backfill

If we consider a level backfill (of well-drained cohensionless soil), the assumed pres-sure diagram is shown in Figure 8-2. The unit pressure intensity p_y in any plane a dis-tance y down from the top is

$$p_y = K_a w_e y$$

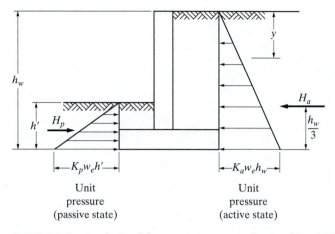

FIGURE 8-2   Analysis of forces acting on walls: level backfill.

Therefore, the total active earth pressure acting on a 1-ft width of wall may be calculated as the product of the average pressure on the total wall height h_w and the area on which this pressure acts:

$$H_a = \tfrac{1}{2}K_a w_e h_w^2 \tag{8-1}$$

where K_a, the *coefficient of active earth pressure*, has been established by both Rankine and Coulomb to be

$$K_a = \frac{1 - \sin \phi}{1 + \sin \phi} = \tan^2\left(45° - \frac{\phi}{2}\right) \tag{8-2}$$

and

$$w_e = \text{unit weight of earth (lb/ft}^3)$$

$$\phi = \text{angle of internal friction (soil on soil)}$$

K_a usually varies from 0.27 to 0.40. The term $K_a w_e$ in Equation (8-1) is generally called an *equivalent fluid weight*, because the resulting pressure is identical to that which would occur in a fluid of that weight (units are lb/ft^3).

In a similar manner, the total passive earth pressure force may be established as

$$H_p = \tfrac{1}{2}K_p w_e (h')^2 \tag{8-3}$$

where h' is the height of earth and K_p is the *coefficient of passive earth pressure*:

$$K_p = \frac{1 + \sin \phi}{1 - \sin \phi} = \tan^2\left(45° + \frac{\phi}{2}\right) = \frac{1}{K_a} \tag{}$$

Note that K_p usually varies from 2.5 to 4.0.

The total force in each case is assumed to act at one-third the height of the triangular pressure distribution, as shown in Figure 8-2.

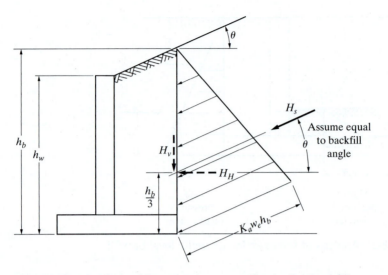

FIGURE 8-3   Analysis of forces acting on walls: sloping backfill.

Sloping Backfill

If we consider a sloping backfill, the assumed active earth pressure distribution is shown in Figure 8-3, where $H_s = \frac{1}{2}K_a w_e h_b^2$, h_b is the height of the backfill at the back of the footing, and K_a is the coefficient of active earth pressure. Thus

$$K_a = \cos\theta\left(\frac{\cos\theta - \sqrt{\cos^2\theta - \cos^2\phi}}{\cos\theta + \sqrt{\cos^2\theta - \cos^2\phi}}\right)$$

where θ is the slope angle of the backfill and ϕ is as previously defined. Note that H_s is shown acting parallel to the slope of the backfill.

For walls approximately 20 ft in height or less, it is recommended that the horizontal force component H_H simply be assumed equal to H_s and be assumed to act at $h_b/3$ above the bottom of the footing, as shown in Figure 8-3. The effect of the vertical force component H_V is neglected. This is a conservative approach.

Assuming a well-drained, cohensionless soil backfill that has a unit weight of 110 lb/ft^3 and an internal friction angle ϕ of 33°40′, values of equivalent fluid weight for sloping backfill may be determined as listed in Table 8-1.

TABLE 8-1 $K_a w_e$ Values for Sloping Backfill

θ (deg)	$K_a w_e$ (lb/ft^3)
0	32
10	33
20	38
30	54

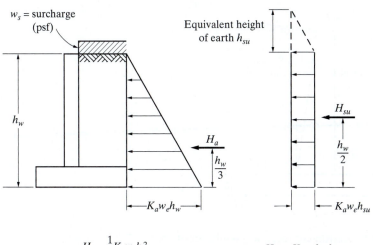

$$H_a = \frac{1}{2}K_a w_e h_w^2$$
(due to level backfill)

$$H_{su} = K_a w_e h_{su} h_w$$
(due to surcharge)

FIGURE 8-4 Forces acting on wall: level backfill and surcharge.

Level Backfill with Surcharge

Loads are often imposed on the backfill surface behind a retaining wall. They may be either live loads or dead loads. These loads are generally termed a *surcharge* and theoretically may be transformed into an equivalent height of earth.

A uniform surcharge over the adjacent area adds the same effect as an additional (equivalent) height of earth. This equivalent height of earth h_{su} may be obtained by

$$h_{su} = \frac{w_s}{w_e}$$

where

w_s = surcharge load (lb/ft^2)

w_e = unit weight of earth (lb/ft^3)

In effect, this adds a rectangle of pressure behind the wall with a total lateral surcharge force assumed acting at its midheight, as shown in Figure 8-4. Surcharge loads far enough removed from the wall cause no additional pressure acting on the wall.

8-3 DESIGN OF REINFORCED CONCRETE CANTILEVER RETAINING WALLS

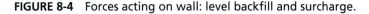

A retaining wall must be stable as a whole, and it must have sufficient strength to resist the forces acting on it. Four possible *modes of failure* will be considered. *Overturning about the toe*, point O, as shown in Figure 8-5, could occur due to lateral

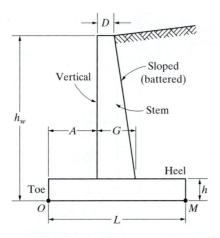

FIGURE 8-5   Cantilever retaining wall proportions.

loads. The stabilizing moment must be sufficiently in excess of the overturning moment so that an adequate factor of safety against overturning is provided. The factor of safety should never be less than 1.5 and should preferably be 2.0 or more. *Sliding on the base of the footing*, surface *OM* in Figure 8-5, could also occur due to lateral loads. The resisting force is based on an assumed coefficient of friction of concrete on earth. The factor of safety against sliding should never be less than 1.5 and should preferably be 2.0 or more. *Excessive soil pressure* under the footing will lead to undesirable settlements and possible rotation of the wall. Actual soil pressures should not be allowed to exceed specified allowable pressures, which depend on the characteristics of the underlying soil. The *structural failure of component parts of the wall* such as stem, toe, and heel, each acting as a cantilever beam, could occur. These must be designed to have sufficient strength to resist all anticipated loads.

A *general design procedure* for specific, known conditions may be summarized as follows:

1.   Establish the general shape of the wall based on the desired height and function.
2.   Establish the site soil conditions, loads, and other design parameters. This includes the determination of allowable soil pressure, earth-fill properties for active and passive pressure calculations, amount of surcharge, and the desired factors of safety.
3.   Establish the tentative proportions of the wall.
4.   Analyze the stability of the wall. Check factors of safety against overturning and sliding and compare actual soil pressure with allowable soil pressure.
5.   Assuming that all previous steps are satisfactory, design the component parts of the cantilever retaining wall, stem, toe, and heel as cantilever beams.

Using a procedure similar to that used for one-way slabs, the analysis and design of cantilever retaining walls is based on a 12-in. (1-ft)-wide strip measured along the length of wall. The tentative proportions of a cantilever retaining wall may be obtained from the following rules of thumb (see Figure 8-5):

1. Footing width L: Use $\frac{1}{2}h_w$ to $\frac{2}{3}h_w$.
2. Footing thickness h: Use $\frac{1}{10}h_w$.
3. Stem thickness G (at top of footing): Use $\frac{1}{12}h_w$.
4. Toe width A: Use $\frac{1}{4}L$ to $\frac{1}{3}L$.
5. Use a minimum wall batter of $\frac{1}{4}$ in./ft to improve the efficiency of the stem as a bending member and to decrease the quantity of concrete required.
6. The top of stem thickness D should not be less than 10 in.

The given rules of thumb will usually result in walls that can reasonably be designed. Depending on the specific conditions, however, dimensions may have to be adjusted somewhat to accommodate such design criteria as reinforcement limits, shear strength, anchorage, and development. One common alternative design approach is to assume a footing thickness and then immediately design the stem thickness for an assumed steel ratio. Once the stem thickness is established, the wall stability can be checked. Whichever procedure is used, adjustment of dimensions during the design is not uncommon.

Example 8-1 —————————————————————————

Design a retaining wall for the conditions shown in Figure 8-6. Use $f_c' = 3000$ psi and $f_y = 60,000$ psi. Other design data are given in step 2. Assume normal-weight concrete.

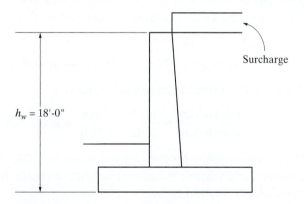

$h_w = 18'\text{-}0''$

Surcharge

FIGURE 8-6   Sketch for Example 8-1.

Solution:

1. The general shape of the wall, as shown, is that of a cantilever wall, because the overall height of 18 ft is within the range in which this type of wall is normally economical.

2.   The design data are unit weight of earth $w_e = 100$ lb/ft^3, allowable soil pressure $= 4000$ psf, equivalent fluid weight $K_a w_e = 30$ lb/ft^3, and surcharge load $w_s = 400$ psf. The desired minimum factor of safety against overturning is 2.0 and against sliding is 1.5.

3.   Establish tentative proportions for the wall.

a.   Footing width:

$$\tfrac{1}{2} \text{ to } \tfrac{2}{3} \text{ of wall height}$$

$$\tfrac{1}{2}(18) \text{ to } \tfrac{2}{3}(18) = 9 \text{ to } 12 \text{ ft}$$

Use 11 ft-0 in.

b.   Footing thickness:

$$\tfrac{1}{10}(18) = 1.8 \text{ ft}$$

Use 1 ft-9 in.

c.   Stem thickness at top of footing:

$$\tfrac{1}{12}(18) = 1.5 \text{ ft}$$

Use 1 ft-6 in.

d.   Toe width:

$$\tfrac{1}{4} \text{ to } \tfrac{1}{3} \text{ footing width}$$

$$\tfrac{1}{4}(11) \text{ to } \tfrac{1}{3}(11) = 2.75 \text{ to } 3.67 \text{ ft}$$

Use 3 ft-0 in.

e.   Use a batter for the rear face of wall approximately $\tfrac{1}{2}$ in./ft.

f.   Top of stem thickness, based on G and a batter of $\tfrac{1}{2}$ in./ft:

$$D = G - (h_w - h)\tfrac{1}{2}$$

$$= 18 \text{ in.} - (18 \text{ ft} - 1.75 \text{ ft})\tfrac{1}{2}\text{in./ft} = 9.88 \text{ in.}$$

Use 10 in. Therefore the calculated batter is

$$\frac{\text{total batter}}{\text{stem height}} = \frac{18 \text{ in.} - 10 \text{ in.}}{18 \text{ ft} - 1.75 \text{ ft}} = 0.492 \text{ in./ft}$$

The preliminary wall proportions are shown in Figure 8-7.

4.   For the stability analysis, use unfactored weights and loads in accordance with the ACI Code, Section 15.2.2.

a.   *Factor of safety against overturning:* The tendency of the wall to overturn is a result of the horizontal loads acting on the wall. An assumption is made that in overturning, the wall will rotate about the toe, and the horizontal loads are said to create an *overturning moment* about the toe. Any vertical loads will tend to create rotation about the toe in the opposite direction and are therefore said to provide a *stabilizing moment*.

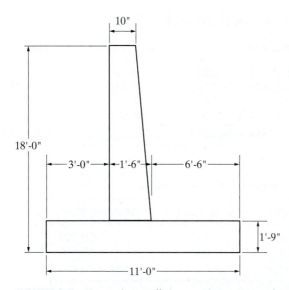

FIGURE 8-7 Tentative wall proportions, Example 8-1.

The factor of safety (FS) against overturning is then expressed as

$$\text{FS} = \frac{\text{stabilizing moment}}{\text{overturning moment}}$$

The required minimum factor of safety against overturning is normally governed by the applicable building code. A minimum of 1.5 is generally considered good practice. The passive earth resistance of the soil in front of the wall is generally neglected in stability computations because of the possibility of its removal by erosion or excavation.

As discussed previously, the surcharge may be converted into an equivalent height of earth,

$$h_{su} = \frac{w_s}{w_e} = \frac{400}{100} = 4 \text{ ft}$$

thus adding a rectangle of earth pressure behind the wall.

The various vertical and horizontal forces and their associated moments are shown in Tables 8-2 and 8-3 (see Figure 8-8). Note that soil on the toe is neglected. Both stabilizing and overturning moments are calculated with respect to point O in Figure 8-8.

In Table 8-2, note that W_2 is the weight of soil and surcharge on the heel from the right-hand edge of the heel to the vertical dashed line in the stem. The weight difference between reinforced concrete and soil (50 lb/ft^3) in the triangular portion that is the back of the stem is represented by W_3.

TABLE 8-2   Stabilizing Moments (Vertical Forces)

Force	Magnitude (lb)	Lever arm (ft)	Moment (ft-lb)
W_1	$0.833(16.25)(150) =$ 2030	3.42	6940
W_2	$7.17(16.25 + 4.0)(100) = 14,520$	7.42	107,700
W_3	$(\frac{1}{2})(16.25)(0.67)(50) =$ 272	4.05	1102
W_4	$11.0(1.75)(150) =$ 2890	5.5	15,900
	$\Sigma W = 19,710$		$\Sigma M = 131,600$

TABLE 8-3   Overturning Moments (Horizontal Forces)

Force	Magnitude (lb)	Lever arm (ft)	Moment (ft-lb)
H_1	$\frac{1}{2}(30)(18)^2 = 4860$	6.0	29,200
H_2	$4(30)(18) = 2160$	9.0	19,440
	$\Sigma H = 7020$		$\Sigma M = 48,600$

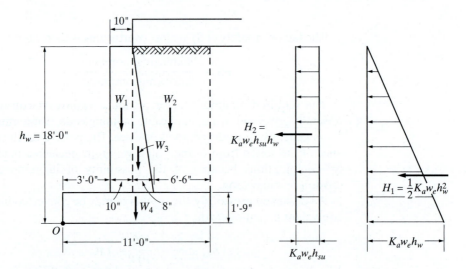

FIGURE 8-8   Stability analysis diagram.

The factor of safety against overturning is

$$\text{FS} = \frac{131,600}{48,600} = 2.71 > 2.0 \qquad \text{(O.K.)}$$

b.   *Factor of safety against sliding:* The tendency of the wall to slide is primarily a result of the horizontal forces, whereas the vertical forces cause the frictional resistance against sliding.

The total frictional force available (or resisting force) may be expressed as

$$F = f(\Sigma W)$$

where f = coefficient of friction between the concrete and soil and ΣW = summation of vertical forces (see Figure 8-8). A typical value for the coefficient of friction is $f = 0.50$. Then the resisting force F can be calculated:

$$F = f(\Sigma W)$$

$$0.50(19,710) = 9860 \text{ lb}$$

The factor of safety against sliding may be expressed as

$$\text{FS} = \frac{\text{resisting force } F}{\text{actual horizontal force } \Sigma H}$$

$$= \frac{9860}{7020} = 1.40$$

The required minimum factor of safety for this problem is 1.5; hence, the resistance against sliding is *inadequate*.

One solution to this problem is to reproportion the wall until the requirements are met. Rather than change the wall, we will use a base shear key to mobilize the passive resistance of the soil and in effect increase the resisting force F and subsequently increase the factor of safety. The design of the base shear key will be one of the last steps in the design of the retaining wall.

c. *Soil pressures and location of resultant force:* The soil pressure under the footing of the wall is a function of the location of the resultant force, which in turn is a function of the vertical forces and horizontal forces. It is generally desirable and usually required that for walls on soil, the resultant of all the forces acting on the wall must lie within the middle third of the base. When this occurs, the resulting pressure distribution could be either triangular, rectangular, or trapezoidal in shape, with the soil in compression under the entire width of the footing. The resulting maximum foundation pressure must not exceed the safe bearing capacity of the soil.

The first step is to locate the point at which the resultant of the vertical forces ΣW and the horizontal forces ΣH intersects the bottom of the footing. This may be visualized with reference to Figure 8-9. Note that ΣW and ΣH are first combined to form the resultant force R. Because R may be moved anywhere along its line of action, it is moved to the bottom of the footing, where it is then resolved back into its components ΣW and ΣH. The

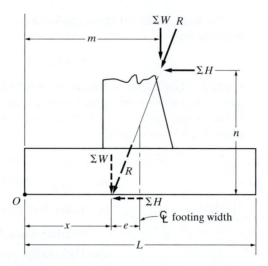

FIGURE 8-9   Resultant of forces acting on wall.

moment about point O due to the resultant force acting at the bottom of the footing must be the same as the moment effect about point O of the components ΣW and ΣH. Therefore, using the components in the plane of the bottom of the footing, the location of the resultant that is a distance x from point O may be determined. Note that the moment due to ΣH at the bottom of the footing is zero:

$$\Sigma W(x) = \Sigma Wm - \Sigma Hn$$

As may be observed, ΣWm is the stabilizing moment of the vertical forces with respect to point O and ΣHn is the overturning moment of the horizontal forces with respect to the same point. Hence, this expression may be rewritten as

$$\Sigma W(x) = \text{stabilizing } M - \text{overturning } M$$

$$x = \frac{\text{stabilizing } M - \text{overturning } M}{\Sigma W}$$

With reference to Figure 8-8,

$$x = \frac{131,600 - 48,600}{19,710} = 4.21 \text{ ft from toe}$$

Therefore the eccentricity e with respect to the centerline of the footing is

$$e = 5.5 - 4.21 = 1.29 \text{ ft}$$

With a footing length of 11 ft, the middle third has a length of $^{11}\!/_3 = 3.667$ ft. Eccentricity e is measured from the centerline of the footing and because

$$e = 1.29 \text{ ft} \le \frac{3.667 \text{ ft}}{2} = 1.834 \text{ ft}$$

the resultant lies in the middle third of the footing, and the resulting pressure distribution is trapezoidal. If the resultant intersected the base at the edge of the middle third (i.e., $e = {}^1\!/_6 = 1.833$ ft), the pressure distribution would have been triangular, and if $e = 0$, the pressure distribution would have been rectangular, indicating a uniform soil pressure distribution.

The pressures may now be calculated considering ΣW applied as an eccentric load on a rectangular section 11 ft long by 1 ft wide (this is a *typical* 1-ft-wide strip of the wall footing). The pressures are obtained by using the basic equations for bending and axial compression:

$$p = \frac{P}{A} \pm \frac{Mc}{I}$$

where

p = unit soil pressure intensity under the footing

P = total vertical load (ΣW)

A = footing cross-sectional area $[(L)(1.0)]$

M = moment due to eccentric load $[(\Sigma W)(e)]$

c = distance from centerline of footing to outside edge $(L/2)$

I = moment of inertia of footing with respect to its centerline $[(1.0)(L)^3/12]$

The expression given previously may be rewritten as

$$p = \frac{\Sigma W}{(L)(1.0)} \pm \frac{(\Sigma W)(e)(L/2)}{1.0(L)^3/12}$$

Simplifying and rearranging yields

$$p = \frac{\Sigma W}{L}\left(1 \pm \frac{6e}{L}\right)$$

Substituting, we obtain

$$p = \frac{19{,}710}{11}\left[1 \pm \frac{6(1.29)}{11}\right]$$

$$= 1792(1 \pm 0.704)$$

$$\text{maximum } p = 1792(1.704) = 3050 \text{ psf}$$

$$\text{minimum } p = 1792(0.296) = 530 \text{ psf}$$

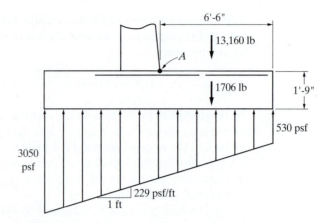

FIGURE 8-10   Pressure distribution under footing.

The resulting pressures are less than 4000 psf and are therefore satisfactory. The resulting pressure distribution beneath the footing is shown in Figure 8-10.

5. Design the component parts.

    a. *Design of heel:* The load on the heel is primarily earth dead load and surcharge, if any, acting vertically downward. With the assumed straight-line pressure distribution under the footing, the downward load is reduced somewhat by the upward-acting pressure. For design purposes, the heel is assumed to be a cantilever beam 1 ft in width and with a span length equal to 6 ft-6 in. fixed at the rear face of the wall (point *A* in Figure 8-10).

    The weight of the footing is

    $$1.75(150)(6.50) = 1706 \text{ lb}$$

    The earth and surcharge weight is

    $$6.50(20.25)(100) = 13{,}160 \text{ lb}$$

    The slope of pressure distribution under the footing is

    $$\frac{3050 - 530}{11} = 229 \text{ psf/linear foot}$$

    Thus far, all analysis has been based on service (unfactored) loads because allowable soil pressures are determined using a certain factor of safety against reaching pressures that will cause unacceptable settlements. Reinforced concrete, however, is designed on the basis of factored loads. The ACI Code, Section 9.2.1, specifies that where lateral earth pressure *H* must be included in design, the strength *U* shall be at least equal to $1.2D + 1.6L + 1.6H$.

A conservative approach to the design of the heel is to use factored loads, as required, and to ignore the relieving effect of the upward pressure under the heel. This is the unlikely condition that would exist if there occurred a lateral force overload (and no associated increased vertical load) causing uplift of the heel.

The *maximum moment* may be obtained by taking a summation of moments about point A (utilizing service loads):

$$M = (13{,}160 + 1706)\frac{6.50}{2} = 48{,}300 \text{ ft-lb}$$

The bending of the heel is such that tension occurs in the top of the footing.

The *maximum shear* may be obtained by taking a summation of vertical forces on the heel side of point A (using service loads):

$$V = 13{,}160 + 1706 = 14{,}870 \text{ lb}$$

The maximum moment and shear must be modified for strength design. As the loads on the heel are predominantly dead load, a load factor of 1.2 is used. This in effect considers the surcharge to be a dead load. This is acceptable, however, because of the conservative nature of the design. Thus

$$M_u = 48{,}300(1.2) = 58{,}000 \text{ ft-lb}$$

$$V_u = 14{,}870(1.2) = 17{,}840 \text{ lb}$$

The *footing size and reinforcement* (for heel) will be determined next. Because it commonly occurs that shear strength will be the controlling factor with respect to footing thickness, the shear will be checked first. The heel effective depth available, assuming 2-in. cover (ACI Code, Section 7.7.1) and No. 8 bars, is

$$d = 21 - 2 - 0.5 = 18.5 \text{ in.}$$

The shear strength ϕV_u of the heel, if no shear reinforcing is provided, is the shear strength of the concrete alone:

$$\phi V_n = \phi V_c = \phi(2\sqrt{f_c'})bd$$
$$= 0.75(2)\sqrt{3000}(12)(18.5)$$
$$\phi V_n = 18{,}240 \text{ lb}$$

Therefore $\phi V_n > V_u$ (O.K.).

The *tensile reinforcement* requirement may now be determined in the normal way. Assuming $\phi = 0.90$:

$$\text{required } \bar{k} = \frac{M_u}{\phi b d^2} = \frac{58.0(12)}{0.9(12)(18.5)^2}$$
$$= 0.1883 \text{ ksi}$$

Therefore, from Table A-8, the required $\rho = 0.0033$, $\epsilon_t > 0.005$, and $\phi = 0.90$ (O.K.). The required steel area is then calculated from

$$\text{required } A_s = \rho bd = 0.0033(12)(18.5) = 0.73 \text{ in.}^2$$

The minimum area of steel required by the ACI Code, Section 10.5.1, may be obtained using Table A-5:

$$A_{s,\min} = 0.0033(12)(18.5) = 0.73 \text{ in.}^2$$

As discussed in Section 2-8, $A_{s,\min}$ must be provided wherever reinforcement is needed, except where such reinforcement is at least one-third greater than that required by analysis (see the ACI Code, Section 10.5.3).

The ACI Code, Section 10.5.4, also permits the use of a minimum reinforcement equal to that required for shrinkage and temperature steel in structural slabs of uniform thickness as furnished in the ACI Code, Section 7.12. This will always be somewhat less than that required by the ACI Code, Section 10.5.1, but not necessarily less than that specified by the ACI Code, Section 10.5.3. For footings, the shrinkage and temperature steel requirement for slabs of uniform thickness will be used as an absolute minimum in this text.

For the heel reinforcement for this retaining wall, because the required steel area is equal to $A_{s,\min}$, we will select No. 7 bars at 9 in. o.c. ($A_s = 0.80 \text{ in.}^2$).

b.   *Design of toe:* The load on the toe is primarily a result of the soil pressure distribution on the bottom of the footing acting in an upward direction. For design purposes, the toe is assumed to be a cantilever beam 1 ft in width and with a span length equal to 3 ft-0 in. fixed at the front face of the wall (point B in Figure 8-11). The soil on the top of the toe is conservatively neglected. Forces and pressures are calculated with reference to Figure 8-11.

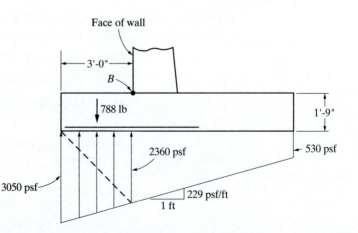

FIGURE 8-11   Pressure distribution for toe design.

As the reinforcing steel will be placed in the bottom of the footing, the effective depth available, assuming a 3-in. cover and No. 8 bars, is

$$d = 21 - 3 - 0.5 = 17.5 \text{ in.}$$

The weight of footing for the toe design is

$$1.75(3.0)(150) = 788 \text{ lb}$$

The soil pressure directly under point B, recalling that the slope of the pressure diagram is 229 psf/ft, is

$$3050 - 3.0(229) = 2360 \text{ psf}$$

The *design moment* M_u may be obtained by a summation of moments about point B in Figure 8-11. Both the moment and the shear must be modified for strength design. According to the ACI Code, Section 9.2.1, a load factor of 1.6 should be used for horizontal earth pressure and for live load, whereas 1.2 should be used for dead load. Because the soil pressure under the toe is largely the result of horizontal earth pressure, however, a conservative procedure of using 1.6 is recommended. In addition, as the weight of the footing reduces the effect of the horizontal earth pressure, a factor of 0.9 is used for the footing dead load, as recommended by the ACI Code (Section 9.2.1). The design moment M_u is then calculated as

$$M_u = 1.6(\tfrac{1}{2})(3050)(3.0)^2(\tfrac{2}{3}) + 1.6(\tfrac{1}{2})(2360)(3.0)^2(\tfrac{1}{3}) - 0.9(788)\left(\frac{3.0}{2}\right)$$

$$= 19{,}240 \text{ ft-lb}$$

The bending of the toe is such that tension occurs in the bottom of the footing.

The *design shear* V_u is obtained by a summation of vertical forces on the toe side of point B and applying load factors, as previously discussed:

$$V_u = 1.6(\tfrac{1}{2})(3050)(3.0) + 1.6(\tfrac{1}{2})(2360)(3.0) - 0.9(788)$$

$$= 12{,}270 \text{ lb}$$

Footing size and reinforcement, based on the requirements of the toe, are treated as they were for the heel. The shear strength ϕV_n of the toe is calculated from

$$\phi V_n = \phi V_c = \phi(2\sqrt{f_c'})bd$$

$$= 0.75(2)\sqrt{3000}(12)(17.5)$$

$$\phi V_n = 17{,}250 \text{ lb}$$

$$\phi V_n > V_u$$

Therefore the thickness of the footing is satisfactory.

Based on the determined M_u, and assuming $\phi = 0.90$:

$$\text{required } \bar{k} = \frac{M_u}{\phi b d^2} = \frac{19.24(12)}{0.9(12)(17.5)^2}$$

$$= 0.0698 \text{ ksi}$$

Therefore, from Table A-8,

$$\text{required } \rho = 0.0012$$

Note that $\epsilon_t > 0.005$. Therefore $\phi = 0.90$. The required steel area is

$$\text{required } A_s = \rho b d = 0.0012(12)(17.5) = 0.25 \text{ in.}^2$$

The minimum area of steel required may be obtained using Table A-5:

$$A_{s,\text{min}} = 0.0033(12)(17.5) = 0.69 \text{ in.}^2$$

Because required $A_s < A_{s,\text{min}}$, other minimum steel criteria should be checked to establish a controlling minimum value:

1. Provide one-third additional reinforcing as outlined in the ACI Code, Section 10.5.3:

$$A_{s,\text{min}} = 1.33(0.25) = 0.33 \text{ in.}^2$$

2. Check the required steel area based on the absolute minimum of shrinkage and temperature steel required for structural slabs of uniform thickness (ACI Code, Section 7.12):

$$\text{required } A_s = 0.0018bh$$

$$= 0.0018(12)(21) = 0.45 \text{ in.}^2$$

As we consider the shrinkage and temperature steel requirement as an absolute minimum, use No. 7 bars at 16 in. o.c. ($A_s = 0.45 \text{ in.}^2$).

c. *Design of stem:* The load on the stem is primarily lateral earth pressure acting (in this problem) horizontally. For design purposes, the stem is assumed to be a vertical cantilever beam 1 ft in width and with a span length equal to 16 ft-3 in. fixed at the top of the footing.

　　The loads acting on the stem from the top of the wall to the top of the footing are depicted in Figure 8-12. The magnitudes of the horizontal forces are

$$H_{s1} = \frac{1}{2}(30)(16.25)^2 = 3960 \text{ lb}$$

$$H_{s2} = 4(30)(16.25) = 1950 \text{ lb}$$

　　The design moment M_u in the stem may be obtained by a summation of moments about the top of the footing. Both shears and moments must be modified for strength design. Because the forces

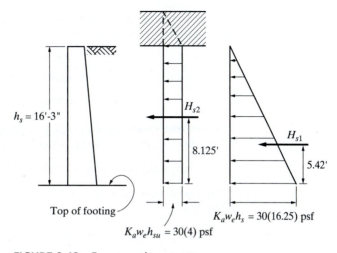

$K_a w_e h_{su} = 30(4)$ psf

$K_a w_e h_s = 30(16.25)$ psf

FIGURE 8-12   Forces acting on stem.

are due to lateral earth pressure, a factor of 1.6 should be used. M_u may then be calculated as

$$M_u = 1.6(3960)(5.42) + 1.6(1950)(8.125)$$

$$= 59{,}700 \text{ ft-lb}$$

The design shear V_u is obtained by a summation of the horizontal forces acting on the stem above the top of the footing:

$$V_u = 1.6(3960) + 1.6(1950)$$

$$= 9460 \text{ lb}$$

The *stem size and reinforcement* requirements will be determined next. Because the stem is assumed to be a cantilever slab, V_u as a limit may equal ϕV_n, where $V_n = V_c$. It is not practical to reinforce the stem for shear; therefore, if shear strength is inadequate, stem thickness must be increased. The effective depth of the stem at the top of the footing, assuming 2-in. cover and No. 8 bars, is

$$d = 18 - 2 - 0.5 = 15.5 \text{ in.}$$

The shear strength ϕV_n of the stem (at the top of the footing), if no shear reinforcement is provided, is the shear strength of the concrete alone:

$$\phi V_n = \phi V_c = \phi 2\sqrt{f_c'}bd$$

$$= 0.75(2)\sqrt{3000}(12)(15.5)$$

$$\phi V_n = 15{,}280 \text{ lb}$$

$$\phi V_n > V_u$$

Thus, the stem thickness need not be increased. Based on M_u at the bottom of the stem and assuming $\phi = 0.90$,

$$\text{required } \bar{k} = \frac{M_u}{\phi b d^2} = \frac{59.7(12)}{0.9(12)(15.5)^2}$$

$$= 0.2761 \text{ ksi}$$

Therefore, from Table A-8, the required $\rho = 0.0049$, and $\epsilon_t > 0.005$, so $\phi = 0.90$. The required steel area is

$$\text{required } A_s = \rho b d = 0.0049(12)(15.5) = 0.91 \text{ in.}^2$$

The minimum area of steel required may be obtained using Table A-5:

$$A_{s,\text{min}} = 0.0033(12)(15.5) = 0.61 \text{ in.}^2$$

Because required $A_s > A_{s,\text{min}}$, the calculated required A_s controls. Use No. 7 bars at 7½ in. o.c. ($A_s = 0.96$ in.^{2}).

The *stem reinforcement pattern* will be investigated more closely. As the moment and shear vary along the height of the wall, it is only logical that the steel requirements also vary. A pattern is usually created whereby some of the stem reinforcement is cut off where it is no longer required. The actual pattern used is the choice of the designer, based on various practical constraints.

A typical approach to the creation of a stem reinforcement pattern is first to draw the complete M_u diagram for the stem and then to determine bar cutoff points using the procedure discussed in Chapter 5. Stem moments due to the external loads will be calculated at 5 ft from the top of the wall and at 10 ft from the top of the wall. These distances are arbitrary and should be chosen so that a sufficient number of points will be available to draw the M_u diagram.

With reference to Figure 8-13, the moment at 5 ft from the top of the wall is calculated by first determining the horizontal forces:

$$H_{s1} = \tfrac{1}{2}(30)(5.0)^2 = 375 \text{ lb}$$

$$H_{s2} = 4(30)(5.0) = 600 \text{ lb}$$

Summing the moments about a plane 5 ft below the top of the wall and introducing the appropriate load factor,

$$M_u = 1.6(375)\left(\frac{5.0}{3}\right) + 1.6(600)\left(\frac{5.0}{2}\right)$$

$$= 3400 \text{ ft-lb}$$

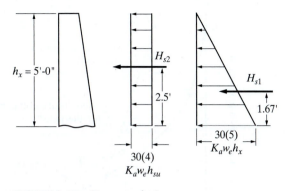

FIGURE 8-13   Stem analysis.

With reference to Figure 8-14, the moment at 10 ft below the top of the wall may be calculated:

$$H_{s1} = \frac{1}{2}(30)(10.0)^2 = 1500 \text{ lb}$$

$$H_{s2} = 4(30)(10.0) = 1200 \text{ lb}$$

and

$$M_u = 1.6(1500)\left(\frac{10.0}{3}\right) + 1.6(1200)\left(\frac{10.0}{2}\right)$$

$$= 17{,}600 \text{ ft-lb}$$

The three M_u values may now be plotted as the solid curve in Figure 8-15. This diagram will reflect both the design moment and moment strength with respect to the distance from the top of the wall and must be plotted to scale.

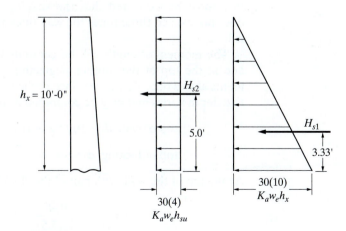

FIGURE 8-14   Stem analysis.

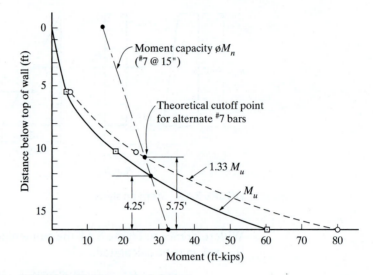

FIGURE 8-15   Stem steel design.

The moment strength ϕM_n of the stem at various locations may be computed from the expression

$$\phi M_n = \phi b d^2 \bar{k}$$

Because we are primarily attempting to establish a cutoff location for some vertical stem reinforcement, the moment strength will be based on a pattern of alternate vertical bars being cut off. With alternate bars cut off, the remaining reinforcement pattern is No. 7 bars at 15 in. o.c., which furnishes an $A_s = 0.48$ in.[2] Caution must be exercised that the spacing of the remaining bars does not exceed three times the wall thickness or 18 in., whichever is less.

The moment strength will be computed at the top of the wall and at the top of the footing, neglecting minimum steel requirements at this time. As the rear face of the wall is battered, the effective depth varies and may be computed as follows:

top of wall: $d = 10 - 2 - 0.5 = 7.5$ in.

top of footing: $d = 18 - 2 - 0.5 = 15.5$ in.

At the top of the wall, with half of the stem bars cut off:

$$\rho = \frac{A_s}{bd} = \frac{0.48}{12(7.5)} = 0.0053$$

Therefore, from Table A-8, $\bar{k} = 0.2982$ ksi. Then

$$\phi M_n = \frac{0.9(12)7.5)^2}{12}(0.2982)$$

$$= 15.10 \text{ ft-kips}$$

At the top of the footing,

$$\rho = \frac{A_s}{bd} = \frac{0.48}{12(15.5)} = 0.0026$$

Therefore, from Table A-8, $\bar{k} = 0.1512$ ksi. Then

$$\phi M_n = \frac{0.9(12)(15.5)^2}{12}(0.1512)$$

$$= 32.7 \text{ ft-kips}$$

These two values (ϕM_n) may now be plotted in Figure 8-15. A straight line will be used to approximate the variation of ϕM_n for No. 7 bars at 15 in. o.c. from the top to the bottom of the wall.

The intersection of the ϕM_n line and the M_u solid curve establishes a theoretical cutoff point for alternate vertical bars based on strength requirements. This location may be determined graphically by scaling on Figure 8-15. The scaled distance is 12.0 ft below the top of the wall or 4 ft-3 in. above the top of the footing. The No. 7 bar must be extended past this theoretical cutoff point a distance equal to the effective depth of the member or 12 bar diameters, whichever is larger.

Because the effective depth of the stem steel at the top of the footing is 15.5 in. and the wall batter is approximately 0.492 in./ft, the effective depth of the stem steel at 4.25 ft above the top of the footing may be calculated as follows:

$$d = 15.5 - 4.25(0.492) = 13.4 \text{ in.}$$

$$12d_b = 12(0.875) = 10.5 \text{ in.}$$

Therefore the required length of bar above the top of the footing would be

$$4.25 + \frac{13.4}{12} = 5.37 \text{ ft}$$

Use 5 ft 5 in. (5.42 ft).

Check the required minimum area of steel, $A_{s,\min}$, at the theoretical cutoff point to ensure that discontinuing half of the steel will not violate this code requirement. Once alternate bars have been terminated, the remaining steel will be No. 7 bars at 15 in.

$(A_s = 0.48 \text{ in.}^2)$. At the theoretical cutoff point 4.25 ft above the top of the footing, the effective depth $d = 13.4$ in. From Table A-5,

$$A_{s,\text{min}} = 0.0033bd$$

$$= 0.0033(12)(13.4) = 0.53 \text{ in.}^2$$

This is unsatisfactory, because the steel area provided is less than $A_{s,\text{min}}$. However, recall from previous discussions that the minimum reinforcing requirement need not be applied if the reinforcing provided is one-third greater than that required by analysis. Therefore, a simple solution is to move the cutoff point upward to where the reinforcing provided is one-third greater than required. This can be accomplished by drawing a $1.33 \times M_u$ curve and reestablishing the theoretical cutoff point for alternate No. 7 bars. At this cutoff point, one-third more steel is provided than is required; therefore the minimum steel requirement does not apply.

Multiplying the previously determined design moments M_u by 1.33, the following moments are obtained:

At 5 ft below the top of the wall:

$$M_u = 1.33(3400) = 4520 \text{ ft-lb}$$

At 10 ft below the top of the wall:

$$M_u = 1.33(17,600) = 23,400 \text{ ft-lb}$$

At the top of the footing:

$$M_u = 1.33(59,700) \text{ ft-lb} = 79,400 \text{ ft-lb}$$

Plotting this moment curve (see the dashed line on Figure 8-15) and scaling the distance above the top of the footing to theoretical cutoff point, we arrive at a value of 5.75 ft., or 5ft-9 in. The bars must extend past this theoretical cutoff point a distance equal to the greater of the effective depth of the member or 12 bar diameters.

$$d = 15.5 - 5.75(0.492) = 12.67 \text{ in.}$$

$$12d_b = 12(0.875) = 10.5 \text{ in.}$$

Adding 12.67 in. to the previously determined theoretical cutoff point above the top of the footing, we have

$$5.75 + \frac{12.67}{12} = 6.81 \text{ ft}$$

Use 6 ft-10 in. (6.83 ft).

Therefore, terminate alternate No. 7 bars at 6 ft-10 in. above the top of the footing.

The ACI Code, Section 12.10.5, stipulates that flexural reinforcement must not be terminated in a tension zone unless one of several

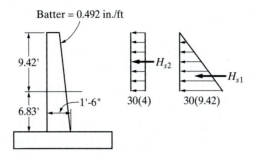

FIGURE 8-16   Shear at bar cutoff point.

conditions is satisfied. One of these conditions is that the shear at the cutoff point does not exceed two-thirds of the shear permitted. Therefore check the shear at the actual cutoff point, 6.83 ft above the top of the footing (9.42 ft below the top of the wall; see Figure 8-16).

For the shear strength calculation, the wall thickness at a height of 6.83 ft above the top of the footing is

$$18.0 - 6.83 (0.492) = 14.64 \text{ in.}$$

from which d may be calculated as

$$d = 14.64 - 2 - 0.5 = 12.14 \text{ in.}$$

$$H_{s1} = \frac{1}{2}(30)(9.42)^2 = 1331 \text{ lb}$$

$$H_{s2} = 4(30)(9.42) = 1130 \text{ lb}$$

$$\text{total} = 2460 \text{ lb}$$

$$V_u = 2460(1.6) = 3940 \text{ lb}$$

$$\tfrac{2}{3}\phi V_n = \tfrac{2}{3}\phi V_c = \tfrac{2}{3}\phi(2\sqrt{f_c'})bd$$

$$= \tfrac{2}{3}(0.75)(2)\sqrt{3000}(12)(12.14)$$

$$\tfrac{2}{3}\phi V_n = 7980 \text{ lb}$$

$$\tfrac{2}{3}\phi V_n > V_u$$

In summary, alternate vertical stem reinforcing bars may be stopped at 6 ft-10 in. above the top of the footing (see the typical wall section shown in Figure 8-20 later in this chapter).

d.   *Additional design details:* Check the anchorage requirements for the stem steel (No. 7 bars at 7½ in. o.c.). Assume all steel is uncoated. Note that anchorage requirements into the stem and into the footing may differ for these bars. Anchorage length required in the stem will be equal to or greater than that required in the footing and will also impact the splice length. Therefore, stem anchorage will be checked.

1. From Table 5-1, $K_D = 82.2$.
2. Establish values for the factors $\psi_t, \psi_e, \psi_s, \lambda$.
   a. $\psi_t = 1.0$ (the bars are not top bars).
   b. The bars are uncoated; $\psi_e = 1.0$.
   c. The bars are No. 7.; $\psi_s = 1.0$.
   d. Normal-weight concrete is used; $\lambda = 1.0$.
3. The product $\psi_t \times \psi_e = 1.0 < 1.7$. (O.K.)
4. Determine c_b. Based on cover (center of bar to nearest concrete surface), consider the clear cover and one-half the diameter of the No. 7 bar:

$$c_b = 2 + \frac{0.875}{2} = 2.44 \text{ in.}$$

Based on bar spacing (one-half the center-to-center distance):

$$c_b = 0.5(7.5) = 3.75 \text{ in.}$$

Therefore use $c_b = 2.44$ in.

5. K_{tr} is taken as zero. There is no transverse steel crossing the potential plane of splitting.
6. Check $(c_b + K_{tr})/d_b \le 2.5$:

$$\frac{c_b + K_{tr}}{d_b} = \frac{2.44 + 0}{0.875} = 2.79 > 2.5$$

Use 2.5.

7. Calculate the excess reinforcement factor:

$$K_{ER} = \frac{A_s \text{ required}}{A_s \text{ provided}} = \frac{0.91}{0.96} = 0.95$$

8. Calculate ℓ_d.
   a. Omitting K_{ER} (this calculated value of ℓ_d will be used shortly for splice length determination):

$$\ell_d = \frac{K_D}{\lambda} \left[\frac{\psi_t \psi_e \psi_s}{\left(\frac{c_b + K_{tr}}{d_b} \right)} \right] d_b$$

$$= \frac{82.2}{1.0} \left(\frac{1.0(1.0)(1.0)}{2.5} \right)(0.875)$$

$$= 28.8 \text{ in.} > 12 \text{ in.} \qquad \text{(O.K.)}$$

b.   Including K_{ER} from step 7, the required anchorage length is

$$\ell_d = 28.8(0.95) = 27.4 \text{ in.} > 12 \text{ in.} \quad \text{(O.K.)}$$

Use $\ell_d = 28$ in. (minimum).

With the footing thickness equal to 21 in. and a minimum of 3 in. of cover required for the steel at the bottom of the footing, the anchorage length available is $21 - 3 = 18$ in. This anchorage length is not adequate. Rather than hook the bars, however, the bars will be extended into a footing base shear key that will be used to increase the sliding resistance of the wall (see the check on factor of safety against sliding).

The *length of splice* required for main stem reinforcing steel (see the typical wall section shown in Figure 8-20 later in this chapter) may be calculated recognizing that the class B splice is applicable for this condition (see Chapter 5, Section 5-7). Therefore the required length of the splice is calculated from

$$1.3(28.8) = 37.4 \text{ in.}$$

Use 38 in. Note that the preceding calculation omits the effect of excess reinforcement.

Stem face steel in the form of horizontal and vertical reinforcement will be provided as per the ACI Code, Section 14.3. The minimum horizontal reinforcement is specified in Section 14.3.3. Although the minimum vertical reinforcement requirement of Section 14.3.2 does not strictly apply to reinforced concrete cantilever retaining walls, it is considered good practice to provide some vertical bars in the exposed face of the wall. The minimum recommended steel for deformed bars not larger than No. 5 with specified yield strength of not less than 60,000 psi is as follows:

horizontal bars (per foot of height of wall): $A_s = 0.0020bt$

vertical bars (per foot of wall horizontally): $A_s = 0.0012bt$

where $t =$ thickness of the wall.

The code also stipulates that walls more than 10-in. thick, except basement walls, must have reinforcement for each direction in each face of the wall. The exposed face must have a minimum of one-half and a maximum of two-thirds the total steel required for each direction. The maximum spacing for the steel must not exceed three times the wall thickness nor 18 in. (ACI Code,

Section 14.3.5) for both vertical bars and horizontal bars. For the front face of the wall (exposed face), use two-thirds of the total steel required and an average stem thickness = 14 in.

For horizontal steel,

$$0.0020(12)(14) = 0.34 \text{ in.}^2/\text{ft of height}$$

$$= 0.34(0.67) = 0.23 \text{ in.}^2$$

Use No. 4 bars at 10 in. ($A_s = 0.24$ in.^{2}).

For vertical steel,

$$0.0012(12)(14) = 0.20 \text{ in.}^2/\text{ft of horizontal length}$$

$$= (0.20)(0.67) = 0.14 \text{ in.}^2$$

Use No. 4 bars at 17 in. ($A_s = 0.14$ in.^{2}) in the front face. No additional vertical steel is needed in the rear face.

Rear face of wall, not exposed, for horizontal steel,

$$0.34(0.33) = 0.11 \text{ in.}^2$$

Use No. 4 bars at 18 in. ($A_s = 0.13$ in.^{2}).

Longitudinal reinforcement in the footing should provide a steel area equal to that required for shrinkage and temperature in slabs. This is a conservative and acceptable approach because temperature and shrinkage exposure are ordinarily less severe for footings than for slabs. These bars serve chiefly as bar supports and spacers to hold the main steel in place during construction. Thus

$$\text{required } A_s = 0.0018(11 \text{ ft})(1.75 \text{ ft})(144 \text{ in.}^2/\text{ft}) = 4.99 \text{ in.}^2$$

Use 12 No. 6 bars ($A_s = 5.28$ in.^{2}), as shown in Figure 8-20 later in this chapter.

Transverse reinforcement in the footing need not run the full width of the footing. Proper anchorage length must be provided from the point of maximum tensile stress, however.

Check anchorage for the top transverse (heel) steel in the footing (No. 7 bars at 9 in. o.c.). This calculation for ℓ_d follows the eight-step procedure presented in Chapter 5, Section 5-2, and is summarized as follows:

1. $K_D = 82.2$
2. $\psi_t = 1.3, \psi_e = 1.0, \psi_s = 1.0, \lambda = 1.0$
3. $\psi_t \times \psi_e = 1.3$ (O.K.)
4. $c_b = 2.44$ in.
5. $K_{tr} = 0$

6. $(c_b + K_{tr})/d_b = 2.79$ in. Use 2.5 in.
7. $K_{ER} = 0.91$
8. $\ell_d = 34$ in.

Check the anchorage of the bottom transverse (toe) steel in the footing (No. 7 bars at 16 in. o.c. As before, this ℓ_d calculation follows the procedure presented in Chapter 5 and is summarized as follows:

1. $K_D = 82.2$
2. $\psi_t = 1.0, \psi_e = 1.0, \psi_s = 1.0, \lambda = 1.0$
3. $\psi_t \times \psi_e = 1.0$ (O.K.)
4. $c_b = 3.44$ in.
5. $K_{tr} = 0$
6. $(c_b + K_{tr})/d_b = 3.93$ Use 2.5
7. $K_{ER} = 1.0$
8. $\ell_d = 28.8$ in. Use 29 in.

These details are included in the typical wall section, Figure 8-20.

The stem and the footing are elements cast at different times and a *shear key* will be used between the two. This is common practice. We will use a depressed key formed by a 2×6 plank (dressed dimensions $1\frac{1}{2} \times 5\frac{1}{2}$), as shown in Figure 8-20. The need for a shear key is questionable, because considerable slip is required to develop the key for purposes of lateral force transfer. It may be considered as an added mechanical factor of safety, however.

The shear-friction design method of the ACI Code Section 11.6 should be used to design for the transfer of the horizontal force between the stem and the footing. This approach eliminates the need for the traditional shear key. The shear-friction approach assumes that all of the horizontal force will be transferred through friction that develops on the contact surface between the two elements. The magnitude of the force that can be so transmitted will depend on the characteristics of the contact surfaces and on the existence of adequate shear-friction reinforcing A_{vf} crossing those surfaces. The angle at which the shear-friction reinforcing crosses the contact surface also plays a role. For our example, we assume the dowels will be placed perpendicular to the top of the footing.

The code allows for two possible contact-surface conditions that could exist in a situation such as between stem and footing in our retaining wall, either not intentionally roughened, though clean and free of laitance, or intentionally roughened. The latter assumption requires the interface to be roughened to a full amplitude of approximately $\frac{1}{4}$ in. This may be accomplished by raking of the fresh concrete or by some other means.

Assuming normal-weight concrete and shear-friction reinforcement placed perpendicular to the interface, the nominal shear strength (or friction force that resists sliding) may be computed from

$$V_n = A_{vf}f_y\mu \qquad \text{[ACI Eq. (11-25)]}$$

where

A_{vf} = area of shear-friction reinforcement

μ = coefficient of friction in accordance with the ACI Code, Section 11.6.4.3; it may be taken as 1.0 for normal-weight concrete placed against hardened concrete roughened as described previously

Note that A_{vf} is steel that is provided for the shear-friction development and that it is in addition to any other steel already provided. This additional steel will be provided in the form of *dowels*. (A *dowel* is defined as a short bar that connects two separately cast sections of concrete.) The dowels will be placed approximately perpendicular to the shear plane.

As a limit,

$$V_u = \phi V_n$$

where V_u is shear force applied at the cracked plane. Substituting for V_n,

$$V_u = \phi A_{vf}f_y\mu$$

Solving for A_{vf},

$$\text{required } A_{vf} = \frac{V_u}{\phi f_y\mu}$$

$$= \frac{9460}{(0.75)(60,000)(1.0)}$$

$$= 0.21 \text{ in.}^2 \text{ per ft}$$

It is desirable to distribute the shear-friction reinforcement across the width of the contact surface. Therefore provide $0.21/2 = 0.11$ in.^{2}/ft in each face. Use No. 4 bars at 18 in. in each face. This will provide 0.13 in.^{2}/ft in each face for a total of 0.26 in.^{2}/ft. Then

$$\text{shear strength } V_n = A_{vf}f_y\mu$$

$$= (0.26)(60,000)(1.0)$$

$$= 15,600 \text{ lb}$$

The ACI Code, Section 11.6.5, stipulates that the maximum V_n for concrete placed monolithically or against hardened concrete intentionally roughened as previously described, shall not exceed the smallest of

$$0.2f'_cA_c, (480 + 0.08f'_c)A_c, \quad \text{and} \quad 1600\,A_c$$

where A_c is the contact area resisting the shear transfer, and for all other cases, V_n shall not exceed the smaller of

$$0.2f'_cA_c \quad \text{and} \quad 800\,A_c$$

Thus, checking the upper limit for V_n:

$$0.2f'_cA_c = 0.2(3000)(12)(18) = 129{,}600 \text{ lb}$$

$$(480 + 0.08f'_c)A_c = [480 + 0.08(3000)](12)(18) = 155{,}500 \text{ lb}$$

$$1600\,A_c = 1600(12)(18) = 346{,}000 \text{ lb}$$

$$\text{Calculated } V_n = 15{,}600 \text{ lb} \ll 129{,}600 \text{ lb} \qquad \text{(O.K.)}$$

The development length for the No. 4 dowels (18 in. o.c.) must be furnished into both the stem and the footing. These are considered to be tension bars. This calculation for ℓ_d follows the eight-step procedure presented in Chapter 5, Section 5-2, and is summarized as follows:

1. $K_D = 82.2$
2. $\psi_t = 1.0, \psi_e = 1.0, \psi_s = 0.8, \lambda = 1.0$
3. $\psi_t \times \psi_e = 1.0$ (O.K.)
4. $c_b = 2.25$ in.
5. $K_{tr} = 0$
6. $(c_b + K_{tr})/d_b = 4.50$ Use 2.5
7. $K_{ER} = 0.808$
8. $\ell_d = 10.6$ in. < 12 in. Use 12 in.

The No. 4 dowels are depicted in Figure 8-17.

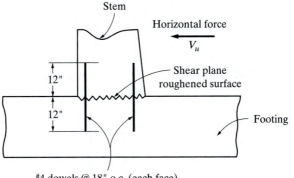

#4 dowels @ 18" o.c. (each face)

FIGURE 8-17   Shear-friction reinforcement.

6. Design the footing base shear key. The *footing base shear key* (sometimes called a *bearing lug*) is primarily used to prevent a sliding failure. The magnitude of the additional resistance to sliding offered by the key is questionable and is a function of the subsoil material. The key, cast in a narrow trench excavated below the bottom of footing elevation, becomes monolithic with the footing. The excavation for a key will generally disturb the subsoil during construction and conceivably will do more harm than good. Hence the use of the key for the purpose intended is controversial.

An acceptable design approach is to use the passive earth resistance H_p in front of the key (from the bottom of the footing to the bottom of the key) as the additional resistance to sliding (see Figure 8-18). This neglects any earth in front of the footing and reflects the case where excavation or scour has removed the earth to the level of the bottom of the footing.

The passive earth resistance may be expressed in terms of an equivalent fluid weight. Because $K_a w_e = 30$ lb/ft^3 and $w_e = 100$ lb/ft^3,

$$K_a = \frac{30}{100} = 0.3$$

The coefficient of passive earth resistance is

$$K_p = \frac{1 + \sin \phi}{1 - \sin \phi} = \frac{1}{K_a} = \frac{1}{0.3} = 3.33$$

and

$$K_p w_e = 3.33(100) = 333 \text{ lb/ft}^3$$

$$H_p = \tfrac{1}{2} K_p w_e h_k^2 = \tfrac{1}{2}(333) h_k^2$$

The unfactored horizontal force $\Sigma H = 7020$ lb. Therefore, the required resistance to sliding that will furnish a factor of safety of 1.5 is

$$1.5(7020) = 10{,}530 \text{ lb}$$

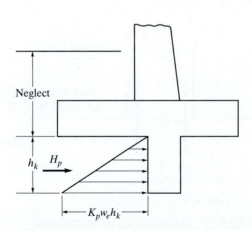

FIGURE 8-18   Base shear key force analysis.

The frictional resistance furnished is 9860 lb (see the discussion of stability analysis). Thus, the resistance that must be furnished by the base shear key and the passive earth resistance is

$$10{,}530 - 9860 = 670 \text{ lb}$$

With reference to Figure 8-18, the height h_k required to furnish this resistance may be obtained by establishing horizontal equilibrium ($\Sigma H = 0$).

$$\tfrac{1}{2}(333)(h_k)^2 = 670$$

$$h_k^2 = 4.02$$

$$\text{required } h_k = 2.01 \text{ ft}$$

Use 2 ft-0 in.

The key must be designed for moment and shear. The worst case would be the situation where excavation had *not* taken place and the full height of earth in front of the wall was available to develop passive pressure. This is shown in Figure 8-19. A 1-ft depth of earth has been assumed on top of the footing. Assuming the key as a vertical cantilever beam, taking a summation of moment about the plane of the bottom of footing, and applying a load factor of 1.6,

$$M_u = 1.6(916)(2.0)(1.0) + 1.6(\tfrac{1}{2})(666)(2.0)(\tfrac{2}{3})(2.0)$$

$$= 4350 \text{ ft-lb}$$

Assume a 10-in. width of key and a No. 8 bar, which provides an effective depth $d = 10 - 3 - 0.5 = 6.5$ in. Assume $\phi = 0.90$.

$$\text{required } \overline{k} = \frac{M_u}{\phi b d^2} = \frac{4350}{0.9(12)(6.5)^2} = 0.1144 \text{ ksi}$$

From Table A-8, the required $\rho = 0.0020$, $\epsilon_t > 0.005$; therefore, $\phi = 0.90$. Now we can calculate the required steel area:

$$\text{required } A_s = \rho b d = 0.0020(12)(6.5) = 0.16 \text{ in.}^2$$

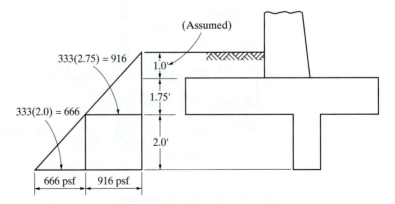

FIGURE 8-19 Base shear key force design.

The minimum area of steel required by the ACI Code, Section 10.5.1, may be obtained using Table A-5:

$$A_{s,min} = 0.0033(12)(6.5) = 0.26 \text{ in.}^2$$

Therefore $A_{s,min}$ of 0.26 in.² controls. This is a very small amount of steel, and it will be more practical to extend some existing stem bars into the key. All stem bars must be extended into the key for anchorage reasons, however. Therefore, all the bars will be extended to within 3 in. of the bottom of the key. This provides 42 in. of anchorage below the bottom of the stem for the No. 7 bars and exceeds the requirement of 28 in. (see typical wall section in Figure 8-20). The steel furnished in the key, therefore, is No. 7 bars at 7½ in. o.c. ($A_s = 0.96$ in.²).

If we conservatively neglect the effect of the greater cover (3.0 in.) in the key, calculations for the required anchorage length at the top of the

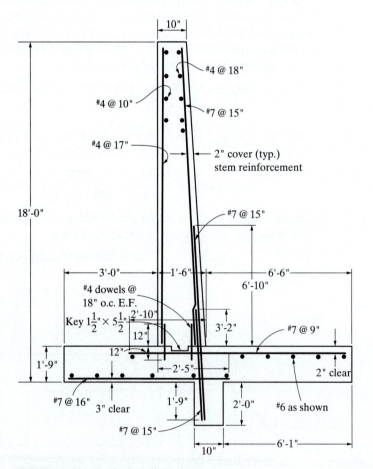

FIGURE 8-20 Typical wall section.

key are identical to those for the stem, except that the excess reinforcement factor $K_{ER} = 0.26/0.96 = 0.27$, from which the final ℓ_d calculation yields

$$\ell_d = 28.8(0.27) = 7.8 \text{ in.} < 12 \text{ in.}$$

Use $\ell_d = 12$ in. The available anchorage length is $24 - 3 = 21$ in., which exceeds the requirements of 12 in.

The shear strength of the key is

$$\phi V_n = \phi V_c$$
$$= \phi(2\lambda\sqrt{f_c'})bd$$
$$= 0.75(2)(1.0)(\sqrt{3000})(12)(6.5)$$
$$= 6410 \text{ lb}$$

The factored shear is

$$V_u = (1.6)(916)(2.0) + 1.6(\tfrac{1}{2})(666)(2) = 4000 \text{ lb}$$

$$\phi V_n > V_u \qquad\qquad\qquad\qquad\qquad\qquad\qquad \text{(O.K.)}$$

8-4 DESIGN CONSIDERATIONS FOR BEARING WALLS

Bearing walls (Figure 8-1g) were briefly described at the beginning of this chapter as those walls that carry vertical load in addition to their own weight. Recommendations for the empirical design of such walls are presented in Chapter 14 of the ACI Code (318-08) and apply primarily to relatively short walls spanning vertically and subject to vertical loads only, such as those resulting from the reactions of floor or roof systems supported on walls. Walls, other than short walls carrying "reasonably concentric" loads, should be designed as compression members for axial load and flexure in accordance with ACI 318-08, Chapter 10. "Reasonably concentric" implies that the resultant factored load falls within the middle third of the cross section.

The design axial load strength or capacity of such a wall will be

$$\phi P_n = 0.55\phi f_c' A_g \left[1 - \left(\frac{k\ell_c}{32h} \right)^2 \right] \qquad \text{[ACI Eq. (14-1)]}$$

where

ϕ = strength-reduction factor corresponding to compression-controlled sections in accordance with ACI Code, Section 9.3.2.2; 0.75 for members with spiral reinforcement and 0.65 for other reinforced members

h = thickness of wall (in.)

ℓ_c = vertical distance between supports (in.)

A_g = gross area of section (in.^{2})

k = effective length factor

The effective length factor k shall be

1. For walls braced top and bottom against lateral translation and
    a. Restrained against rotation at one or both ends
        (top and/or bottom)   0.8
    b. Unrestrained against rotation at both ends   1.0
2. For walls not braced against lateral translation   2.0

Where the wall is subject to concentrated loads, the effective length of wall for each concentration must not exceed the center-to-center distance between loads nor exceed the width of bearing plus four times the wall thickness.

The following requirements applicable to bearing walls are prescribed, among others, by the ACI Code, Chapter 14:

1. Reinforced concrete bearing walls must have a thickness of at least ¹⁄₂₅ of the unsupported height or width, whichever is shorter, and not less than 4 in.
2. Thickness of nonbearing walls shall not be less than 4 in. nor less than ¹⁄₃₀ times the least distance between members that provide lateral support.
3. The area of horizontal reinforcement must be at least 0.0025 times the area of the wall ($0.0025bh$ per foot) and the area of vertical reinforcement not less than 0.0015 times the area of the wall ($0.0015bh$ per foot), where $b = 12$ in. and h is the wall thickness. These values may be reduced to 0.002 and 0.0012, respectively, if the reinforcement is deformed bars not larger than No. 5 with f_y not less than 60,000 psi or if the reinforcement is welded wire reinforcement larger than W31 or D31.
4. Exterior basement walls and foundation walls must not be less than 7½-in. thick.
5. Reinforced concrete walls must be anchored to intersecting elements such as floors and roofs or to columns, pilasters, buttresses, intersecting walls, and to footings.
6. Walls more than 10-in. thick, except for basement walls, must have reinforcement in each direction for each face. The exterior surface shall have a minimum of one-half and a maximum of two-thirds of the total steel required, with the interior surface having the balance of the reinforcement.
7. Vertical reinforcement must be enclosed by lateral ties if in excess of 0.01 times the gross concrete area or when it is required as compression reinforcement.

Additionally, from the ACI Code, Section 10.14, the design-bearing strength of concrete under a bearing plate may be taken as ϕ ($0.85 f_c' A_1$), where A_1 is the loaded area. An exception occurs when the supporting surface is wider on all sides than the loaded area. In that case, the foregoing expression for bearing strength may be multiplied by $\sqrt{A_2/A_1} \leq 2.0$, where A_2 is a concentric and geometrically similar support area that is the lower base of a frustum (upper base of which is A_1) of a pyramid having 1:2 sloping sides and fully contained within the support.

Example 8-2 _____

Design a reinforced concrete bearing wall to support a series of steel wide-flange beams at 8 ft-0 in. o.c. Each beam rests on a bearing plate 6 in. × 12 in. The wall is braced top and bottom against lateral translation. Assume the bottom end fixed against rotation. The wall height is 15 ft and the design (factored) load P_u from each beam is 115 kips. Use $f'_c = 3000$ psi and $f_y = 60,000$ psi. See Figure 8-21.

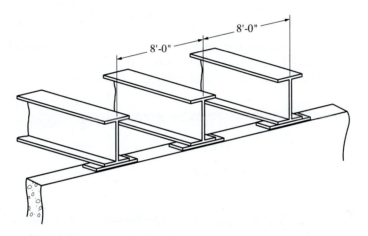

FIGURE 8-21   Bearing wall.

Solution:

1. Assume an 8-in.-thick wall with full concentric bearing.
2. From our previous discussion, the bearing strength of the concrete under the bearing plate, neglecting the $\sqrt{A_2/A_1}$ multiplier, is

$$\phi(0.85)f'_c A_1 = 0.65(0.85)(3000)(6)(12)$$

$$= 119,300 \text{ lb}$$

$$\text{factored bearing load } P_u = 115,000 \text{ lb}$$

$$119,300 > 115,000 \qquad\qquad \text{(O.K.)}$$

3. The effective length of the wall (ACI Code, Section 14.2.4) must not exceed the center-to-center distance between loads nor the width of bearing plus four times the wall thickness. Beam spacing = 96 in. Thus

$$12 + 4(8) = 44 \text{ in.}$$

Therefore, use 44 in.

4.   The minimum thickness required is $\frac{1}{25}$ times the shorter of the unsupported height or width. Assume, for our case, that width does not control. Then

$$h_{\min} = \frac{\ell_c}{25} = \frac{15(12)}{25} = 7.2 \text{ in.}$$

Also, $h_{\min} = 4$ in. Therefore, the 8-in. wall is satisfactory.

5.   The capacity of wall is calculated from

$$\phi P_n = 0.55\phi f_c' A_g \left[1 - \left(\frac{k\ell_c}{32h}\right)^2\right]$$

The strength-reduction factor ϕ for compression-controlled sections is discussed in Section 2-9 of this text. In this case, because the wall will not contain spiral lateral reinforcing, ϕ is taken as 0.65.

$$\phi P_n = 0.55(0.65)(3)(44)(8)\left\{1 - \left[\frac{0.8(15)(12)^2}{32(8)}\right]\right\}$$

$$= 378(1 - 0.316)$$

$$= 258 \text{ kips}$$

$$\phi P_n > P_u \qquad\qquad\qquad\qquad\qquad\qquad\qquad \text{(O.K.)}$$

6.   The reinforcing steel (ACI Code, Section 14.3), assuming No. 5 bars or smaller, can be found: For the vertical reinforcement per foot of wall length,

$$\text{required } A_s = 0.0012bh$$

$$= 0.0012(12)(8) = 0.12 \text{ in.}^2$$

For the horizontal reinforcement per foot of wall height,

$$\text{required } A_s = 0.002bh$$

$$= 0.002(12)(8) = 0.19 \text{ in.}^2$$

The maximum spacing of reinforcement must not exceed three times the wall thickness nor 18 in. (ACI Code, Section 14.3.5). Thus

$$3(8) = 24 \text{ in.}$$

Use 18 in.

To select reinforcing,

vertical steel: Use No. 4 bars at 18 in. ($A_s = 0.13$ in.^{2})

horizontal steel: Use No. 4 bars at 12 in. ($A_s = 0.20$ in.^{2})

The steel may be placed in one layer, as the wall is less than 10-in. thick (see Figure 8-22).

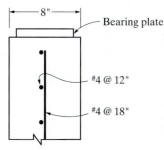

FIGURE 8-22   Section of wall.

8-5   DESIGN CONSIDERATIONS FOR BASEMENT WALLS

A basement wall is a type of retaining wall in which there is lateral support assumed to be provided at bottom and top by the basement floor slab and first-floor construction, respectively. As previously mentioned, the wall would be designed as a simply supported member with a loading diagram and moment diagram as shown in Figure 8-23.

If the wall is part of a bearing wall, the vertical load will relieve some of the tension in the vertical reinforcement. This may be neglected because its effect may be small compared with the uncertainties in the assumption of loads. If the vertical load is of a permanent nature and of significant magnitude, its effect should be considered in the design.

When a part of the basement wall is above ground, the lateral bending moment may be small and may be computed as shown in Figure 8-24. This assumes that the wall is spanning in a vertical direction. Depending on the type of construction, the basement wall may also span in a horizontal direction and may behave as a slab reinforced in either one or two directions. If the wall design assumes two horizontal reactions, as shown, caution must be exercised that the two supports are in place prior to backfilling behind the wall.

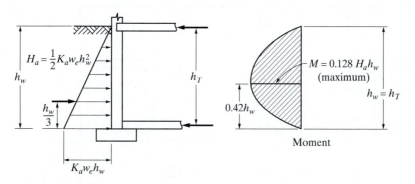

FIGURE 8-23   Basement wall with full height backfill: forces and moment diagram.

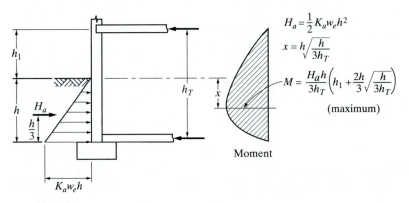

$$H_a = \frac{1}{2} K_a w_e h^2$$

$$x = h\sqrt{\frac{h}{3h_T}}$$

$$M = \frac{H_a h}{3h_T}\left(h_1 + \frac{2h}{3}\sqrt{\frac{h}{3h_T}}\right)$$

(maximum)

Moment

FIGURE 8-24 Basement wall with partial backfill: forces and moment diagram.

8-6   SHEAR WALLS

Concrete or masonry walls fixed at their base are used to resist lateral wind and seismic loads in building structures parallel to the plane of the shear wall, in addition to supporting gravity loads. These lateral load resisting elements may consist of single walls located internally within the building or on the exterior face of the building, or they could be the stair and elevator core walls. Shear walls are very efficient lateral load resisting elements that resist lateral loads by acting as a vertical cantilever. Sometimes, the walls may be perforated by door or window openings or corridors, and when these walls are joined together with deep beams spanning the openings or corridor, they are called "coupled shear" walls. The strength of a coupled shear wall depends on the stiffness of the coupled walls and the stiffness of the coupling beams and lies between the strength of a moment frame and that of an unperforated shear wall. Figures 8-25 and 8-26 show some typical shear wall layout and typical shear

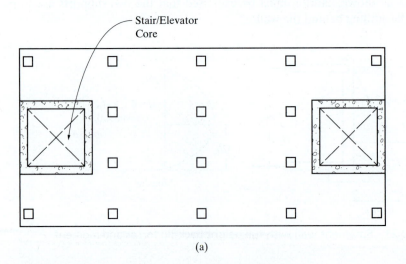

(a)

FIGURE 8-25 Typical shear wall layout in buildings.

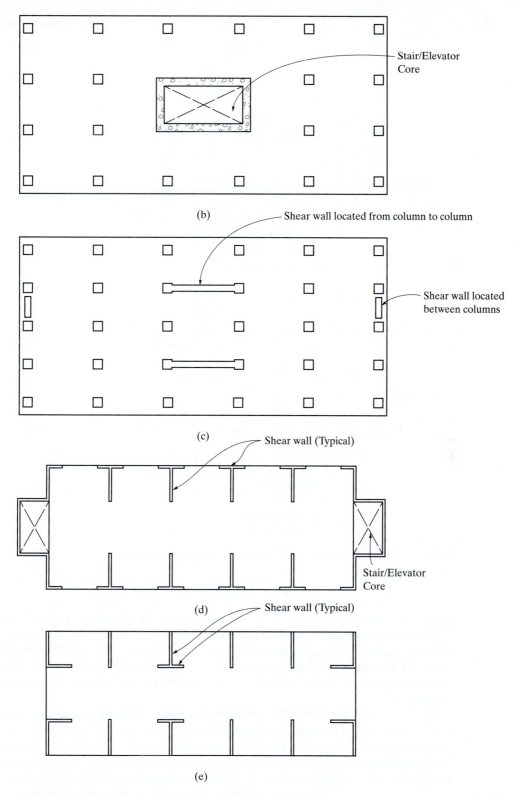

(b)

Stair/Elevator Core

Shear wall located from column to column

(c)

Shear wall located between columns

(d)

Shear wall (Typical)

Stair/Elevator Core

(e)

Shear wall (Typical)

FIGURE 8-25 (CONT.) Typical shear wall layout in buildings.

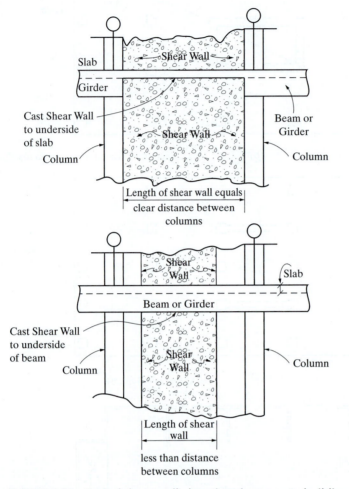

FIGURE 8-26   Typical shear wall elevations in concrete buildings.

wall elevations, respectively, in building structures. Single shear walls could be laid out between column lines as shown in Figure 8-26. Where the shear wall extends to and is built integrally with the columns, the column at both ends of the wall will serve as the boundary members for the shear wall and the vertical reinforcement in the columns will serve as the vertical end reinforcement in the shear wall.

Typical concrete buildings have floor and roof slabs that can be considered rigid in the horizontal plane, and thus the floors and roofs can be modeled as rigid diaphragms. For buildings with rigid diaphragms, the lateral load acting on the building is distributed to the lateral force resisting elements parallel to the lateral load in proportion to the stiffness of the lateral force resisting element. For shear walls, the stiffness, K, consists of a flexural stiffness component and a shear stiffness component, and is calculated as

$$K = \frac{3EI}{h_w^3} + \frac{GA}{1.2h_w}$$

If the total lateral force on the building at level x is F_x, the lateral force at level x is distributed to each shear wall that is parallel to the lateral load as follows:

$$F_{\text{shear wall},x} = \frac{K_{\text{wall}}}{\Sigma K_{\text{wall}}} F_x$$

where

h_w = overall or total height of shear wall from top of footing to the top of the wall (see Figure 8-27a)

E = modulus of elasticity

I = moment of inertia of the wall about the strong axis $= \dfrac{h\ell_w^3}{12}$

ℓ_w = overall length of shear wall (see Figure 8-27a)

h = wall thickness (minimum practical wall thickness is 8 in.; see Figure 8-27b)

A = gross cross-sectional area of the wall $= h\ell_w$

G = shear modulus of elasticity $= \dfrac{E}{2(1 + \nu)}$

ν = Poisson's ratio for concrete ≈ 0.20.

K_{wall} = stiffness of the shear wall being considered

ΣK_{wall} = sum of the stiffnesses of all the shear walls *parallel* to the lateral load

The following practical considerations should be taken into account when laying out shear walls in concrete buildings:

- Locate shear walls to minimize the effect on architectural features in the building such as doors and windows.
- Utilize stair and elevator shaft walls as shear walls. Shear walls can also be located on the outer perimeter of a building, but this may reduce the number of available windows in a building and therefore lead to a reduction in natural light and exterior views.
- Locate shear walls in each orthogonal direction as symmetrically as possible to minimize twisting or torsional deformations of the building from lateral loads. If a symmetrical arrangement is *not* feasible because of architectural or other constraints, the building should be analyzed for the resulting inplane *twisting forces,* and these would lead to additional lateral forces in the shear walls.
- Shear walls or other forms of lateral force resisting systems are required in both orthogonal directions of the building.

Shear Wall Design Considerations (ACI 11.9)

The shear wall design considerations in this chapter pertain to ordinary reinforced concrete shear walls as presented in ACI Code, Section 11.9. The design considerations for ductile or special shear walls used in highly seismic zones are presented in ACI Code, Section 21.9. The following load effects should be considered in the design of shear walls:

- The varying horizontal *shear* force that is maximum at the base of the wall.
- The *bending moment* that is maximum at the base of the wall. This produces compression at the end zone at one end of the wall and tension at the end zone at the opposite end. The location of the tension and compression forces will change depending on the direction of the lateral load.
- The *gravity* or *vertical loads* (i.e., roof and floor dead and live loads) that cause compression on the wall.

For low-rise concrete buildings, the effect of the interaction of the gravity loads with the shear or bending moment capacity of the wall will be minimal, and therefore, it is practical to consider the load effects just listed separately for low-rise buildings. However, for high-rise buildings where the gravity load on the wall could be substantial, the interactions between the axial load and bending moment and shear capacities of the wall have to be considered, and in that case, the shear wall is usually designed as a "column" with combined axial load and moment, and a column-type interaction diagram is used.

Reinforcement in Shear Walls

The reinforcement in shear walls consists of distributed horizontal reinforcement used to resist shear forces in the wall, distributed vertical reinforcement used to resist gravity loads and to control shrinkage and cracking, and concentrated vertical end reinforcement used to resist the bending moment due to lateral loads. The typical shear wall reinforcement is shown in the wall elevation and section in Figure 8-27, and the typical reinforcement details at the corners and at the ends of shear walls are shown in Figure 8-28.

Minimum Reinforcement in Shear Walls

The minimum ratio of distributed *transverse* or *horizontal* reinforcement, ρ_t in the wall to the gross cross-sectional area of the wall perpendicular to the reinforcement is given in ACI 11.9.9.2:

$$\rho_t = \frac{A_v}{sh} \geq 0.0025$$

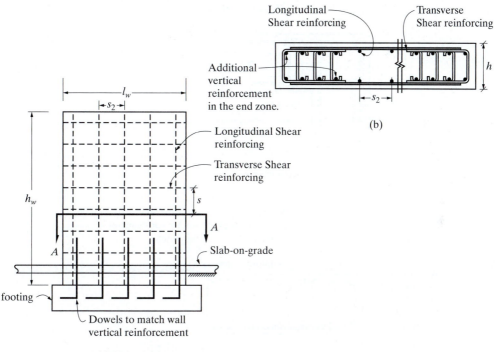

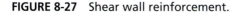

(a) Elevation

FIGURE 8-27   Shear wall reinforcement.

The minimum ratio of distributed *vertical* or *longitudinal* reinforcement, ρ_ℓ in the wall to the gross cross-sectional area of the wall perpendicular to the reinforcement is given in ACI 11.9.9.4:

$$\rho_\ell = \frac{A_v}{s_2 h} = 0.0025 + 0.5\left(2.5 - \frac{h_w}{\ell_w}\right)(\rho_t - 0.0025) \geq 0.0025$$

but need not be greater than ρ_t.

where

A_v = area of distributed transverse reinforcement

s = center-to-center spacing of the transverse or horizontal reinforcement

s_2 = center-to-center spacing of the vertical reinforcement

and h, ℓ_w, and h_w are as previously defined.

The size of the distributed horizontal and vertical reinforcement is usually No. 4 or larger bars. The maximum spacing of the *horizontal* reinforcement allowed by the Code (*s* maximum) is the smallest of $\ell_w/5$, $3h$, or 18 in, and the maximum spacing of the *vertical* reinforcement allowed by the Code (*s* maximum) is the smallest of $\ell_w/3$, $3h$, or 18 in.

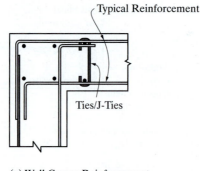

Typical Reinforcement

Ties/J-Ties

(a) Wall Corner Reinforcement

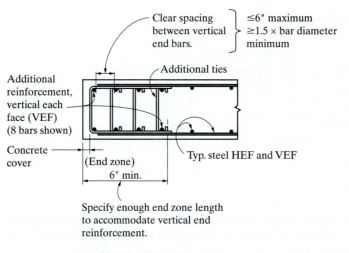

Clear spacing
between vertical
end bars.

≤6" maximum
≥1.5 × bar diameter
minimum

Additional
reinforcement,
vertical each
face (VEF)
(8 bars shown)

Additional ties

Concrete
cover

(End zone)
6" min.

Typ. steel HEF and VEF

Specify enough end zone length
to accommodate vertical end
reinforcement.

(b) Wall end Reinforcement

FIGURE 8-28   Typical added reinforcement at the corners and ends of shear walls.

For 8-in.- or 10-in.-thick walls, although ACI Code, Section 14.3.4 implies that the horizontal and vertical reinforcement be placed at the center of the wall, it is common in design practice to use two layers of reinforcement (i.e., reinforcement on both faces of the wall) for 8-in.- and 10-in.-thick shear walls.

Strength of Shear Walls

In the limit states design for shear, the ACI Code requires the design shear strength, ϕV_n to be greater than or equal to the required shear strength or factored shear, V_u. That is,

$$\phi V_n \geq V_u$$

where

$\phi = 0.75$

V_n = nominal shear strength = $V_c + V_s \leq 10\sqrt{f'_c}hd$ (ACI 11.9.3)

V_c = concrete shear strength = $2\lambda\sqrt{f'_c}hd$

V_s = shear strength of shear reinforcing

d = effective depth = $0.8\ell_w$ (ACI 11.9.4)

λ = lightweight concrete modification factor defined in Chapter 1 and is 1.0 for normal-weight concrete

When $V_u \leq \phi V_c$, minimum *horizontal* reinforcing should be provided in the shear wall per ACI Sections 11.9.9.2 and 11.9.9.3.

When $V_u > \phi V_c$, the required *horizontal* reinforcing is calculated as follows:

The limit states design equation for shear requires that

$$\phi V_n = \phi V_c + \phi V_s \geq V_u$$

Therefore, $\phi V_s \geq V_u - \phi V_c$, and as a limit

$$V_s = \frac{V_u - \phi V_c}{\phi}$$

From Chapter 4, the strength of any shear reinforcing is determined from

$$V_s = \frac{A_v f_y d}{s}$$

Therefore, by substitution,

$$\frac{V_u - \phi V_c}{\phi} = \frac{A_v f_y d}{s}$$

and

$$\frac{A_v}{s} = \frac{V_u - \phi V_c}{\phi f_y d}$$

After selecting the size of the horizontal reinforcement (usually No. 4 or larger bars), the cross-sectional area of the horizontal reinforcement, A_v, can be obtained as well as the required spacing of the reinforcement using the above equation.

The end zone vertical reinforcement due to the bending moment in the shear wall is determined using the rectangular beam design procedure in Chapter 2.

Example 8-3 _____

Shear Wall Design

The shear wall layout for a three-story building and the unfactored north–south lateral seismic loads acting on the building are shown in Figure 8-29. Design the north–south shear walls for the seismic lateral loads shown. Assume normal-weight concrete; f'_c is 4000 psi and f_y is 60,000 psi. Assume all shear walls are of equal thickness.

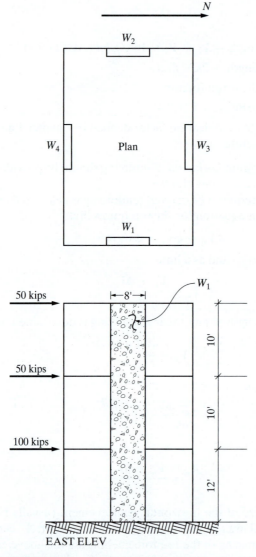

FIGURE 8-29 Shear wall layout and unfactored lateral loads for Example 8-3.

Solution:

1.   The maximum load factor for seismic loads from ACI Section 9.2.1 is 1.0 (1.6 for wind loads). Because this is a low-rise building, the effect of the vertical loads on the moment capacity of the shear walls will be small; thus, the interaction between the gravity loads and the moment is assumed to be negligible and hence the load effects on the wall will be considered separately.

2. As the length, thickness, and total height of the north–south shear walls, W1 and W2, are equal, their stiffness, K, will also be equal. The factored seismic lateral forces for shear walls W1 or W2, including the seismic load factor, are calculated as follows:

$$\text{Roof: } F_r = 1.0 \times \frac{K}{2K}(50 \text{ kips}) = 25 \text{ kips}$$

$$\text{Third floor: } F_3 = 1.0 \times \frac{K}{2K}(50 \text{ kips}) = 25 \text{ kips}$$

$$\text{Second floor: } F_2 = 1.0 \times \frac{K}{2K}(100 \text{ kips}) = 50 \text{ kips}$$

For high-rise buildings, the wall reinforcement is typically specified or designed for two or three story lifts, but for this low-rise building, only the reinforcement at the base of the wall will be designed and this reinforcement will be used throughout the full height of the shear wall. The factored shear wall lateral loads and load effects are shown in Figure 8-30.

3. Select wall thickness: Assume $h = 8$ in. (the reinforcement will be placed on both faces of the wall).

4. Check the maximum allowed shear strength of the wall. The effective depth, $d = 0.8 \times$ length of shear wall $= (0.8)(8 \text{ ft})(12 \text{ in./ft}) = 76.8$ in.

    Total maximum allowable shear strength $= 10\phi\sqrt{f_c'}hd$

    $$= 10(0.75)\sqrt{4000}(8 \text{ in.})(76.8 \text{ in.}) = 291.4 \text{ kips}$$

    Required shear strength V_u at the base of the shear wall
    $$= 100 \text{ kips} < 291.4 \text{ kips} \qquad \text{(O.K.)}$$

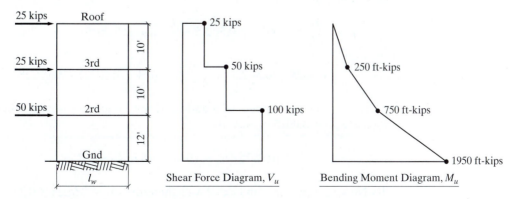

FIGURE 8-30   Shear wall lateral loads and load effects.

5. Calculate shear strength provided by concrete alone.

$$\phi V_c = 2\phi \sqrt{f'_c} hd = 2(0.75)\sqrt{4000}\,(8\text{ in.})(76.8\text{ in.}) = 58.3\text{ kips}$$

Because $V_u = 100$ kips $> \phi V_c$, shear reinforcement therefore is required in the wall.

6. Determine the required *horizontal* shear reinforcement, A_v:

$$\frac{A_v}{s} = \frac{V_u - \phi V_c}{\phi f_y d} = \frac{100\text{ kips} - 58.3\text{ kips}}{(0.75)(60\text{ ksi})(76.8\text{ in.})} = 0.0121$$

Try No. 4 horizontal bars on both faces of the wall; therefore, $A_v = 2$ faces $\times$ 0.2 in.^{2} = 0.4 in.^{2}. The required spacing of the horizontal shear reinforcement is

$$s\text{ required} = \frac{0.4}{0.0121} = 33\text{ in.}$$

The maximum spacing of the horizontal reinforcement that is allowed by the Code (s maximum) is the smallest of the following:

-   $\dfrac{\ell_w}{5} = \dfrac{96\text{ in.}}{5} = 19.2\text{ in.}$
-   $3h = (3)(8\text{ in.}) = 24\text{ in.}$
-   18 in. (Controls)

Therefore, Try $s = 18$ in.

The corresponding horizontal reinforcement ratio provided is

$$\rho_t = \frac{A_v}{sh} = \frac{0.4\text{ in.}^2}{(18\text{ in.})(8\text{ in.})} = 0.0028 \geq 0.0025 \qquad \text{(O.K.)}$$

For the distributed horizontal reinforcement, provide No. 4 HEF @ 18 in. o.c. (HEF = horizontal each face of wall.)

7. Determine distributed vertical shear reinforcement. The required vertical reinforcement ratio is

$$\rho_\ell = \frac{A_v}{s_2 h} = 0.0025 + 0.5\left(2.5 - \frac{h_w}{\ell_w}\right)(\rho_t - 0.0025)$$

$$= 0.0025 + 0.5\left(2.5 - \frac{32\text{ ft}}{8\text{ ft}}\right)(0.0028 - 0.0025) = 0.00228$$

ρ_ℓ must not be less than 0.0025 and need not be greater than ρ_t. Therefore, use $\rho_\ell = 0.0025 = A_v / s_2 h$.

$$\frac{A_v}{s_2} = (0.0025)(h) = (0.0025)(8\text{ in.}) = 0.02\text{ in.}$$

Try No. 4 vertical bar both face of wall; therefore, $A_v = 2$ faces $\times$ 0.2 in.^{2} = 0.4 in.^{2}

Therefore, the required spacing of the vertical shear reinforcement is

$$s_2 \text{ required} = \frac{0.4}{0.02} = 20 \text{ in.}$$

The maximum spacing vertical reinforcement allowed by the Code (s_2 maximum) is the smallest of the following:

- $\dfrac{\ell_w}{3} = \dfrac{96 \text{ in.}}{3} = 32 \text{ in.}$
- $3h = (3)(8 \text{ in.}) = 24 \text{ in.}$
- 18 in. (Controls)

Because maximum spacing = 18 in. $< s_2$ required = 20 in.; therefore, use $s_2 = 18$ in.

For the distributed vertical reinforcement, provide No. 4 VEF @ 18 in. o.c. (VEF = vertical each face of wall.)

8. Design the shear wall for flexure or bending and determine the end zone vertical reinforcement. The maximum factored bending moment at the base of the wall due to the factored seismic lateral load is

$$M_u = 1950 \text{ ft-kips}$$

The limit states design equation for flexure requires that $\phi M_n = \phi h d^2 \bar{k} \geq M_u$. Initially, assume $\phi = 0.9$ (this will be checked later after ϵ_t is determined) and then calculate the required $\bar{k}$ as follows:

$$\bar{k} = \frac{M_u}{\phi h d^2} = \frac{1950(12)}{(0.9)(8 \text{ in.})(76.8 \text{ in.})^2} = 0.55$$

From Table A-10, we obtain $\rho = 0.0101$ and $\epsilon_t \gg 0.005$; therefore, $\phi = 0.9$, as initially assumed. The concentrated vertical reinforcement required at each end zone of the shear wall is

$$A_s \text{ required} = \rho h d = (0.0101)(8 \text{ in.})(76.8 \text{ in.}) = 6.21 \text{ in.}^2$$

The minimum area of concentrated steel required for bending at the ends of the shear wall is

$$A_{s,\min} = \frac{3\sqrt{f_c'}}{f_y} h d \geq \frac{200}{f_y} h d$$

$$= \frac{3\sqrt{4000}}{60,000} (8)(76.8) = 1.94 \text{ in.}^2$$

$$\geq \frac{200}{60,000} (8)(76.8) = 2.05 \text{ in.}^2$$

Because A_s required = 6.21 in.^{2} > $A_{s,\min}$ = 2.05 in.^{2}, A_s required = 6.21 in.^{2}

Try 8 No. 8 vertical reinforcement at each end of the wall (i.e., 4 No. 8 VEF) Total area of steel provided at each end of the wall = 6.32 in.^{2} > A_s required (O.K.).

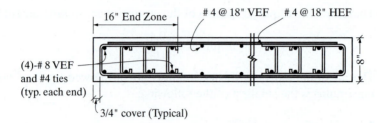

FIGURE 8-31 Shear wall reinforcement detail.

A plan detail showing the reinforcement provided in the shear wall is shown in Figure 8-31. An end zone length of 16 in. has been assumed. This will be checked below. The concentrated vertical reinforcement must be provided at both ends of the shear wall as the lateral load can reverse direction and each end of the shear wall will be subjected to tension or compression forces depending on the direction of the lateral load. From Figure 8-31, the clear spacing between the concentrated vertical bars is

$$\frac{16 \text{ in.} - 0.75 \text{ in. cover} - 0.5 \text{ in. tie} - 0.5 \text{ in. tie} - 4 \text{ bars}(1 \text{ in. diameter})}{3 \text{ spaces}}$$

$$= 3.42 \text{ in.} > 1.5(1 \text{ in. diameter}) = 1.5 \text{ in.} \quad \text{(O.K.)}$$

Therefore, the assumed end zone length of 16 in. is adequate.

PROBLEMS

8-1. Compute the active earth pressure horizontal force on the wall shown for the following conditions. Use $w_e = 100 \text{ lb/ft}^3$.

Case	ϕ	θ	w_s (psf)	h_w (ft)
(a)	25	0	400	15
(b)	28	10	0	18
(c)	30	0	200	20
(d)	33	20	0	25

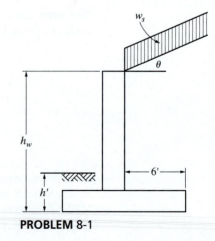

PROBLEM 8-1

8-2. Find the passive earth pressure force in front of the wall for Problem 8-1(a), if $h' = 4$ ft.

8-3. For the wall shown, determine the factors of safety against overturning and sliding and determine the soil pressures under the footing. Use $K_a = 0.3$ and $w_e = 100$ lb/ft^3. The coefficient of friction $f = 0.50$.

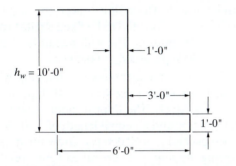

PROBLEM 8-3

8-4. Same as Problem 8-3, but the toe is 2 ft-0 in.

8-5. In Problem 8-1(c), assume a footing depth of 2 ft. Design the stem steel for M_u at the top of the footing. Check the anchorage into the footing. Disregard other stem steel details. Use $f_c' = 4000$ psi and $f_y = 60,000$ psi. Use thickness at top of stem = 12 in. and thickness at bottom of stem = 1 ft-9 in.

8-6. Completely design the cantilever retaining wall shown. The height of wall h_w is 22 ft, the backfill is level, the surcharge is 600 psf, $w_e = 100$ lb/ft^3, $h' = 4$ ft, $K_a = 0.30$, and the allowable soil pressure is 4 ksf. Use a coefficient of friction of concrete on soil of 0.5, $f_c' = 3000$ psi, and $f_y = 60,000$ psi. The required factor of safety against overturning is 2.0 and against sliding is 1.5. The wall batter should be between ¼ and ½ in./ft. The design is to be in accordance with the ACI Code (318-08).

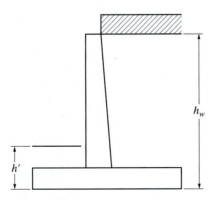

PROBLEM 8-6

8-7. Develop a spreadsheet application that will determine the location of the theoretical cutoff point for alternate bars in the stem of Example 8-1 (refer to Figure 8-15). Set up the spreadsheet so that a user could input any typical design data for a reinforced concrete cantilever retaining wall.

8-8. Design a reinforced concrete bearing wall to support a series of precast single tees spaced 7 ft-6 in. on centers. The stem of each tee section is 8-in. wide and bears on the full thickness of wall. The wall is braced top and bottom against lateral translation. Assume the bottom end to be fixed against rotation and the top to be unrestrained against rotation. The wall height is 14 ft, and P_u from each tee is 65 kips. Use $f'_c = 4000$ psi and $f_y = 60,000$ psi.

8-9. Design the first story shear wall for a three-story building with unfactored north–south lateral wind loads of 20 kips, 40 kips, 40 kips at the roof, third floor, and second floor, respectively. The floor-to-floor height is 12 ft and all shear walls have equal thickness and an equal length of 15 ft. Assume normal-weight concrete; f'_c is 3000 psi and f_y is 60,000 psi. Draw the detail of the shear wall reinforcement showing the required vertical end zone reinforcement and the distributed vertical and horizontal face reinforcement.

Columns

9-1   INTRODUCTION

The main vertical load-carrying members in buildings are called *columns*. The ACI Code defines a column as a member used primarily to support axial compressive loads and with a height at least three times its least lateral dimension. The code further defines a *pedestal* as an upright compression member having a ratio of unsupported height to least lateral dimension of 3 or less. The code definition for columns

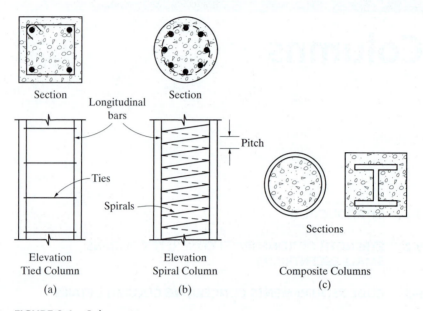

Section   Section

Longitudinal
bars

Pitch

Ties

Spirals

Elevation   Elevation   Composite Columns
Tied Column   Spiral Column   (c)
(a)   (b)

FIGURE 9-1 Column types.

will be extended to include members subjected to combined axial compression and
bending moment (in other words, eccentrically applied compressive loads), because,
for all practical purposes, no column is truly axially loaded.

The three basic types of reinforced concrete columns are shown in Figure 9-1.
Tied columns (Figure 9-1a) are reinforced with longitudinal bars enclosed by horizontal, or lateral, ties placed at specified spacings. *Spiral columns* (Figure 9-1b) are
reinforced with longitudinal bars enclosed by a continuous, rather closely spaced,
steel spiral. The spiral is made up of either wire or bar and is formed in the shape of
a helix. A third type of reinforced concrete column, a *composite column*, is shown in
Figure 9-1c. This type of column encompasses compression members reinforced lon-
gitudinally with structural steel shapes, pipes, or tubes with or without longitudinal
bars. Code requirements for composite compression members are found in Section
10.13. Our discussion is limited to the first two types: tied and spiral columns. Tied
columns are generally square or rectangular, whereas spiral columns are normally
circular. This is not a hard-and-fast rule, however, as square, spirally reinforced
columns, and circular tied columns do exist, as do other shapes, such as octagonal
and L-shaped columns.

We initially discuss the analysis and design of columns that are *short*. A column is
said to be short when its length is such that lateral buckling need not be considered.
The ACI Code does, however, require that the length of columns be a design consideration. It is recognized that as length increases, the usable strength of a given cross
section is decreased because of the buckling problem. By their very nature, concrete
columns are more massive and therefore stiffer than their structural steel counterparts. For this reason, slenderness is less of a problem in reinforced concrete

PHOTO 9-1   Foundation columns, turbine generator building. Seabrook Station, New Hampshire.

columns. It has been estimated that more than 90% of typical reinforced concrete columns existing in braced frame buildings may be classified as short columns, and slenderness effects may be neglected.

9-2   STRENGTH OF REINFORCED CONCRETE COLUMNS: SMALL ECCENTRICITY

If a compressive load P is applied coincident with the longitudinal axis of a symmetrical column, it theoretically induces a uniform compressive stress over the cross-sectional area. If the compressive load is applied a small distance e away from the longitudinal axis, however, there is a tendency for the column to bend due to the moment $M = Pe$. This distance e is called the *eccentricity*. Unlike the zero eccentricity condition, the compressive stress is not uniformly distributed over the cross section but is greater on one side than the other. This is analogous to our previously discussed eccentricity of applied loads with respect to retaining wall footings, where the eccentrically applied load resulted in a nonuniform soil pressure under the footing.

We consider an *axial load* to be a load that acts parallel to the longitudinal axis of a member but *need not* be applied at any particular point on the cross section, such as a centroid or a geometric center. The column that is loaded with a compressive axial load at zero eccentricity is probably nonexistent, and even the axial load/small eccentricity (axial load/small moment) combination is relatively rare. Nevertheless, we first consider the case of columns that are loaded with compressive axial loads at *small eccentricities*, further defining this situation as that in which the induced moments, although they are present, are so small that they are of little significance. Earlier codes have defined small eccentricity as follows:

For spirally reinforced columns: $e/h \le 0.05$.
For tied columns: $e/h \le 0.10$.

where h is the column dimension perpendicular to the axis of bending.

The fundamental assumptions for the calculation of column axial load strength (small eccentricities) are that at nominal strength the concrete is stressed to $0.85f_c'$ and the steel is stressed to f_y. For the cross sections shown in Figure 9-1a and b, the nominal axial load strength at small eccentricity is a straightforward sum of the forces existing in the concrete and longitudinal steel when each of the materials is stressed to its maximum. The following ACI notation will be used:

A_g = gross area of the column section (in.^{2})

A_{st} = total area of *longitudinal* reinforcement (in.^{2})

P_0 = nominal, or theoretical, axial load strength at zero eccentricity

P_n = nominal, or theoretical, axial load strength at given eccentricity

P_u = factored applied axial load at given eccentricity

For convenience, we will use the following longitudinal steel reinforcement ratio:

ρ_g = ratio of total longitudinal reinforcement area to cross-sectional area of column (A_{st}/A_g)

The nominal, or theoretical, axial load strength for the special case of zero eccentricity may be written as

$$P_0 = 0.85f_c'(A_g - A_{st}) + f_y A_{st}$$

This theoretical strength must be further reduced to a maximum usable axial load strength using two different strength reduction factors.

Extensive testing has shown that spiral columns are tougher than tied columns, as depicted in Figure 9-2. Both types behave similarly up to the column yield point, at which time the outer shell spalls off. At this point, the tied column fails through crushing and shearing of the concrete and through outward buckling of the bars between the ties. The spiral column, however, has a core area within the spiral that is effectively laterally supported and continues to withstand load. Failure occurs only

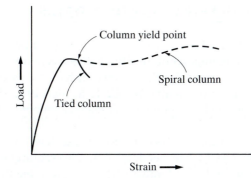

FIGURE 9-2   Load–strain relationship for columns.

when the spiral steel yields following large deformation of the column. Naturally, the size and spacing of the spiral steel will affect the final load at failure. Although both columns have exceeded their usable strengths once the outer shell spalls off, the ACI Code recognizes the greater tenacity of the spiral column in Section 9.3.2, where it directs that a strength-reduction factor of 0.75 be used for a spiral column, whereas the strength-reduction factor for the tied column is 0.65.

The code directs that the basic load–strength relationship be

$$\phi P_n \geq P_u$$

where P_n is the nominal axial load strength at a given eccentricity and ϕP_n is designated as the *design axial load strength*. Logically, for the case of zero eccentricity, if it could exist, P_n would equal P_0. The ACI Code recognizes that no practical column can be loaded with zero eccentricity, however. Therefore, in addition to imposing the strength reduction factor ϕ, the code directs that the nominal strengths be further reduced by factors of 0.80 and 0.85 for tied and spiral columns, respectively. This results in the following expressions for usable axial load strengths.

For spiral columns,

$$\phi P_{n(\text{max})} = 0.85\phi[0.85f_c'(A_g - A_{st}) + f_y A_{st}] \qquad \text{[ACI Eq. (10-1)]}$$

For tied columns,

$$\phi P_{n(\text{max})} = 0.80\phi[0.85f_c'(A_g - A_{st}) + f_y A_{st}] \qquad \text{[ACI Eq. (10-2)]}$$

These expressions provide the magnitude of the maximum design axial load strength that may be realized from any column cross section. This will be the design axial load strength at small eccentricity. Should the eccentricity (and the associated moment) become larger, ϕP_n will have to be reduced, as shown in Section 9-9. It may be recognized that the code equations for $\phi P_{n(\text{max})}$ provide for an extra margin of axial load strength. This will, in effect, provide some reserve strength to carry small moments.

9-3   CODE REQUIREMENTS CONCERNING COLUMN DETAILS

Main (longitudinal) reinforcing should have a cross-sectional area so that ρ_g will be between 0.01 and 0.08. The minimum number of longitudinal bars is four within rectangular or circular ties, three within triangular ties, and six for bars enclosed by spirals. The foregoing requirements are stated in the ACI Code, Section 10.9. Although not mentioned in the present code, the 1963 code recommended a minimum bar size of No. 5.

The clear distance between longitudinal bars must not be less than 1.5 times the nominal bar diameter nor $1\frac{1}{2}$ in. (ACI Code, Section 7.6.3). This requirement also holds true where bars are spliced. Table A-14 may be used to determine the maximum number of bars allowed in one row around the periphery of circular or square columns.

Cover shall be $1\frac{1}{2}$ in. minimum over primary reinforcement, ties, or spirals (ACI Code, Section 7.7.1).

Tie requirements are discussed in detail in the ACI Code, Section 7.10.5. The minimum size is No. 3 for longitudinal bars No. 10 and smaller; otherwise, minimum tie size is No. 4 (see Table A-14 for a suggested tie size). Usually, No. 5 is a maximum. The center-to-center spacing of ties should not exceed the smaller of 16 longitudinal bar diameters, 48 tie-bar diameters, or the least column dimension. Furthermore, the ties shall be arranged so that every corner and alternate longitudinal bar will have lateral support provided by the corner of a tie having an included angle of not more than 135°, and no bar shall be farther than 6 in. clear on each side from such a laterally supported bar. See Figure 9-3 for typical tie arrangements.

Spiral requirements are discussed in the ACI Code, Sections 7.10.4. and 10.9.3. The minimum spiral size is $\frac{3}{8}$ in. in diameter for cast-in-place construction ($\frac{5}{8}$ in. is usually maximum). Clear space between spirals must not exceed 3 in. or be less than 1 in. The *spiral steel ratio* ρ_s must not be less than the value given by

$$\rho_{s(\min)} = 0.45\left(\frac{A_g}{A_{ch}} - 1\right)\frac{f_c'}{f_{yt}} \qquad \text{[ACI Eq. (10-5)]}$$

where

$$\rho_s = \frac{\text{volume of spiral steel in one turn}}{\text{volume of column core in height } (s)}$$

s = center-to-center spacing of spiral (in.) (sometimes called the *pitch*)

A_g = gross cross-sectional area of the column (in.²)

A_{ch} = cross-sectional area of the core (in.²) (out-to-out of spiral)

f_{yt} = *spiral* steel yield point (psi) ≤ 60,000 psi

f_c' = compressive strength of concrete (psi)

This particular spiral steel ratio will result in a spiral that will make up the strength lost due to the spalling of the outer shell (see Figure 9-2).

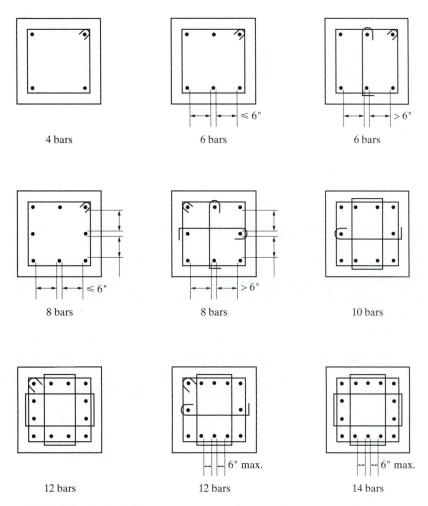

4 bars   6 bars   6 bars

8 bars   8 bars   10 bars

12 bars   12 bars   14 bars

FIGURE 9-3   Typical tie arrangements.

An approximate formula for the calculated spiral steel ratio in terms of physical properties of the column cross section may be derived from the preceding definition of ρ_s. In Figure 9-4, we denote the overall core diameter (out-to-out of spiral) as D_{ch} and the spiral diameter (center to center) as D_s. The cross-sectional area of the spiral bar or wire is denoted A_{sp}. From the definition of ρ_s:

$$\text{calculated } \rho_s = \frac{A_{sp}\pi D_s}{(\pi D_{ch}^2/4)(s)}$$

If the small difference between D_{ch} and D_s is neglected, then in terms of D_{ch},

$$\text{calculated } \rho_s = \frac{4A_{sp}}{D_{ch}s}$$

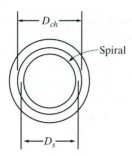

FIGURE 9-4   Definition of D_{ch} and D_s.

9-4   ANALYSIS OF SHORT COLUMNS: SMALL ECCENTRICITY

The analysis of short columns carrying axial loads that have small eccentricities involves checking the maximum design axial load strength and the various details of the reinforcing. The procedure is summarized in Section 9-6.

Example 9-1 ⸻

Find the maximum design axial load strength for the tied column of cross section shown in Figure 9-5. Check the ties. Assume a short column. Use $f'_c = 4000$ psi and $f_y = 60{,}000$ psi for both longitudinal steel and ties.

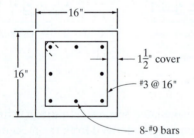

FIGURE 9-5   Sketch for Example 9-1.

Solution:

1. Check the steel ratio for the longitudinal steel:

$$\rho_g = \frac{A_{st}}{A_g} = \frac{8.00}{(16)^2} = 0.0313$$

$$0.01 < 0.0313 < 0.08 \qquad\qquad\qquad \text{(O.K.)}$$

2. From Table A-14, using a 13-in. core (column size less cover on each side), the maximum number of No. 9 bars is eight. Therefore the number of longitudinal bars is satisfactory.

3. The maximum design axial load strength may now be calculated:

$$\phi P_{n(\text{max})} = 0.80\phi[0.85f_c'(A_g - A_{st}) + f_y A_{st}]$$

$$= 0.80(0.65)[0.85(4)(256 - 8) + 60(8)]$$

$$= 688 \text{ kips}$$

4. Check the ties. Tie size of No. 3 is acceptable for longitudinal bar size up to No. 10. The spacing of the ties must not exceed the smaller of

$$48 \text{ tie-bar diameters} = 48(\tfrac{3}{8}) = 18 \text{ in.}$$

$$16 \text{ longitudinal-bar diameters} = 16(1.128) = 18 \text{ in.}$$

$$\text{least column dimension} = 16 \text{ in.}$$

Therefore the tie spacing is O.K. The tie arrangement for this column may be checked by ensuring that the clear distance between longitudinal bars does not exceed 6 in. Clear space in excess of 6 in. would require additional ties in accordance with the ACI Code, Section 7.10.5.3. Thus

$$\text{clear distance} = \frac{16 - 2(1\tfrac{1}{2}) - 2(\tfrac{3}{8}) - 3(1.128)}{2}$$

$$= 4.4 \text{ in.} < 6 \text{ in.}$$

Therefore no extra ties are needed.

Example 9-2 _____

Determine whether the spiral column of cross section shown in Figure 9-6 is adequate to carry a factored axial load (P_u) of 540 kips. Assume small eccentricity. Check the spiral. Use $f_c' = 4000$ psi and $f_y = 60,000$ psi.

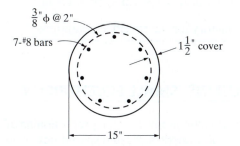

$\tfrac{3}{8}" \phi @ 2"$

7-#8 bars

$1\tfrac{1}{2}"$ cover

15"

FIGURE 9-6   Sketch for Example 9-2.

Solution:

1. From Table A-2, $A_{st} = 5.53$ in.^{2}, and from Table A-14, a diameter of 15 in. results in a circular area $A_g = 176.7$ in.^{2}. Therefore

$$\rho_g = \frac{5.53}{176.7} = 0.0313$$

$$0.01 < 0.0313 < 0.08 \qquad \text{(O.K.)}$$

2. From Table A-14, using a 12-in. core, seven No. 8 bars are satisfactory.

3. Find the maximum design axial load strength:

$$\phi P_{n(\text{max})} = 0.85\phi[0.85f_c'(A_g - A_{st}) + f_y A_{st}]$$

$$= 0.85(0.75)[0.85(4)(176.7 - 5.53) + 60(5.53)]$$

$$= 583 \text{ kips}$$

The strength is O.K. because 583 kips > 540 kips.

4. Check the spiral steel. The ⅜-in. diameter is O.K. (ACI Code, Section 7.10.4.2; and Table A-14). The minimum ρ_s is calculated using Table A-14 for the value of A_{ch}:

$$\rho_{s(\text{min})} = 0.45\left(\frac{A_g}{A_{ch}} - 1\right)\frac{f_c'}{f_{yt}}$$

$$= 0.45\left(\frac{176.7}{113.1} - 1\right)\frac{4}{60} = 0.0169$$

$$\text{actual } \rho_s = \frac{4A_{sp}}{D_{ch}s} = \frac{4(0.11)}{12(2)} = 0.0183$$

$$0.0183 > 0.0169 \qquad \text{(O.K.)}$$

The clear distance between spirals must not be in excess of 3 in. nor less than 1 in.:

$$\text{clear distance} = 2.0 - \tfrac{3}{8} = 1\tfrac{5}{8} \text{ in.} \qquad \text{(O.K.)}$$

The column is satisfactory for the specified conditions.

9-5   DESIGN OF SHORT COLUMNS: SMALL ECCENTRICITY

The design of reinforced concrete columns involves the proportioning of the steel and concrete areas and the selection of properly sized and spaced ties or spirals. Because the ratio of steel to concrete area must fall within a given range ($0.01 \le \rho_g \le 0.08$), the strength equation given in Section 9-2 is modified to include this term. For a tied column,

$$\phi P_{n(max)} = 0.80\phi[0.85f_c'(A_g - A_{st}) + f_y(A_{st})]$$

$$\rho_g = \frac{A_{st}}{A_g}$$

from which

$$A_{st} = \rho_g A_g$$

Therefore

$$\phi P_{n(max)} = 0.80\phi[0.85f_c'(A_g - \rho_g A_g) + f_y\rho_g A_g]$$
$$= 0.80\phi A_g[0.85f_c'(1 - \rho_g) + f_y\rho_g]$$

As

$$P_u \leq \phi P_{n(max)}$$

an expression can be written for required A_g in terms of the material strengths, P_u and ρ_g. For tied columns,

$$\text{required } A_g = \frac{P_u}{0.80\phi[0.85f_c'(1 - \rho_g) + f_y\rho_g]}$$

Similarly for spiral columns,

$$\text{required } A_g = \frac{P_u}{0.85\phi[0.85f_c'(1 - \rho_g) + f_y\rho_g]}$$

It should be recognized that there can be many valid choices for the size of column that will provide the necessary strength to carry any load P_u. A low ρ_g will result in a larger required A_g and vice versa. Other considerations will normally affect the practical choice of column size. Among them are architectural requirements and the desirability of maintaining column size from floor to floor so that forms may be reused.

The procedure for the design of short columns for loads at small eccentricities is summarized in Section 9-6.

Example 9-3

Design a square tied column to carry axial service loads of 320 kips dead load and 190 kips live load. There is no identified applied moment. Assume that the column is short. Use ρ_g about 0.03, $f_c' = 4000$ psi, and $f_y = 60,000$ psi.

Solution:

1. Material strengths and approximate ρ_g are given.
2. The factored axial load is

$$P_u = 1.6(190) + 1.2(320) = 688 \text{ kips}$$

3. The required gross column area is

$$\text{required } A_g = \frac{P_u}{0.80\phi[0.85f_c'(1 - \rho_g) + f_y\rho_g]}$$

$$= \frac{688}{0.80(0.65)[0.85(4)(1 - 0.03) + 60(0.03)]}$$

$$= 260 \text{ in.}^2$$

4. The required size of the square column would be

$$\sqrt{260} = 16.1 \text{ in.}$$

Use a 16-in.-square column. This choice will require that the actual ρ_g be slightly in excess of 0.03:

$$\text{actual } A_g = (16 \text{ in.})^2 = 256 \text{ in.}^2$$

5. The load on the concrete area (this is approximate since ρ_g will increase slightly) is

$$\text{load on concrete} = 0.80\phi(0.85f_c')A_g(1 - \rho_g)$$

$$= 0.80(0.65)(0.85)(4)(256)(1 - 0.03)$$

$$= 439 \text{ kips}$$

Therefore the load to be carried by the steel is

$$688 - 439 = 249 \text{ kips}$$

Because the maximum design axial load strength of the steel is $(0.80\phi A_{st}f_y)$, the required steel area may be calculated as

$$\text{required } A_{st} = \frac{249}{0.80(0.65)(60)} = 7.98 \text{ in.}^2$$

We will distribute bars of the same size evenly around the perimeter of the column and must therefore select bars in multiples of four. Use eight No. 9 bars ($A_{st} = 8.0 \text{ in.}^2$). Table A-14 indicates a maximum of eight No. 9 bars for a 13-in. core (O.K.).

6. Design the ties. From Table A-14, select a No. 3 tie. The spacing must not be greater than

$$48 \text{ tie-bar diameters} = 48(\tfrac{3}{8}) = 18 \text{ in.}$$

$$16 \text{ longitudinal-bar diameters} = 16(1.128) = 18.0 \text{ in.}$$

$$\text{least column dimension} = 16 \text{ in.}$$

Use No. 3 ties spaced 16 in. o.c. Check the arrangement with reference to Figure 9-7. The clear space between adjacent bars in the same face is

$$\frac{16 - 3 - 0.75 - 3(1.13)}{2} = 4.43 \text{ in.} < 6.0 \text{ in.}$$

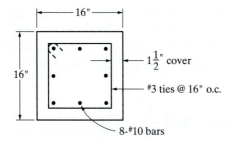

FIGURE 9-7   Design sketch for Example 9-3.

Therefore no additional ties are required by the ACI Code, Section 7.10.5.3.

7.   The design sketch is shown in Figure 9-7.

Example 9-4 _____

Redesign the column of Example 9-3 as a circular, spirally reinforced column.

Solution:

1.   Use $f'_c = 4000$ psi, $f_y = 60,000$ psi, and ρ_g approximately 0.03.

2. $P_u = 688$ kips, as in Example 9-3.

3.
$$\text{required } A_g = \frac{P_u}{0.85\phi[0.85f'_c(1 - \rho_g) + f_y\rho_g]}$$

$$= \frac{688}{0.85(0.75)[0.85(4)(1 - 0.03) + 60(0.03)]}$$

$$= 212 \text{ in.}^2$$

4.   From Table A-14, use an 18-in.-diameter column. Thus

$$A_g = 254.5 \text{ in.}^2$$

5.
$$\text{load on concrete} = 0.85\phi(0.85)f'_cA_g(1 - \rho_g)$$

$$= 0.85(0.75)(0.85)(4)(254.5)(1 - 0.03)$$

$$= 535 \text{ kips}$$

$$\text{load on steel} = 688 - 535 = 153 \text{ kips}$$

$$\text{required } A_{st} = \frac{153}{0.85\phi f_y}$$

$$= \frac{153}{0.85(0.75)60} = 4.00 \text{ in.}^2$$

Use seven No. 7 bars (A_{st} = 4.20 in.²). Table A-14 indicates a maximum of 13 No. 7 bars for a circular core 15 in. in diameter (O.K.).

6.   Design the spiral. From Table A-14, select a ⅜-in.-diameter spiral. The spacing will be based on the required spiral steel ratio. Here A_{ch} is from Table A-14. Thus

$$\rho_{s(\text{min})} = 0.45\left(\frac{A_g}{A_{ch}} - 1\right)\frac{f_c'}{f_{yt}}$$

$$= 0.45\left(\frac{254.5}{176.7} - 1\right)\frac{4}{60} = 0.0132$$

The maximum spiral spacing may be found by setting the calculated ρ_s equal to $\rho_{s(\text{min})}$:

$$\text{calculated } \rho_s = \frac{4A_{sp}}{D_{ch}s}$$

from which

$$S_{\text{max}} = \frac{4A_{sp}}{D_{ch}\rho_{s(\text{min})}}$$

$$= \frac{4(0.11)}{15(0.0132)} = 2.22 \text{ in.}$$

Use a spiral spacing of 2 in. The clear space between spirals must not be less than 1 in. nor more than 3 in.:

$$\text{clear space} = 2 - \tfrac{3}{8} = 1\tfrac{5}{8} \text{ in.} \qquad\qquad (\text{O.K.})$$

7.   The design sketch is shown in Figure 9-8.

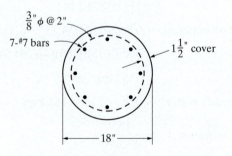

$\tfrac{3}{8}" \phi$ @ 2"

7-#7 bars

$1\tfrac{1}{2}"$ cover

18"

FIGURE 9-8   Design sketch for Example 9-4.

9-6   SUMMARY OF PROCEDURE FOR ANALYSIS AND DESIGN OF SHORT COLUMNS WITH SMALL ECCENTRICITIES

Analysis

1.   Check ρ_g within acceptable limits:

$$0.01 \leq \rho_g \leq 0.08$$

2.   Check the number of bars within acceptable limits for the clear space (see Table A-14). The minimum number is four for bars with rectangular or circular ties and six for bars enclosed by spirals.
3.   Calculate the maximum design axial load strength $\phi P_{n(\max)}$. See Section 9-2.
4.   Check the lateral reinforcing. For ties, check size, spacing, and arrangement. For spirals, check size, ρ_s, and clear distance.

Design

1.   Establish the material strengths. Establish the desired ρ_g (if any).
2.   Establish the factored axial load P_u.
3.   Determine the required gross column area A_g.
4.   Select the column dimensions. Use full-inch increments.
5.   Find the load carried by the concrete and the load required to be carried by the longitudinal steel. Determine the required longitudinal steel area. Select the longitudinal steel.
6.   Design the lateral reinforcing (ties or spiral).
7.   Sketch the design.

9-7   THE LOAD–MOMENT RELATIONSHIP

The equivalency between an eccentrically applied load and an axial load–moment combination is shown in Figure 9-9. Assume that a force P_u is applied to a cross

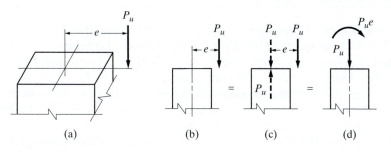

FIGURE 9-9   Load–moment–eccentricity relationship.

section at a distance e (eccentricity) from the centroid, as shown in Figure 9-9a and b. Add equal and opposite forces P_u at the centroid of the cross section (Figure 9-9c). The original eccentric force P_u may now be combined with the upward force P_u to form a couple, $P_u e$, that is a pure moment. This will leave remaining one force, P_u, acting downward at the centroid of the cross section. It can therefore be seen that if a force P_u is applied with an eccentricity e, the situation that results is identical to the case where an axial load of P_u at the centroid and a moment of $P_u e$ are simultaneously applied (Figure 9-9d). If we define M_u as the factored moment to be applied on a *compression* member along with a factored axial load of P_u at the centroid, the relationship between the two is

$$e = \frac{M_u}{P_u}$$

We have previously spoken of a column's nominal axial load strength at zero eccentricity P_0. Temporarily neglecting the strength reduction factors that must be applied to P_0, let us assume that we may apply a load P_u with zero eccentricity on a column of nominal axial load strength P_0 (where $P_u = P_0$). If we now move the load P_u away from the zero eccentricity position a distance of e, the column must resist the load P_u and, in addition, a moment $P_u e$. Because $P_u = P_0$ (in our hypothetical case), it is clear that there is no additional strength to carry the moment and that the column is overloaded. For this particular column to not be overloaded when subjected to load at eccentricity e, we must reduce P_u to the point where the column can carry both P_u and $P_u e$. The amount of the required decrease in P_u will depend on the magnitude of the eccentricity.

The preceding discussion must be modified because the ACI Code imposes $\phi P_{n(\max)}$ as the upper limit of axial load strength for any column. Nevertheless, the strength of any column cross section is such that it will support a broad spectrum of load and moment (or load and eccentricity) combinations. In other words, we may think of a column cross section as having many different axial load strengths, each with its own related moment strength.

9-8   COLUMNS SUBJECTED TO AXIAL LOAD AT LARGE ECCENTRICITY

At one time the ACI Code stipulated that compression members be designed for an eccentricity e of not less than $0.05h$ for spirally reinforced columns or $0.10h$ for tied columns, but at least 1 in. in any case. Here h is defined as the overall dimension of the column. These specified minimum eccentricities were originally intended to serve as a means of reducing the axial load design strength of a section in pure compression. The effect of the minimum eccentricity requirement was to limit the *maximum* axial load strength of a compression member.

Under the 2008 ACI Code, as we have discussed, the maximum design axial load strength $\phi P_{n(\max)}$ is given by ACI Equations (10-1) and (10-2). These two equations

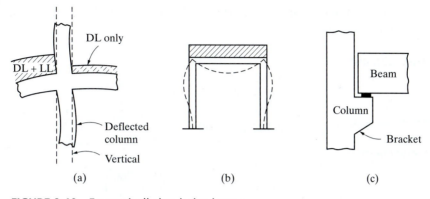

FIGURE 9-10   Eccentrically loaded columns.

apply when eccentricities are not in excess of *approximately* the 0.10h and 0.05h minimum eccentricity limits previously discussed. Therefore *small eccentricities* may be considered as those eccentricities up to about 0.10h and 0.05h for tied and spiral columns, respectively. We will consider cases of *large eccentricities* as those in which ACI Equations (10-1) or (10-2) no longer apply and where ϕP_n must be reduced below $\phi P_{n(\max)}$.

The occurrence of columns subjected to eccentricities sufficiently large so that moment must be a design consideration is common. Even interior columns supporting beams of equal spans will receive unequal loads from the beams due to applied live load patterns. These unequal loads could mean that the column must carry both load and moment, as shown in Figure 9-10a, and the resulting eccentricity of the loads could be appreciably in excess of our definition of small eccentricity. Another example of a column carrying both load and moment is shown in Figure 9-10b. In both cases, the rigidity of the joint will require the column to rotate along with the end of the beam that it is supporting. The rotation will induce moment in the column. A third and very practical example can be found in precast work (Figure 9-10c), where the beam reaction can clearly be seen to be eccentrically applied on the column through the column bracket.

9-9 ϕ FACTOR CONSIDERATIONS

Columns discussed so far have had strength-reduction factors applied in a straightforward manner. That is, $\phi = 0.75$ for spiral columns, and $\phi = 0.65$ for tied columns. These ϕ factors correspond to the compression-controlled strain limit or a net tensile strain in the extreme tension reinforcement, $\epsilon_t \leq 0.002$. Eccentrically loaded columns, however, carry both axial load and moment. For values of ϵ_t larger than 0.002, the ϕ equations from ACI Code, Section 9.3.2, discussed in

Chapter 2 will give higher values than indicated above. The ϕ equations are repeated here as follows:

Tied columns:

$$\phi = 0.65 + (\epsilon_t - 0.002)\left(\frac{250}{3}\right)$$

$$0.65 \le \phi \le 0.90$$

Spiral columns:

$$\phi = 0.75 + (\epsilon_t - 0.002)\left(\frac{200}{3}\right)$$

$$0.75 \le \phi \le 0.90$$

9-10   ANALYSIS OF SHORT COLUMNS: LARGE ECCENTRICITY

The first step in our investigation of short columns carrying loads at large eccentricity is to determine the strength of a given column cross section that carries loads at various eccentricities. This may be thought of as an analysis process. For this development, we will find the design axial load strength ϕP_n, where P_n is defined as the nominal axial load strength at a given eccentricity.

Example 9-5

Find the design axial load strength ϕP_n for the tied column for the following conditions: (a) small eccentricity ($e = 0$ to $0.10h$); (b) $e = 5$ in.; (c) the balanced strain condition or compression-controlled strain limit, $\epsilon_t = 0.002$; (d) $\epsilon_t = 0.004$; (e) the tension-controlled strain limit, $\epsilon_t = 0.005$; and (f) pure moment. The column cross section is shown in Figure 9-11. Assume a short column. Bending is about the Y–Y axis. Use $f_c' = 4000$ psi and $f_y = 60{,}000$ psi.

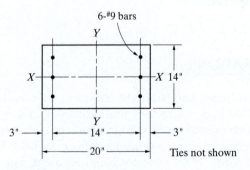

FIGURE 9-11   Column cross section for Example 9-5.

Solution:

a. The analysis of the small eccentricity condition is similar to the analyses of Examples 9-1 and 9-2. We can calculate the design axial load strength from

$$\phi P_n = \phi P_{n(\text{max})}$$

$$= 0.80\phi[0.85f'_c(A_g - A_{st}) + f_y A_{st}]$$

$$= 0.80(0.65)[0.85(4)(280 - 6) + 60(6)]$$

$$= 672 \text{ kips}$$

The corresponding maximum moment:

$$\phi M_n = \phi P_n e = 672(0.10)\left(\frac{20}{12}\right) = 112 \text{ ft-kips}$$

b. The situation of $e = 5$ in. is shown in Figure 9-12. In part (a) of this example, all steel was in compression. As eccentricity increases, the steel on the side of the column away from the load is subjected to less compression. Therefore there is some value of eccentricity at which this steel will change from compression to tension. Because this value of eccentricity is not known, the strain situation shown in Figure 9-13 is *assumed* and will be verified (or disproved) by calculation.

   The assumptions at nominal strength are

1. Maximum concrete strain = 0.003.
2. $\epsilon'_s > \epsilon_y$. Therefore $f'_s = f_y$.
3. ϵ_s is tensile.
4. $\epsilon_s < \epsilon_y$. Therefore $f_s < f_y$.

The unknown quantities are P_n and c.

   Using basic units of kips and inches, the tensile and compressive forces are evaluated. Force C_2 is the force in the compressive steel

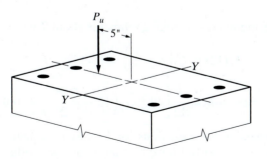

FIGURE 9-12   Example 9-5b, $e = 5$ in.

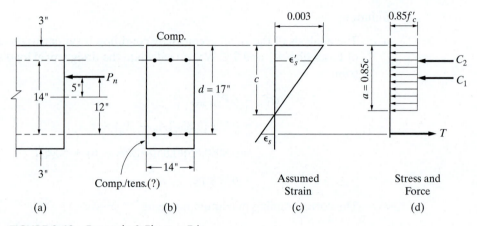

FIGURE 9-13   Example 9-5b, e = 5 in.

accounting for the concrete displaced by the steel. With reference to Figure 9-13c,

$$C_1 = 0.85f'_c ab = 0.85(4)(0.85c)(14) = 40.46c$$

$$C_2 = f_y A'_s - 0.85f'_c A'_s = A'_s(f_y - 0.85f'_c)$$

$$= 3[60 - 0.85(4)] = 169.8$$

$$T = f_s A_s = \epsilon_s E_s A_s = 87\left(\frac{d-c}{c}\right)A_s$$

$$= 87\left(\frac{17-c}{c}\right)3 = 261\left(\frac{17-c}{c}\right)$$

From Σ forces = 0 in Figure 9-13,

$$P_n = C_1 + C_2 - T$$

$$= 40.46c + 169.8 - 261\left(\frac{17-c}{c}\right)$$

From Σ moments = 0, taking moments about T in Figure 9-13,

$$P_n(12) = C_1\left(d - \frac{a}{2}\right) + C_2(14)$$

$$P_n = \frac{1}{12}\left[40.46(c)\left(17 - \frac{0.85c}{2}\right) + 169.8(14)\right]$$

The preceding two equations for P_n may be equated and the resulting cubic equation solved for c by trial or by some other iterative method. The solution will yield c = 14.86 in., which will result in a value for P_n

(from either equation) of 733 kips. The net tensile strain in the extreme tension reinforcement may be calculated from

$$\epsilon_t = 0.003\left(\frac{d-c}{c}\right) = 0.003\left(\frac{17-14.86}{14.86}\right) = 0.00043 < 0.002$$

For $\epsilon_t \le 0.002$, the corresponding tied column strength-reduction factor ϕ is 0.65. Therefore,

$$\phi P_n = 0.65(733)$$

$$= 476 \text{ kips}$$

Check the assumptions that were made:

$$\epsilon_s' = \left(\frac{14.86-3}{14.86}\right)(0.003) = 0.0024$$

$$\epsilon_y = 0.00207$$

Because $\epsilon_s' > \epsilon_y$,

$$f_s' = f_y \qquad \text{(O.K.)}$$

Based on the location of the neutral axis, the steel away from the load *is* in tension and

$$f_s = 87\left(\frac{17-14.86}{14.86}\right) = 12.53 \text{ ksi} < 60 \text{ ksi} \qquad \text{(O.K.)}$$

All assumptions are verified.

We may also determine the design moment strength for an eccentricity of 5 in. as follows:

$$\phi P_n e = \frac{476(5)}{12}$$

$$= 198 \text{ ft-kips}$$

Therefore the given column has a design load–moment combination strength of 476 kips axial load and 198 ft-kips moment. This assumes that the moment is applied about the Y–Y axis.

c.   The compression-controlled strain limit (balanced condition) exists when the concrete reaches a strain of 0.003 at the same time the extreme tension steel reaches a strain of 0.002, as shown in Figure 9-14c. Here P_b is defined as nominal axial load strength at the balanced condition, e_b is the associated eccentricity, and c_b is the distance from the compression face to the balanced neutral axis.

Using the strain diagram in Figure 9-14, we may calculate the value of c_b:

$$\frac{0.003}{c_b} = \frac{0.002}{17-c_b}$$

from which $c_b = 0.6(17) = 10.2$ in.

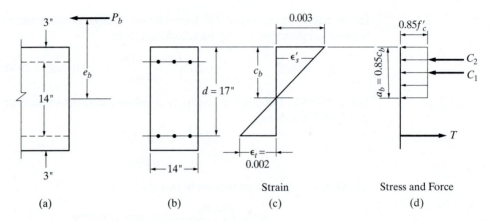

FIGURE 9-14 Example 9-5c, balanced condition.

For $\epsilon_t = 0.002$, the tied column strength-reduction factor ϕ is 0.65. We then may determine ϵ_s':

$$\epsilon_s' = \frac{7.2}{10.2}(0.003) = 0.0021$$

Because $0.0021 > 0.00207$, the compression steel has yielded and $f_s' = f_y = 60$ ksi.

Summarize the forces in Figure 9-14d and let C_2 account for the force in the concrete displaced by the three No. 9 bars:

$$C_1 = 0.85(4)(0.85)(10.20)(14) = 413 \text{ kips}$$

$$C_2 = 60(3) - 0.85(4)(3) = 170 \text{ kips}$$

$$T = 60(3) = 180 \text{ kips}$$

$$P_b = C_1 + C_2 - T = 413 + 170 - 180$$

$$= 403 \text{ kips}$$

The value of e_b may be established by summing moments about T:

$$P_b(e_b + 7) = C_1\left(d - \frac{0.85c_b}{2}\right) + C_2(14)$$

$$403(e_b + 7) = 413\left[17 - \frac{0.85(10.20)}{2}\right] + 170(14)$$

from which $e_b = 11.88$ in. Therefore, at the balanced condition,

$$\phi P_b = 0.65(403) = 262 \text{ kips}$$

$$\phi P_b e_b = \frac{262}{12}(11.88) = 259 \text{ ft-kips}$$

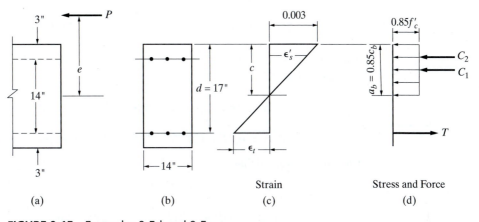

Strain   Stress and Force

(a)   (b)   (c)   (d)

FIGURE 9-15 Examples 9-5d and 9-5e.

d. Net tensile strain in the extreme tension steel, $\epsilon_t = 0.004$ and a corresponding compression strain of 0.003 in the compression face.

Using the strain diagram in Figure 9-15, the neutral axis depth, c is calculated as follows:

$$\frac{0.003}{c} = \frac{0.004}{17 - c}$$

from which $c = 0.429(17) = 7.29$ in.
We may then determine ϵ_s':

$$\epsilon_s' = 0.003\frac{(7.29 - 3)}{7.29} = 0.00177 < \epsilon_y$$

$$f_s' = E\,\epsilon_s' = 29{,}000\,(0.00177) = 51.2 \text{ ksi}$$

Using similar equations from case (b), the forces in the concrete, compression steel, and tension steel are as follows:

$$C_1 = 0.85(4)(0.85)(7.29)(14) = 295 \text{ kips}$$

$$C_2 = 51.2(3) - (0.85)(4)(3) = 143 \text{ kips}$$

$$T = 60(3) = 180 \text{ kips}$$

From Σ forces $= 0$ in Figure 9-15,

$$P_n = C_1 + C_2 - T$$

$$= 295 + 143 - 180 = 258 \text{ kips}$$

From Σ moments $= 0$, taking moments about T in Figure 9-15,

$$P_n(e + 7) = C_1\left(d - \frac{0.85c}{2}\right) + C_2(14)$$

$$258(e + 7) = 295\left(d - \frac{0.85(7.29)}{2}\right) + 143(14)$$

from which $e = 16.7$ in.
  For $\epsilon_t = 0.004$,

$$\phi = 0.65 + (0.004 - 0.002)\left(\frac{250}{3}\right) = 0.82$$

$$0.65 < 0.82 < 0.9 \qquad\qquad\qquad\qquad\qquad \text{(O.K.)}$$

Therefore, at $\epsilon_t = 0.004$,

$$\phi P_n = 0.82(258) = 212 \text{ kips}$$

$$\phi M_n = \phi P_n e = 212\left(\frac{16.7}{12}\right) = 295 \text{ ft-kips}$$

e. Net tensile strain in the extreme tension steel ϵ_t is 0.005, and a correspon-
    ding compression strain of 0.003 exists in the compression face.
      Using the strain diagram in Figure 9-15, the neutral axis depth, c, is
    calculated as follows:

$$\frac{0.003}{c} = \frac{0.005}{17 - c}$$

from which $c = 0.375(17) = 6.38$ in.
  We may then determine ϵ'_s:

$$\epsilon'_s = 0.003\left(\frac{6.38 - 3}{6.38}\right) = 0.00159 < \epsilon_y$$

$$f'_s = E\,\epsilon'_s = 29{,}000\,(0.00159) = 46.1 \text{ ksi}$$

Using similar equations from case (b), the forces in the concrete, compression steel, and tension steel are as follows:

$$C_1 = 0.85(4)(0.85)(6.38)(14) = 258 \text{ kips}$$

$$C_2 = 46.1(3) - (0.85)(4)(3) = 128 \text{ kips}$$

$$T = 60(3) = 180 \text{ kips}$$

From Σ forces $= 0$ in Figure 9-15,

$$P_n = C_1 + C_2 - T$$

$$= 258 + 128 - 180 = 206 \text{ kips}$$

From Σ moments $= 0$, taking moments about T in Figure 9-15,

$$P_n(e + 7) = C_1\left(d - \frac{0.85c}{2}\right) + C_2(14)$$

$$206(e + 7) = 258\left(d - \frac{0.85(6.38)}{2}\right) + 128(14)$$

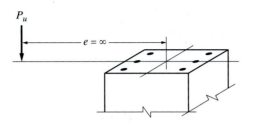

FIGURE 9-16   Column of Example 9-5 loaded with pure moment.

from which $e = 19.6$ in.
   For $\epsilon_t \geq 0.005, \phi = 0.90$. Therefore, at $\epsilon_t = 0.005$,

$$\phi P_n = 0.90(206) = 185 \text{ kips}$$

$$\phi M_n = \phi P_n e = 185 \left(\frac{19.6}{12} \right) = 302 \text{ ft-kips}$$

f.   The analysis of the pure moment condition is similar to the analysis of
the case where eccentricity e is infinite, shown in Figure 9-16. We will find
the design moment strength ϕM_n, because P_u and ϕP_n will both be zero.
      With reference to Figure 9-17d, notice that for pure moment the bars
on the load side of the column are in compression, whereas the bars on
the side away from the load are in tension. The total tensile and compressive forces must be equal to each other. Since $A_s = A'_s$, A'_s *must* be at a
stress less than yield. Assume that A_s is at yield stress. Then

$$C_1 = \text{concrete compressive force}$$

$$C_2 = \text{steel compressive force}$$

$$T = \text{steel tensile force}$$

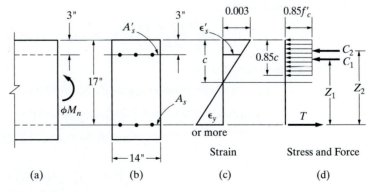

FIGURE 9-17   Example 9-5f, $e = \infty$.

Referring to the compressive strain diagram portion of Figure 9-17c and noting that the basic units are kips and inches,

$$\epsilon'_s = 0.003 \frac{(c - 3)}{c}$$

Because

$$f'_s = E_s \epsilon'_s$$

substituting yields

$$f'_s = 29,000(0.003) \frac{(c - 3)}{c}$$

$$= 87 \frac{(c - 3)}{c}$$

For equilibrium in Figure 9-17d,

$$C_1 + C_2 = T$$

Substituting into the foregoing and accounting for the concrete displaced by the compression steel, we obtain

$$(0.85f'_c)(0.85c)(b) + f'_s A'_s - 0.85f'_c A'_s = f_y A_s$$

$$(0.85)(4)(0.85c)(14) + 87\left(\frac{c - 3}{c}\right)(3) - 0.85(4)(3) = 3(60)$$

Solving the preceding equation for the one unknown quantity c yields

$$c = 3.62 \text{ in.}$$

The net tensile strain in the extreme tension reinforcement may be calculated from

$$\epsilon_t = 0.003\left(\frac{d - c}{c}\right) = 0.003\left(\frac{17 - 3.62}{3.62}\right) = 0.011$$

For $\epsilon_t \geq 0.005$, the corresponding strength reduction factor ϕ is 0.90. Therefore

$$f'_s = 87\left(\frac{3.62 - 3}{3.62}\right) = 14.90 \text{ ksi} \qquad \text{(compression)}$$

Summarizing the forces,

$$C_1 = 0.85f'_c(0.85)cb = 0.85(4)(0.85)(3.62)(14) = \quad 146.5 \text{ kips}$$

$$\text{displaced concrete} = 0.85f'_c A'_s = 0.85(4)(3) = -10.2 \text{ kips}$$

$$C_2 = f'_s A'_s = 14.90(3) = \quad \underline{44.7 \text{ kips}}$$

$$181.0 \text{ kips}$$

$$T = f_y A_s = 60(3.0) = \quad 180 \quad \text{kips}$$

The slight error between T and $(C_1 + C_2)$ will be neglected.

Summarizing the internal couples,

$$M_{n1} = C_1 Z_1$$

$$= \frac{146.5}{12}\left[17 - \frac{0.85(3.62)}{2}\right]$$

$$= 188.8 \text{ ft-kips}$$

$$M_{n2} = C_2 Z_2$$

$$= (44.7 - 10.2)\left(\frac{14}{12}\right)$$

$$= 40.3 \text{ ft-kips}$$

$$M_n = M_{n1} + M_{n2}$$

$$= 188.8 + 40.3$$

$$= 229 \text{ ft-kips}$$

The design moment strength becomes

$$\phi M_n = 0.90(229)$$

$$= 206 \text{ ft-kips}$$

The results of the six parts of Example 9-5 are tabulated (see Table 9-1) and plotted in Figure 9-18. All design axial load strengths are denoted ϕP_n, and all design moment strengths are denoted $\phi P_n e$. This plot is commonly called an *interaction diagram*. It applies *only* to the column analyzed, but it is a representation of *all* combinations of axial load and moment strengths for that column cross section.

In Figure 9-18, any point *on* the solid line represents an allowable combination of load and moment. Any point *within* the solid line represents a load–moment combination that is also allowable, but for which this column is *overdesigned*. Any point

TABLE 9-1   Column Axial Load–Moment Interaction for Example 9-5

Eccentricity, e	Net tensile strain in extreme tension steel, ϵ_t	Strength-reduction factor ϕ	Axial load strength, (ϕP_n, kips)	Moment strength, ($\phi P_n e$, ft-kips)
Small		0.65	672	112
(i.e., 0 to 0.10h)5″	0.00043	0.65	476	198
11.88″ (balanced)	0.002	0.65	262	259
16.7″	0.004	0.82	212	295
19.6″	0.005	0.90	185	302
Infinite (pure moment)	$\gg 0.005$	0.90	0	206

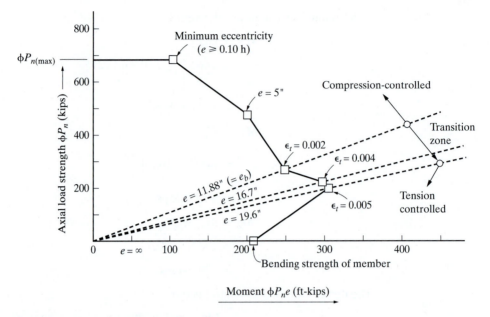

FIGURE 9-18   Column interaction diagram.

outside the solid line represents an unacceptable load–moment combination or a load–moment combination for which this column is *underdesigned*. The value of $\phi P_{n(max)}$, which we calculated in part (a), is superimposed on the plot as the horizontal line.

Radial lines from the origin represent various eccentricities. (Actually, the slopes of the radial lines are equal to $\phi P_n / \phi P_n e$ or $1/e$.) The intersection of the $e = e_b$ line with the solid line represents the balanced condition. Any eccentricity *less* than e_b will result in compression controlling the column. For eccentricities between e_b (11.88 in.) and 19.6 in., the column will be in the transition zone $(0.002 \leq \epsilon_t \leq 0.005)$. For eccentricities greater than 19.6 in., the column will be tension-controlled.

The calculations involved with column loads at large eccentricities are involved and tedious. The previous examples were analysis examples. Design of a cross section using the calculation approach would be a trial-and-error method and would become exceedingly tedious. Therefore design and analysis aids have been developed that shorten the process to a great extent. These aids may be found in the form of tables and charts. A chart approach is developed in ACI Publication SP-17(97), *ACI Design Handbook* [1]. The design aids are based on the assumptions of ACI 318-95 and on the principles of static equilibrium; they are developed in a fashion similar to what was done in Example 9-5. No ϕ factors are incorporated into the diagrams. Eight interaction diagrams are included in Appendix A (Diagrams A-15 through A-22.)

The diagrams take on the general form of Figure 9-18 but are generalized to be applicable to more situations. Referring to Diagram A-15, which corresponds to our Figure 9-18, the following definitions will be useful:

$$\rho_g = \frac{A_{st}}{A_g}$$

h = column dimension perpendicular to the bending axis (see the sketch included on Diagram A-15)

γ = ratio of distance between centroids of outer rows of bars and column dimension perpendicular to the bending axis

Note that the vertical axis and the horizontal axis are in general terms of K_n and R_n, where

$$K_n = \frac{P_n}{f'_c A_g}$$

$$R_n = \frac{P_n e}{f'_c A_g h}$$

Note also that P_n and $P_n e$ are nominal axial load strength and nominal moment strength. The slope of a radial line from the origin can be represented as follows:

$$\text{slope} = \frac{\text{rise}}{\text{run}} = \frac{\left(\dfrac{P_n}{f'_c A_g}\right)}{\left(\dfrac{P_n e}{f'_c A_g h}\right)} = \frac{h}{e}$$

Curves are shown for the range of allowable ρ_g values from 0.01 to 0.08. A line near the horizontal axis labeled $\epsilon_t = 0.0050$ indicates the limit for tension-controlled sections. Columns with load–moment–strength combinations below this line are tension-controlled ($\phi = 0.90$). The line labeled $f_s/f_y = 1.0$ indicates the balanced condition. Columns with load–moment–strength combinations above this line are compression-controlled ($\phi = 0.65$ for tied columns; 0.75 for spiral columns). Columns with load–moment–strength combinations between these two lines are in the transition zone. The line labeled $K_{\max}$ indicates the maximum allowable nominal load strength $[\phi P_{n(\max)}]$ for columns loaded with small eccentricities. A horizontal line drawn through the intersection of the $K_{\max}$ line and a ρ_g curve corresponds to the horizontal line near the top of the interaction diagram in Figure 9-18.

The following three examples illustrate the use of the ACI interaction diagrams for analysis and design of short reinforced concrete columns.

Example 9-6 _____

Using the interaction diagrams of Appendix A, find the axial load strength ϕP_n and the moment strength ϕM_n for the column cross section with six No. 9 bars,

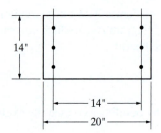

FIGURE 9-19   Sketch for Example 9-6.

as shown in Figure 9-19. Eccentricity $e = 5$ in., and use $f'_c = 4000$ psi and $f_y = 60,000$ psi. Compare the results with Example 9-5b.

Solution:

First, determine which interaction diagram to use, based on the type of cross section, the material strengths, and the factor γ.

$$\gamma h = 14 \text{ in.}$$

$$\gamma = \frac{14}{20} = 0.7$$

Therefore, use interaction Diagram A-15.

$$\rho_g = \frac{6.00}{14(20)} = 0.0214$$

$$0.01 \le 0.0214 \le 0.08 \qquad\qquad \text{(O.K.)}$$

Next, calculate the slope of the radial line from the origin, which relates h and e:

$$\text{slope} = \frac{h}{e} = \frac{20}{5} = 4$$

A straight edge and some convenient values (e.g., $K_n = 1.0$ and $R_n = 0.25$) may be used to intersect this radial line with an estimated $\rho_g = 0.0214$ curve. At this intersection, we read $K_n \approx 0.64$ and $R_n \approx 0.16$. Because this combination of load and moment is above the $f_s/f_y = 1.0$ line, this is a compression-controlled section and $\phi = 0.65$.

$$\phi P_n = \phi K_n f'_c A_g$$

$$= 0.65\,(0.64)(4)(20)(14) = 466 \text{ kips}$$

$$\phi M_n = \phi R_n f'_c A_g h$$

$$= \frac{0.65(0.160)(4)(20)(14)(20)}{12 \text{ in./ft}} = 194 \text{ ft-kips}$$

or

$$\phi M_n = \phi P_n e = \frac{466 \text{ ft-kips (5 in.)}}{12 \text{ in./ft}} = 194 \text{ ft-kips}$$

This compares reasonably well with the results of Example 9-5b: $\phi P_n = 476$ kips and $\phi M_n = 198$ ft-kips.

Example 9-7

Design a circular spirally reinforced concrete column to support a design load $P_u = 1100$ kips and a design moment $M_u = 285$ ft-kips. Use $f_c' = 4000$ psi and $f_y = 60,000$ psi.

Solution:

Estimate the column size required based on $\rho_g = 1\%$ and axial load only.

$$\text{required } A_g = \frac{P_u}{0.85\phi[0.85f_c'(1 - \rho_g) + f_y\rho_g]}$$

$$= \frac{1100}{0.85 \, (0.75)[0.85(4)(0.99) + 60(0.01)]}$$

$$= 435 \text{ in.}^2$$

Try a 24-in.-diameter column ($A_g = 452$ in.^{2}).

If No. 9 bars are eventually chosen (refer to Figure 9-20),

$$\gamma h = 24 - 2(1\tfrac{1}{2}) - 2\left(\frac{3}{8}\right) - 1.13 = 19.12 \text{ in.}$$

$$\gamma = \frac{19.12}{h} = \frac{19.12}{24} = 0.797$$

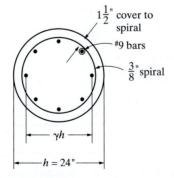

FIGURE 9-20   Sketch for Example 9-7.

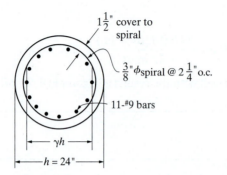

FIGURE 9-21   Design sketch for Example 9-7.

Therefore, use Diagram A-21 from Appendix A (ACI Interaction Diagram C4-60.8). Next, determine the required ρ_g. Assume that this column will be compression-controlled ($\phi = 0.75$) subject to later check.

Recognizing that required $P_n = P_u/\phi$ and required $P_n e = M_u/\phi$, we can calculate the values of required K_n and R_n:

$$\text{required } K_n = \frac{P_u}{\phi f_c' A_g} = \frac{1100}{0.75(4)(452)} = 0.811$$

$$\text{required } R_n = \frac{M_u}{\phi f_c' A_g h} = \frac{285(12)}{0.75(4)(452)(24)} = 0.105$$

From Diagram A-21, $\rho_g = 0.024$. Note that this is well above the $f_s/f_y = 1.0$ line; therefore, the column is compression-controlled and the assumption that $\phi = 0.75$ is O.K.

$$\text{required } A_s = \rho_g A_g = 0.024(452) = 10.85 \text{ in.}^2$$

Select 11 No. 9 bars ($A_s = 11.00 \text{ in.}^2$). Check the maximum number of No. 9 bars from Table A-14: 15 (O.K.).

Design the spiral. Use a ⅜-in.-diameter spiral.

The concrete core diameter is $D_{ch} = 24 - 2(1\frac{1}{2}) = 21$ in.

$$\text{required } \rho_s = 0.45\left(\frac{A_g}{A_{ch}} - 1\right)\frac{f_c'}{f_{yt}} = 0.45\left(\frac{452}{346} - 1\right)\frac{4}{60} = 0.0092$$

$$\text{required } s = \frac{4A_{sp}}{D_{ch}\rho_s} = \frac{4(0.11)}{21(0.0092)} = 2.27$$

Use 2¼-in. spacing. The design is shown in Figure 9-21.

Example 9-8 _____

Design a square-tied reinforced concrete column to support a design load $P_u = 1300$ kips and a design moment $M_u = 550$ ft-kips. Use $f_c' = 4000$ psi and $f_y = 60,000$ psi.

Solution:

Estimate the column size required based on $\rho_g = 1\%$ and axial load only.

$$\text{required } A_g = \frac{P_u}{0.80\phi[0.85 f_c' (1 - \rho_g) + f_y\rho_g]}$$

$$= \frac{1300}{0.80 (0.65)[0.85(4)(0.99) + 60(0.01)]}$$

$$= 631 \text{ in.}^2$$

Try a 26-in.-square column ($A_g = 676$ in.^{2}).

If No. 9 bars are eventually chosen (refer to Figure 9-22):

$$\gamma h = 26 - 2(1\tfrac{1}{2}) - 2\left(\frac{3}{8}\right) - 1.13 = 21.12 \text{ in.}$$

$$\gamma = \frac{21.12}{h} = \frac{21.12}{26} = 0.812$$

Therefore, use Diagram A-18 from Appendix A (ACI Interaction Diagram R4-60.8).

Next, determine the required ρ_g. Assume that this column will be compression-controlled ($\phi = 0.65$) subject to later check.

Recognizing that required $P_n = P_u/\phi$ and required $P_u e = M_u/\phi$, we can calculate the values of required K_n and R_n:

$$\text{required } K_n = \frac{P_u}{\phi f_c' A_g} = \frac{1300}{0.65(4)(676)} = 0.740$$

$$\text{required } R_n = \frac{M_u}{\phi f_c' A_g h} = \frac{550\,(12)}{0.65(4)(676)(26)} = 0.144$$

From Diagram A-18, $\rho_g \approx 0.023$. Note that this is well above the $f_s/f_y = 1.0$ line; therefore, the column is compression-controlled and the assumption that $\phi = 0.65$ is O.K.

$$\text{required } A_s = \rho_g A_g = 0.023(676) = 15.55 \text{ in.}^2$$

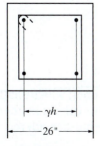

FIGURE 9-22 Sketch for Example 9-8.

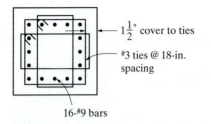

FIGURE 9-23   Design sketch for Example 9-8.

Select 16 No. 9 bars ($A_s = 16.00$ in.^{2}). Check the maximum number of No. 9 bars from Table A-14: 20 (O.K.).

Design the ties. Use a ⅜-in.-diameter tie, because the vertical bar size (No. 9 bar) is not greater than a No. 10.

The maximum tie spacing is the smallest of the following:

$$16 \text{ (bar diameter)} = 16 \times 1.13 = 18 \text{ in.}$$

$$48 \text{ (tie diameter)} = 48 \times \frac{3}{8} = 18 \text{ in.}$$

least column dimension = 26 in.

Therefore, use No. 3 ties at 18-in. spacing. The design is shown in Figure 9-23.

9-11   THE SLENDER COLUMN

Thus far, our design and analysis have been limited to short columns that require no consideration of necessary strength reduction due to the possibility of buckling. All compression members will experience the buckling phenomenon as they become longer and more flexible. These are sometimes termed *slender columns*. A column may be categorized as *slender* if its cross-sectional dimensions are small in comparison to its unsupported length. The degree of slenderness may be expressed in terms of the *slenderness ratio*

$$\frac{k\ell_u}{r}$$

where

k = effective length factor for compression members

ℓ_u = the unsupported length of a compression member, which shall be taken as the clear distance between floor slabs, beams, or other members capable of providing lateral support in the direction being considered (ACI Code, Section 10.1.1) r = radius of gyration of the cross section of the compression members, which may be taken as 0.30h, where h is the overall dimension of a rectangular column in the direction of the moment, or 0.25D, where D is the diameter of a circular column (ACI Code, Section 10.1.2)

The numerator $k\ell_u$ is termed the *effective length*. It is a function not only of the unsupported length and end conditions of the column but also a function of whether or not *sidesway* exists. Sidesway may be described as a kind of deformation whereby one end of a member moves laterally with respect to the other. Sidesway is also termed *lateral drift*.

The ACI Code, Section 10.6.3, states that for compression members braced against sidesway, k may be taken as 1.0. This is conservative. The ACI Code, Section 10.7.2, states that for compression members not braced against sidesway, the effective length must be greater than 1.0. Therefore, as a rule, compression members free to buckle in a sidesway mode are appreciably weaker than when braced against sidesway.

A simple example is a column fixed at one end and entirely free at the other (cantilever column or flagpole). Such a column will buckle, as shown in Figure 9-24. The upper end would move laterally with respect to the lower end. This lateral movement is the sidesway (or lateral drift). In reinforced concrete structures, it is common to deal with indeterminate rigid frames, such as illustrated by the simple portal frame in Figure 9-25. The upper end of the frame can move sideways as it is unbraced. This type of frame is sometimes termed a *sway frame*, and it depends on the rigidity of the joints for stability. The lower ends of the columns may be theoretically pin corrected, fully restrained, or somewhere in between.

As an example of how the effective length of a column is influenced by sidesway, consider the simple case of a single member, as shown in Figure 9-26. The member braced against sidesway (Figure 9-26a) has an effective length half that of the

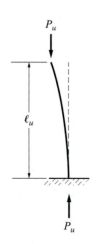

FIGURE 9-24   Fixed-free column.

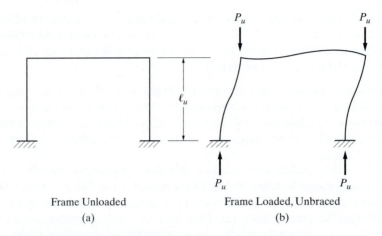

FIGURE 9-25   Sidesway on portal frame.

member without sidesway bracing (Figure 9-26b) and has four times the axial-load capacity based on the Euler critical column load theory.

If we consider the column shown in Figure 9-26b to be part of a frame and give the sidesway the notation Δ as shown in Figure 9-27, it is seen that the axial load now acts eccentrically and creates end moments of $P_u\Delta$. This is referred to as the *P-delta effect*. These moments are also referred to as "second-order end moments" because

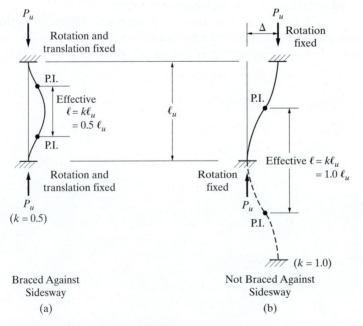

FIGURE 9-26   Sidesway and effective length.

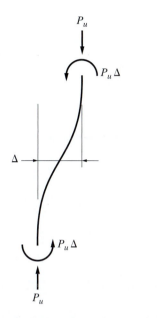

FIGURE 9-27   Column subjected to sidesway.

they are in addition to any primary (first-order) end moments that result from applied loads on the frame with no consideration of geometry change (sidesway).

Actual structures are rarely completely braced (nonsway) or completely unbraced (sway). Sidesway may be minimized in various ways. The common approach is to use walls or partitions sufficiently strong and rigid in their own planes to prevent the horizontal displacement. Another method is to use a rigid central core that is capable of resisting lateral loads and lateral displacements due to unsymmetrical loading conditions. For those cases when it is not readily apparent whether a structure is braced or unbraced, the ACI Code (Sections 10.10.1 and 10.10.5) provides analytical methods to aid in the decision.

For braced columns, slenderness effects may be neglected when

$$\frac{k\ell_u}{r} \leq 34 - 12\left(\frac{M_1}{M_2}\right) \qquad \text{[ACI Eq. (10-7)]}$$

where M_1 is the smaller end moment and M_2 is the larger end moment, both obtained by an elastic frame analysis. The ratio M_1/M_2 is positive if the column is bent in single curvature, is negative if bent in double curvature (see Figure 9-28), and the term $[34 - 12M_1/M_2]$ shall not be taken greater than 40. For columns in sway frames (not braced against sidesway), slenderness effects may be neglected when $k\ell_u/r$ is less than 22 (ACI 318-08, Section 10.10.1).

Fortunately, for ordinary beam and column sizes and typical story heights of concrete framing systems, effects of slenderness may be neglected in more than 90% of columns in braced (nonsway) frames and in about 40% of columns in unbraced (sway) frames [2].

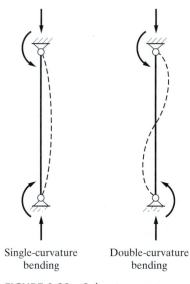

Single-curvature   Double-curvature
   bending   bending

FIGURE 9-28   Column curvature.

In cases where slenderness must be considered, the ACI Code gives the methods that can be used. These range from an approximate method (ACI 318-08, Section 10.10.5) in which moments are magnified to account for slenderness to a nonlinear second-order analysis. The approximate method uses a moment magnification factor that amplifies the factored moment computed from a conventional elastic analysis. In turn, the combination of the magnified factored moment and the factored axial load is used in the design of the compression member. Although the approximate analyses have been the traditional design methods of the past, more exact analyses have become possible and practical with the increased availability of sophisticated computer methods.

The design of slender reinforced concrete columns is one of the more complex aspects of reinforced concrete design and is not within the intended scope of this book. For the more rigorous theoretical background and applications relative to slender columns, the reader is referred to other comprehensive texts on reinforced concrete design that may offer a different emphasis and orientation.

REFERENCES

[1] *ACI Design Handbook*, "Design of Structural Reinforced Concrete Elements in Accordance with Strength Design Method of ACI 318-95," Publication SP-17(97). American Concrete Institute, P.O. Box 9094, Farmington Hills, MI 48333.

[2] "Notes on ACI 318-02, Building Code Requirements for Structural Concrete, with Design Applications." Portland Cement Association, 5420 Old Orchard Road, Skokie, IL 60077-1083, 2002.

PROBLEMS

9-1. Compute the maximum design axial load strength of the tied columns shown. Assume that the columns are short. Check the tie size and spacing. Use $f'_c = 4000$ psi and $f_y = 60,000$ psi.

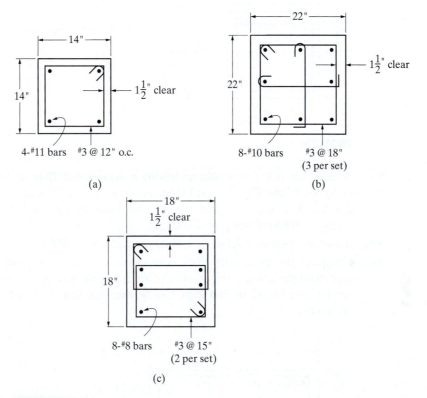

(a)

(b)

(c)

PROBLEM 9-1

9-2. Compute the maximum design axial load strength for the tied column shown. The column is short. Check the tie size and spacing. Use $f'_c = 4000$ psi and $f_y = 60,000$ psi.

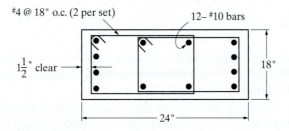

PROBLEM 9-2

9-3. Find the maximum axial compressive service loads that the column of cross section shown can carry. The column is short. Assume that the service dead load and live load are equal. Check the ties. Use f'_c = 4000 psi and f_y = 60,000 psi.

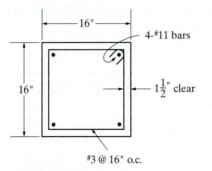

16"

4-#11 bars

16"

$1\frac{1}{2}$" clear

#3 @ 16" o.c.

PROBLEM 9-3

9-4. A short, circular spiral column having a diameter of 18 in. is reinforced with eight No. 9 bars. The cover is 1½ in., and the spiral is ⅜ in. in diameter spaced 2 in. o.c. Find the maximum design axial load strength and check the spiral. Use f'_c = 3000 psi and f_y = 40,000 psi.

9-5. Same as Problem 9-4, but f'_c = 4000 psi and f_y = 60,000 psi.

9-6. Compute the maximum axial compressive service live load that may be placed on the column shown. The column is short and is subjected to an axial service dead load of 200 kips. Check the ties. Use f'_c = 3000 psi and f_y = 40,000 psi.

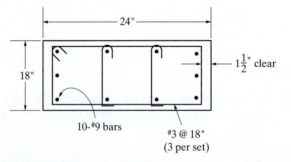

24"

$1\frac{1}{2}$" clear

18"

10-#9 bars

#3 @ 18"
(3 per set)

PROBLEM 9-6

9-7. Design a short, square tied column to carry a total factored design load P_u of 905 kips. Space and practical limitations require a column size of 18 in. × 18 in. Use f'_c = 4000 psi and f_y = 60,000 psi.

9-8. Design a short, square tied column for service loads of 205 kips dead load and 165 kips live load. Use ρ_g of about 0.04, f'_c = 3000 psi, and f_y = 60,000 psi. Assume that eccentricity is small.

9-9. Same as Problem 9-8, but use $f'_c = 4000$ psi, $f_y = 60,000$ psi, and a ρ_g of about 0.03.

9-10. Design a short, circular spiral column for service loads of 175 kips dead load and 325 kips live load. Assume that the eccentricity is small. Use $f'_c = 4000$ psi and $f_y = 60,000$ psi. Make ρ_g about 3%.

9-11. For the short column of cross section shown, find ϕP_n and e_b at the balanced condition using basic principles. Use $f'_c = 4000$ psi and $f_y = 60,000$ psi. Bending is about the strong axis.

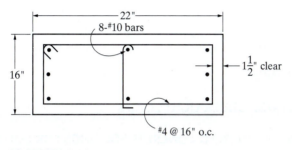

PROBLEM 9-11

9-12. Design a short square tied column to carry a factored axial design load P_u of 890 kips and a factored design moment M_u of 390 ft-kips. Place the longitudinal reinforcing uniformly in the four faces. Use $f'_c = 4000$ psi and $f_y = 60,000$ psi.

9-13. Same as Problem 9-12, but design a circular spiral column.

9-14. For the short tied column of cross section shown, find the design axial load strength ϕP_n for an eccentricity of 14 in. Use $f'_c = 4000$ psi and $f_y = 60,000$ psi. Assume that the ties and bracket design are adequate.

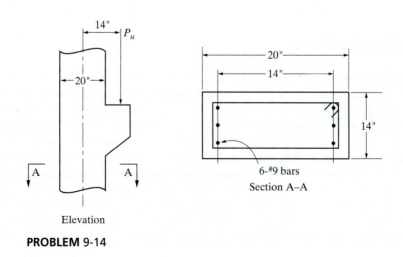

PROBLEM 9-14

Chapter **10**

Footings

10-1 INTRODUCTION

The purpose of the structural portion of every building is to transmit applied loads safely from one part of the structure to another. The loads pass from their point of application into the *superstructure*, then to the *foundation*, and then into the underlying supporting material. We have discussed the superstructure and foundation walls to some extent. The foundation is generally considered the entire lowermost supporting part of the structure. Normally, a *footing* is the last, or nearly the last, structural element of the foundation through which the loads pass. A footing has as its function the requirement of spreading out the superimposed load so as not to exceed the safe capacity of the underlying material, usually soil, to which it delivers the load. Additionally, the design of footings must take into account certain practical and, at times, legal considerations. Our discussion is

342

concerned only with spread footings depicted in Figure 10-1. Footings supported on piles will not be discussed.

The more common types of footings may be categorized as follows:

1.   *Individual column footings* (Figure 10-1a) are often termed *isolated spread footings* and are generally square. If space limitations exist, however, the footing may be rectangular in shape.

2.   *Wall footings* support walls that may be either bearing or nonbearing walls (Figure 10-1b).

3.   *Combined footings* support two or more columns and may be either rectangular or trapezoidal in shape (Figure 10-1c and d). If two isolated footings are joined by a *strap beam*, the footing is sometimes called a *cantilever footing* (Figure 10-1e).

4.   *Mat foundations* are large continuous footings that support all columns and walls of a structure. They are commonly used where undesirable soil conditions prevail (Figure 10-1f).

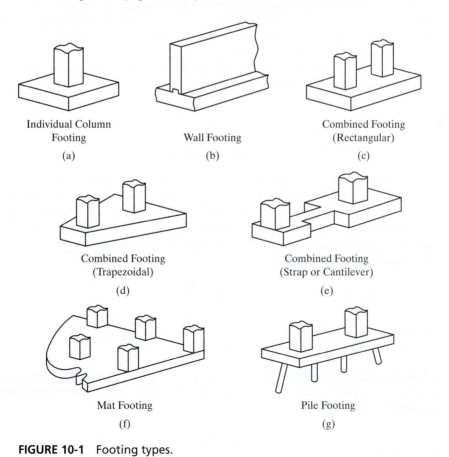

Individual Column
Footing
(a)

Wall Footing
(b)

Combined Footing
(Rectangular)
(c)

Combined Footing
(Trapezoidal)
(d)

Combined Footing
(Strap or Cantilever)
(e)

Mat Footing
(f)

Pile Footing
(g)

FIGURE 10-1   Footing types.

5.   *Pile caps or pile footings* serve to transmit column loads to a group of piles, which will in turn transmit the loads to the supporting soil through friction or to underlying rock in bearing (Figure 10-1g).

10-2   WALL FOOTINGS

Wall footings are commonly required to support direct concentric loads. An exception to this is the footing for a retaining wall. A wall footing may be of either plain or reinforced concrete. Because it has bending in only one direction, it is generally designed in much the same manner as a one-way slab, by considering a typical 12-in.-wide strip along the length of wall. Footings carrying relatively light loads on well-drained cohesionless soil are often made of plain concrete— that is, concrete without reinforcing. This material is referred to as *structural plain concrete*.

A wall footing under concentric load behaves similarly to a cantilever beam, where the cantilever extends out from the wall and is loaded in an upward direction by the soil pressure. The flexural tensile stresses induced in the bottom of the footing are acceptable for a plain concrete footing.

From the ACI Code, Section 22.5, the nominal flexural design strength of a plain concrete cross section is calculated from

$$M_n = 5\lambda \sqrt{f_c'} S_m$$

if tension controls and

$$M_n = 0.85\, f_c' S_m$$

if compression controls. The quantity S_m is the elastic section modulus of the section. Lamda (λ) is the modification factor reflecting the lower tensile strength of lightweight concrete relative to normal-weight concrete and is described in Chapter 1. For normal-weight concrete, $\lambda = 1.0$. In this chapter we will consider only normal-weight concrete; therefore, λ will be omitted in the examples. These formulas are based on the flexure formula,

$$f_b = \frac{Mc}{I}$$

rewritten

$$M = f_b S$$

where $5\lambda \sqrt{f_c'}$ and $0.85 f_c'$ are limiting stress f_b values in tension and compression, respectively. Similarly, nominal shear strength for beam action for a plain concrete member is calculated from

$$V_n = \frac{4}{3}\lambda \sqrt{f_c'} bh$$

PHOTO 10-1   Column and wall footings. The Health, Physical Education and Recreation Complex, Hudson Valley Community College, Troy, New York.

This is based on the general shear formula for a rectangular section:

$$f_v = \frac{3V}{2bh}$$

which is rewritten

$$V = \frac{2}{3}f_v bh$$

where the limiting shear stress value is $2\lambda\sqrt{f_c'}$, a familiar value from our previous discussions of shear in flexural members and $\lambda = 1.0$ for normal-weight concrete.

In each case, the basis for the design must be

$$\phi M_n \geq M_u \quad \text{and} \quad \phi V_n \geq V_u$$

as applicable. The strength-reduction factor for plain concrete is 0.60 (ACI Code Section 9.3.5).

In a *reinforced* concrete wall footing, the behavior is identical to that just described. Reinforcing steel is placed in the bottom of the footing in a direction perpendicular to the wall, however, thereby resisting the induced flexural tension, similar to a reinforced concrete beam or slab.

In either case, the cantilever action is based on the maximum bending moment occurring at the face of the wall if the footing supports a concrete wall or at a point halfway between the middle of the wall and the face of the wall if the footing supports a masonry wall. This difference is primarily because a masonry wall is somewhat less rigid than a concrete wall.

For each type of wall, the critical section for shear in the footing may be taken at a distance from the face of the wall equal to the effective depth of the footing.

Example 10-1_____

Design a plain normal-weight concrete wall footing to carry a 12-in. concrete block masonry wall, as shown in Figure 10-2. The service loading may be taken as 10 kips/ft dead load (which includes the weight of the wall) and 20 kips/ft live load. Use $f_c' = 3000$ psi. The gross allowable soil pressure is 5000 psf (5.0 ksf), and the weight of earth $w_e = 100$ lb/ft^3.

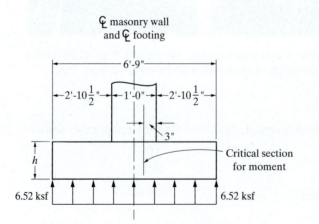

FIGURE 10-2 Plain concrete wall footing for Example 10-1.

Solution:

1. Compute the factored load:

$$w_u = 1.2w_{DL} + 1.6w_{LL}$$
$$= 1.2(10) + 1.6(20)$$
$$= 44 \text{ kips/ft}$$

2. Assume a footing thickness of 3 ft 0 in. This will be checked later in the design. The footing weight is

$$0.150(3) = 0.450 \text{ ksf}$$

Assuming the bottom of the footing to be 4 ft-0 in. below the finished ground line, the weight of the soil on top of the footing is

$$(1)(100) = 100 \text{ psf} = 0.100 \text{ ksf}$$

Therefore the net allowable soil pressure for superimposed service loads is

$$5.00 - 0.45 - 0.100 = 4.45 \text{ ksf}$$

Sometimes, the geotechnical engineering report for a project may specify the net allowable soil pressure directly, thus obviating the need for the calculations in this step.

3. The maximum allowable soil pressure for strength design must now be found. It must be modified in a manner consistent with the modification of the service loads. This may be accomplished by multiplying by the ratio of the total design load (44 kips) to the total service load (30 kips). We then obtain a maximum allowable soil pressure solely for use in the strength design of the footing. This soil pressure should not be construed as an actual allowable soil pressure. Thus

$$4.45\left(\frac{44}{30}\right) = 6.53 \text{ ksf}$$

4. We may now determine the required footing width:

$$\frac{44.0}{6.53} = 6.74 \text{ ft}$$

Use 6 ft-9 in.

5. Determine the factored soil pressure to be used for the footing design if the footing width is 6 ft-9 in.:

$$\frac{44.0}{6.75} = 6.52 \text{ ksf} < 6.53 \text{ ksf} \qquad \text{(O.K.)}$$

6. With the factored soil pressure known, the bending moment in the footing may be calculated. For concrete block masonry walls, the critical section for moment should be taken at the quarter-point of wall thickness (ACI Code, Section 15.4.2b).

With reference to Figure 10-2, the factored moment is determined as follows:

$$M_u = \frac{6.52(3.125)^2}{2} = 31.8 \text{ ft-kips}$$

7. Find the required footing thickness based on the required moment strength. Assuming that tension controls:

$$\phi M_n = \phi 5\sqrt{f_c'}S_m = \phi 5\sqrt{f_c'}\frac{bh^2}{6}$$

Setting $\phi M_n = M_u$ and considering a typical 12-in.-wide strip:

$$\phi 5\sqrt{f_c'}\frac{(12)(h^2)}{6} = 31.8 \text{ ft-kips}$$

from which

$$\text{required } h = \sqrt{\frac{(31.8 \text{ ft-kips})(12 \text{ in./ft})(1000 \text{ lb/kip})(6)}{0.60(5)\sqrt{3000}(12 \text{ in.})}} = 34.1 \text{ in.}$$

8. It is common practice to assume that the bottom 1 or 2 in. of concrete placed against the ground may be of poor quality and therefore may be neglected for strength purposes. The total required footing thickness may then be determined:

$$34.1 + 2.0 = 36.1 \text{ in.}$$

We will use $h = 38$ in. This checks closely with the assumed thickness of 36 in. No revision of the calculations is warranted.

9. Shear is generally of little significance in plain concrete footings because of the large concrete footing thickness. With reference to Figure 10-3, if the critical section for shear is considered a distance equal to the depth of the member h from the face of the wall and if the depth is taken as 34.1 in.

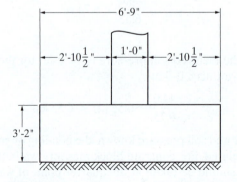

FIGURE 10-3   Plain concrete wall footing for Example 10-1.

(the required h), it may be observed that the critical section is outside the edge of the footing. Therefore the shear check may be neglected.

10.   It is common practice to use some *longitudinal* steel in continuous wall footings whether or not *transverse* steel is present. This will somewhat enhance the structural integrity by limiting differential movement between parts of the footing should transverse cracking occur. It will also lend some flexural strength in the longitudinal direction. The rationale for this stems from the many uncertainties that exist in both the supporting soil and the applied loads.

As a guide, this longitudinal steel may be computed in a manner similar to that for temperature and shrinkage steel in a one-way slab, where

$$\text{required } A_s = 0.0018bh \qquad (\text{using } f_y = 60{,}000 \text{ psi})$$

The total longitudinal steel required for the 6-ft-9-in. footing width is

$$A_s = 0.0018(6.75)(12 \text{ in./ft})(38) = 5.54 \text{ in.}^2$$

Use 13 No. 6 bars ($A_s = 5.72$ in.^{2}). This would satisfy any longitudinal steel requirement. It is obvious, however, that this large steel requirement does not lend itself to an economical footing design, but rather leads us to conclude that we should question the use of plain concrete wall footings when superimposed loads are heavy.

A redesign using a reinforced concrete wall footing is accomplished in the following problem.

Example 10-2_____

Design a normal-weight reinforced concrete wall footing to carry a 12-in. concrete block masonry wall, as shown in Figure 10-4. The service loading is 10 kips/ft dead load (which includes the weight of the wall) and 20 kips/ft live load. Use $f_c' = 3000$ psi, $f_y = 60{,}000$ psi, weight of earth $= 100$ lb/ft^3, and gross allowable soil pressure $= 5000$ psf (5.0 ksf). The bottom of the footing is to be 4 ft-0 in. below the finished ground line.

Solution:

1.   Compute the factored load:

$$w_u = 1.2w_{\text{DL}} + 1.6w_{\text{LL}}$$
$$= 1.2(10) + 1.6(20)$$
$$= 44 \text{ kips/ft}$$

2.   Assume a total footing thickness $= 18$ in. Therefore the footing weight is

$$0.150(1.5) = 0.225 \text{ ksf}$$

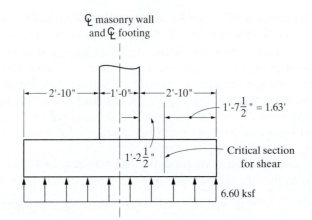

FIGURE 10-4   Reinforced concrete wall footing for Example 10-2.

Because the bottom of the footing is to be 4 ft below the finished ground line, there will be 30 in. of earth on top of the footing. This depth of earth has a weight of

$$\frac{30(100)}{12} = 250 \text{ psf} = 0.250 \text{ ksf}$$

The net allowable soil pressure for superimposed service loads is

$$5.00 - 0.225 - 0.250 = 4.53 \text{ ksf}$$

3. As in Example 10-1, we use the ratio of factored load to service load to determine the soil pressure for strength design:

$$\frac{44(4.53)}{30} = 6.64 \text{ ksf}$$

4. The required footing width is

$$\frac{44.0}{6.64} = 6.63 \text{ ft}$$

Use 6 ft-8 in.

5. The factored soil pressure to be used for the footing design is

$$\frac{44.0}{6.67} = 6.60 \text{ ksf}$$

6. The assumed effective depth d for the footing is determined by subtracting the concrete cover [see the ACI Code, Section 7.7.1(a)] and one-half of the bar diameter (No. 8 assumed) from the total thickness:

$$d = 18 - 3 - 0.5 = 14.5 \text{ in.}$$

7. Because required thicknesses of reinforced concrete footings are generally controlled by shear requirements, the shear should be checked first.

With reference to Figure 10-4, as the wall footing carries shear in a manner similar to that of a one-way slab or beam, the critical section for shear will be taken at a distance equal to the effective depth of the footing (14.5 in.) from the face of the wall (ACI Code, Section 11.1.3):

$$V_u = 1.63(1)(6.60) = 10.75 \text{ kips/ft of wall}$$

The total nominal shear strength V_n is the sum of the shear strength of the concrete V_c and the shear strength of any shear reinforcing V_s.

$$V_n = V_c + V_s$$

$$\phi V_n = \phi V_c + \phi V_s$$

Assuming no shear reinforcing,

$$\phi V_n = \phi V_c$$

In footings, shear reinforcing is not required if

$$\phi V_c > V_u$$

Computing ϕV_c,

$$\phi V_c = \phi 2\sqrt{f_c'}bd$$

$$= 0.75(2)\sqrt{3000}(12)(14.5)$$

$$= 14.30 \text{ kips/ft of wall}$$

$$14.30 \text{ kips} > 10.75 \text{ kips}$$

Therefore

$$\phi V_c > V_u$$

Thus the assumed thickness of footing is satisfactory for shear, and no revisions are necessary with respect to footing weight.

8. With reference to Figure 10-5, the critical section for moment is taken at the quarter-point of wall thickness (ACI Code, Section 15.4.2b). The maximum factored moment, assuming the footing to be a cantilever beam, is

$$M_u = \frac{6.60(3.08)^2}{2} = 31.3 \text{ ft-kips}$$

9. The required area of tension steel is then determined in the normal way, using $d = 14.5$ in. and $b = 12$ in., and assuming $\phi = 0.90$:

$$\text{required } \overline{k} = \frac{M_u}{\phi bd^2} = \frac{31.3(12)}{0.9(12)(14.5)^2}$$

$$= 0.1645 \text{ ksi}$$

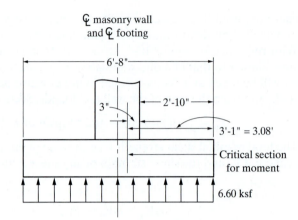

FIGURE 10-5   Reinforced concrete wall footing for Example 10-2.

From Table A-8, the required $\rho = 0.0029$, and $\epsilon_t > 0.005$; therefore $\phi = 0.90$ (O.K.).

$$\text{required } A_s = \rho bd$$

$$= 0.0029(12)(14.5)$$

$$= 0.50 \text{ in.}^2/\text{ft of wall}$$

We will use the ACI Code minimum reinforcement requirement for beams as being applicable for footings. From Table A-5,

$$A_{s,\text{min}} = 0.0033(12)(14.5)$$

$$= 0.57 \text{ in.}^2/\text{ft of wall}$$

Finally, as discussed in Chapter 8, the provisions of the ACI Code, Section 10.5.4, for minimum reinforcement in structural slabs of uniform thickness may be considered applicable for footings such as this that transmit vertical loads to the underlying soil. In this text, this will be used only as an absolute minimum. Checking the provision of the code for the minimum reinforcement required for grade 60 steel gives us

$$\text{required } A_s = 0.0018bh = 0.0018(12)(18)$$

$$= 0.39 \text{ in.}^2/\text{ft of wall}$$

Therefore, considering the minimum reinforcement ratio for beams as applicable, use required $A_s = 0.57 \text{ in.}^2$ per ft of wall and use No. 6 bars at 9 in. o.c. ($A_s = 0.59 \text{ in.}^2$).

10.   The development length should be checked for the bars selected. Assume uncoated bars. This calculation for ℓ_d follows the eight-step procedure presented in Chapter 5, Section 5-2, and is summarized as follows:

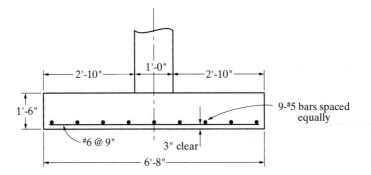

FIGURE 10-6   Design sketch for Example 10-2.

1. $K_D = 82.2$
2. $\psi_t = 1.0, \psi_e = 1.0, \psi_s = 0.8, \lambda = 1.0$
3. $\psi_t \times \psi_e = 1.0$ (O.K.)
4. $c_b = 3.38$ in.
5. $K_{tr} = 0$
6. $(c_b + K_{tr})/d_b = 4.51$ in. Use 2.5 in.
7. $K_{ER} = 0.97$
8. $\ell_d = 19.1$ in. > 12 in. (O.K.)

The development length provided, measured from the critical section for moment and allowing for 3-in. end cover, is 34 in. Because 34 in. > 19.1 in., the development length provided is adequate.

11. Although not specifically required in footings by the ACI Code, longitudinal steel will be provided on the same basis as for one-way slabs (Section 7.12). Thus

$$\text{required } A_s = 0.0018bh$$

$$= 0.0018(6.67 \text{ ft})(12 \text{ in./ft})(18 \text{ in.}) = 2.59 \text{ in.}^2$$

Use nine No. 5 bars ($A_s = 2.79$ in.^{2}) spaced equally. The footing design is shown in Figure 10-6.

10-3   WALL FOOTINGS UNDER LIGHT LOADS

A relatively common situation is one in which a lightly loaded wall is supported on average soil. As previously indicated in this chapter, a design would result in a very small footing thickness and width.

In such a situation, experience has shown that for footings carrying plain concrete or block masonry walls, the minimum recommended dimensions shown in

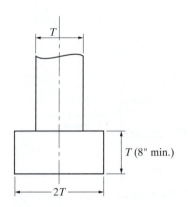

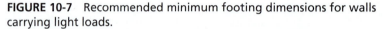

FIGURE 10-7 Recommended minimum footing dimensions for walls carrying light loads.

Figure 10-7 should be used. The minimum depth or thickness of footing should be 8 in. but not less than the wall thickness. The minimum width of footing should equal twice the wall thickness.

10-4 INDIVIDUAL REINFORCED CONCRETE FOOTINGS FOR COLUMNS

An individual reinforced concrete footing for a column, also termed an *isolated spread footing*, is probably the most common, simplest, and most economical of the various types of footings used for structures. Individual column footings are generally square in plan. Rectangular shapes are sometimes used where dimensional limitations exist, however. The footing is a slab that directly supports a column. At times, a pedestal is placed between a column and a footing so that the base of the column need not be set below grade.

The footing behavior under concentric load is that of two-way cantilever action extending out from the column or pedestal. The footing is loaded in an upward direction by the soil pressure. Tensile stresses are induced in each direction in the bottom of the footing. Therefore the footing is reinforced by two layers of steel perpendicular to each other and parallel to the edges. The required footing–soil contact area is a function of, and determined by, the allowable soil bearing pressure and the column loads being applied to the footing.

Shear

Because the footing is subject to two-way action, two different types of shear strength must be considered: two-way shear and one-way shear. The footing thickness (depth) is generally established by the shear requirements. The two-way shear is commonly termed *punching shear*, because the column or pedestal tends to punch through the footing, inducing stresses around the perimeter of the column or pedestal. Tests have verified that, if failure occurs, the fracture takes the form of a truncated pyramid with sides sloping away from the face of the column or pedestal. The critical section for this two-way shear is taken perpendicular to the plane of the footing and located so that its perimeter, b_0, is a minimum but does not come closer to the edge of the column or pedestal than one-half the effective depth of the footing (ACI Code, Section 11.11.1.2).

The design of the footing for two-way action is based on a shear strength V_n, which is not to be taken greater than V_c unless shear reinforcement is provided. V_c may be determined from the ACI Code, Section 11.11.2, and shall be the smallest of

a.
$$V_c = \left(2 + \frac{4}{\beta_c}\right)\lambda \sqrt{f_c'}\, b_0 d \qquad \text{[ACI Eq. (11-31)]}$$

b.
$$V_c = \left(\frac{\alpha_s d}{b_0} + 2\right)\lambda \sqrt{f_c'}\, b_0 d \qquad \text{[ACI Eq. (11-32)]}$$

c.
$$V_c = 4\lambda \sqrt{f_c'}\, b_0 d \qquad \text{[ACI Eq. (11-33)]}$$

where

β_c = ratio of the long side to the short side of the concentrated load or reaction area (loaded area)

b_0 = perimeter of critical section for two-way shear action in the footing

α_s = 40 for interior columns, 30 for edge columns, and 20 for corner columns

and V_c, f_c', λ, and d are as previously defined. Note that the terms *interior*, *edge*, and *corner* columns in ACI 11.11.2.1(b) refer to the location of the column relative to the edges of the spread footing. Therefore, *interior*, *edge*, and *corner columns* will have four-, three-, and two-sided critical sections, respectively.

The introduction of shear reinforcement in footings is impractical and undesirable purely on an economic basis. It is general practice to design footings based solely on the shear strength of the concrete.

The one-way (or beam) shear may be compared with the shear in a beam or one-way slab. The critical section for this one-way shear is taken on a vertical plane extending across the entire width of the footing and located at a distance equal to the effective depth of the footing from the face of the concentrated load or reaction area (ACI Code, Section 11.1). As in a beam or one-way slab, the shear strength provided by the footing concrete may be taken as

$$V_c = 2\lambda \sqrt{f_c'}\, b_w d \qquad \text{[ACI Eq. (11-3)]}$$

For both one- and two-way action, if we assume no shear reinforcement, the basis for the shear design will be $\phi V_n > V_u$, where $V_n = V_c$.

Moment and Development of Bars

The size and spacing of the footing reinforcing steel is primarily a function of the bending moment induced by the net upward soil pressure. The footing behaves as a cantilever beam in two directions. It is loaded by the soil pressure. The fixed end, or critical section for the bending moment, is located as follows (ACI Code, Section 15.4.2):

1. At the face of the column or pedestal, for a footing supporting a concrete column or pedestal (see Figure 10-8a).
2. Halfway between the face of the column and the edge of a steel base plate, for a footing supporting a column with a steel base plate (see Figure 10-8b).

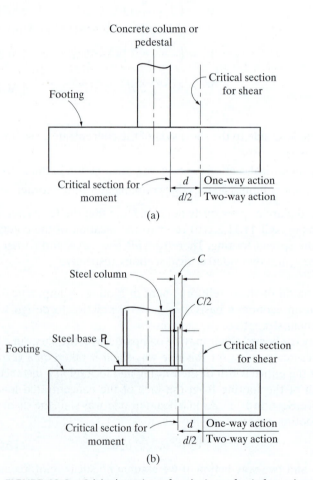

(a)

(b)

FIGURE 10-8 Critical sections for design of reinforced concrete footings supporting columns or pedestals.

The ACI Code, Section 15.6.3, stipulates that the critical section for development length of footing reinforcement shall be assumed to be at the same location as the critical section for bending moment.

Transfer of Load from Column into Footing

All loads applied to a column must be transferred to the top of the footing (through a pedestal, if there is one) by compression in the concrete, by reinforcement, or by both.

The bearing strength of the concrete contact area of supporting and supported member cannot exceed $\phi(0.85f_c'A_1)$ as directed by the ACI Code, Section 10.14.1. When the supporting surface is wider on all sides than is the loaded area, the design-bearing strength on the loaded area may be multiplied by $\sqrt{A_2/A_1} \leq 2.0$, as discussed in Chapter 8, Section 8-4. Therefore in no case can the design-bearing strength for the loaded area be in excess of

$$\phi(0.85f_c'A_1)(2)$$

where $\phi = 0.65$ for bearing on concrete and f_c' is as previously defined.

It is common for the footing concrete to be of a lower strength (f_c') than the supported column concrete. This suggests that both supporting and supported members should be considered in determining load transfer.

Where a reinforced concrete column cannot transfer the load entirely by bearing, the excess load must be transferred by reinforcement where the required $A_s = $ (excess load)$/f_y$. This may be accomplished by furnishing dowels, one per column bar if necessary but not larger than No. 11 (ACI Code, Section 15.8.2.3).

To provide a positive connection between a reinforced concrete column and footing (whether dowels are required or not), the ACI Code, Section 15.8.2.1, requires a minimum area of reinforcement crossing the bearing surface of 0.005 of the column cross-sectional area. It is generally recommended that a minimum of four bars be used. These four bars should preferably be dowels for the four corner bars of a square column.

The development length of the dowels must be sufficient on both sides of the bearing surface to provide the necessary development length for bars in compression (see Chapter 5).

When the dowel carries excess load into the footing, it must be spliced to the column bar using the necessary compression splice. The same procedure applies where a column rests on a pedestal and where a pedestal rests on a footing.

Where structural steel columns and column base plates are used, the total load is usually transferred entirely by bearing on the concrete contact area. The design-bearing strength as stipulated previously also applies in this case. Where a column base detail is inadequate to transfer the total load, adjustments may be made as follows:

1. Increase column base plate dimensions.
2. Use higher-strength concrete (f_c') for the pedestal or footing.
3. Increase the supporting area with respect to the base plate area until the ratio reaches the maximum allowed by the ACI Code.

In building design, it is common practice to use a concrete pedestal between the footing and the column. The pedestal, in effect, distributes the column load over a larger area of the footing, thereby contributing to a more economical footing design. Pedestals may be either plain or reinforced. If the ratio of height to least lateral dimension is in excess of 3, the member is by definition a column and must be designed and reinforced as a column (see Chapter 9). If the ratio is less than 3, it is categorized as a pedestal and theoretically may not require any reinforcement.

The cross-sectional area of a pedestal is usually established by the concrete-bearing strength as stipulated in ACI Code, Section 10.14, by the size of a steel column base plate, or by the desire to distribute the column load over a larger footing area. It is common practice to design a pedestal in a manner similar to a column using a minimum of four corner bars (for a square or rectangular cross section) anchored into the footing and extending up through the pedestal. Ties should be provided in pedestals according to the same requirements as in columns.

10-5   SQUARE REINFORCED CONCRETE FOOTINGS

In isolated square footings, the reinforcement should be uniformly distributed over the width of the footing in each direction. Because the bending moment is the same in each direction, the reinforcing bar size and spacing should be the same in each direction. In reality, the effective depth is not the same in both directions. It is common practice to use the same average effective depth for design computations for both directions, however.

It is also common practice to assume that the minimum tensile reinforcement for beams is applicable to two-way footings for each of the two directions, unless the reinforcement provided is one-third greater than required. As discussed in Section 2-8 (and in the ACI Code, Section 10.5.1), the minimum tensile reinforcement is determined from

$$A_{s,\min} = \frac{3\sqrt{f_c'}}{f_y}b_w d \geq \frac{200}{f_y}b_w d$$

The use of this minimum is conservative for footings. The ACI Code, Section 10.5.4, permits the use of a minimum reinforcement equal to that required for shrinkage and temperature steel in structural slabs of uniform thickness. This will always be somewhat less than that required by $A_{s,\min}$, but not necessarily less than that specified by the ACI Code, Section 10.5.3. In this book the criteria of the ACI Code, Section 10.5.4, will be used in isolated footing cases only as an absolute minimum of steel area to be provided.

Example 10-3_____

Design a square reinforced concrete footing to support an 18-in.-square tied concrete column, as shown in Figure 10-9. The column is a typical interior column in a building. Assume normal-weight concrete.

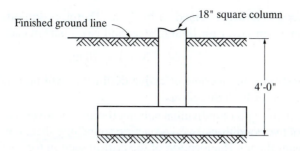

FIGURE 10-9   Sketch for Example 10.3.

Solution:

1. The design data are as follows: service dead load = 225 kips, service live load = 175 kips, allowable soil pressure = 5000 psf (5.00 ksf), f'_c for the column = 4000 psi and for the footing = 3000 psi, f_y for all steel = 60,000 psi, and longitudinal column steel consists of No. 8 bars. The weight of earth = 100 lb/ft³.

2. Assume a total footing thickness of 24 in. The footing weight may then be calculated as

    $$0.150(2.0) = 0.300 \text{ ksf}$$

    Because the bottom of the footing is to be 4 ft below the finished ground line, there will be 24 in. of earth on top of the footing. This depth of earth has a weight of

    $$^{24}\!/_{12}(0.100) = 0.200 \text{ ksf}$$

    Therefore, the *net allowable* soil pressure for the superimposed loads becomes

    $$5.00 - 0.300 - 0.200 = 4.50 \text{ ksf}$$

    The required area of footing may be determined using service loads and allowable soil pressure or by modifying both the service loads and the allowable soil pressure with the ACI load factors. Using the service loads,

    $$\text{required } A = \frac{225 + 175}{4.50} = 88.9 \text{ ft}^2$$

    Use a 9-ft 6-in. square footing. This furnishes an actual area A of 90.3 ft².

3. The factored soil pressure from superimposed loads to be used for the footing design may now be calculated:

    $$p_u = \frac{P_u}{A} = \frac{1.2(225) + 1.6(175)}{90.3} = 6.09 \text{ ksf}$$

4. The footing thickness is usually determined by shear strength requirements. Therefore, using the assumed footing thickness, check the shear strength.

The thickness h was assumed to be 24 in. Therefore the effective depth, based on a 3-in. cover for bottom steel and No. 8 bars in each direction, is

$$d = 24 - 3 - 1 = 20 \text{ in.}$$

This constitutes an average effective depth that will be used for design calculations for both directions.

5.   The shear strength of individual column footings is governed by the more severe of two conditions: two-way action (punching shear) or one-way action (beam shear). The location of the critical section for each type of behavior is depicted in Figure 10-10. *For two-way action* (Figure 10-10a),

$$B = \text{column width} + \left(\frac{d}{2}\right)2$$

$$= 18 + 20 = 38 \text{ in.} = 3.17 \text{ ft}$$

The total factored shear acting on the critical section is

$$V_u = p_u(W^2 - B^2)$$
$$= 6.09(9.5^2 - 3.17^2)$$
$$= 488 \text{ kips}$$

The shear strength of the concrete is taken as the smallest of

a.

$$V_c = \left(2 + \frac{4}{\beta_c}\right)\sqrt{f'_c}b_0d$$

$$= \left(2 + \frac{4}{1}\right)\sqrt{3000}\,(38)(4)(20) = 999{,}000 \text{ lb}$$

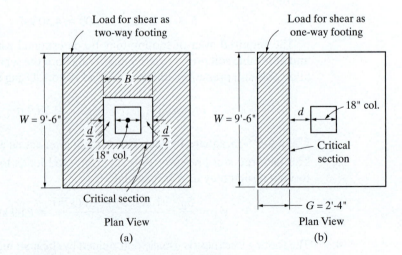

FIGURE 10-10   Footing shear analysis.

b.   With $\alpha_s = 40$ for an interior column,

$$V_c = \left(\frac{\alpha_s d}{b_0} + 2\right)\sqrt{f_c'}\,b_0 d$$

$$= \left[\frac{40(20)}{38(4)} + 2\right]\sqrt{3000}\,(38)(4)(20) = 1{,}209{,}000 \text{ lb}$$

c. $\qquad V_c = 4\sqrt{f_c'}\,b_0 d = 4\sqrt{3000}\,(4)(38)(20) = 666{,}000 \text{ lb}$

from which $V_c = 666{,}000$ lb $= 666$ kips. Thus

$$\phi V_n = \phi V_c = 0.75(666) = 500 \text{ kips}$$

Therefore

$$\phi V_n > V_u \qquad\qquad\qquad\text{(O.K.)}$$

When V_u is relatively close to the shear strength of the concrete (ϕV_c), it indicates that the assumed footing thickness is approximately equal to that required for shear. If these two values were significantly different, the assumed footing thickness should be modified.

For one-way action, the total factored shear acting on the critical section is distance d from the face of the column:

$$V_u = p_u W G$$

$$= 6.09(9.5)(2.33)$$

$$= 134.8 \text{ kips}$$

The shear strength of the concrete is

$$V_c = 2\sqrt{f_c'}\,b_w d$$

$$= 2\sqrt{3000}\,(9.5)(12)(20)$$

$$= 250{,}000 \text{ lb}$$

$$= 250 \text{ kips}$$

$$\phi V_n = \phi V_c = 0.75(250) = 187.5 \text{ kips}$$

Therefore

$$\phi V_n > V_u \qquad\qquad\qquad\text{(O.K.)}$$

The 24-in.-deep footing is satisfactory with respect to shear. Our assumption of step 2 with regard to weight of the footing and the soil on the footing is satisfactory.

6.   The critical section for bending moment may be taken at the face of the column, as depicted in Figure 10-11. Using the factored soil pressure and

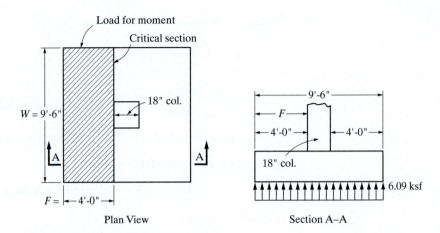

FIGURE 10-11 Footing moment analysis.

assuming the footing to act as a wide cantilever beam in both directions, the design moment may be computed:

$$M_u = p_u F \left(\frac{F}{2} \right) (W)$$

$$= 6.09(4) \left(\frac{4}{2} \right) (9.5)$$

$$= 463 \text{ ft-kips}$$

7. Assume $\phi = 0.90$ and design the tension steel as follows:

$$\text{required } \bar{k} = \frac{M_u}{\phi b d^2} = \frac{463(12)}{0.9(9.5)(12)(20)^2}$$

$$= 0.1354 \text{ ksi}$$

From Table A-8, the required $\rho = 0.0024$, $\epsilon_t > 0.005$, and $\phi = 0.90$. Therefore

$$\text{required } A_s = \rho b d$$

$$= 0.0024(9.5)(12)(20)$$

$$= 5.47 \text{ in.}^2$$

Check the ACI Code minimum reinforcement requirement. From Table A-5,

$$A_{s,\text{min}} = 0.0033(9.5)(12)(20)$$

$$= 7.52 \text{ in.}^2$$

Of the two steel areas, the larger (7.52 in.^{2}) controls.

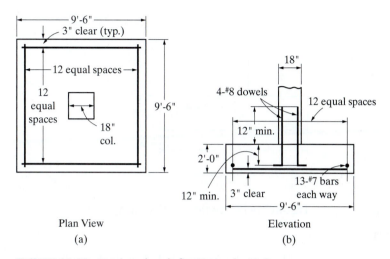

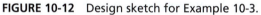

FIGURE 10-12   Design sketch for Example 10-3.

Because the footing is square and an average effective depth was used, the steel requirements in the other direction may be assumed to be identical. Therefore use 13 No. 7 bars each way ($A_s = 7.80$ in.^{2} in each direction) and distribute the bars uniformly across the footing in each direction, as shown in Figure 10-12a.

Check the development length for the No. 7 bars.

This calculation for ℓ_d follows the eight-step procedure presented in Chapter 5, Section 5-2, and is summarized as follows:

1. $K_D = 82.2$
2. $\psi_t = 1.0, \psi_e = 1.0, \psi_s = 1.0, \lambda = 1.0$
3. $\psi_t \times \psi_e = 1.0$ (O.K.)
4. $c_b = 3.44$ in.
5. $K_{tr} = 0$
6. $(c_b + K_{tr})/d_b = 3.93$ in. Use 2.5 in.
7. $K_{ER} = 0.96$
8. $\ell_d = 27.6$ in. > 12 in. (O.K.)

Use $\ell_d = 27.6$ in. (minimum). The development length provided is 45 in., which is in excess of that required (O.K.).

8. The concrete bearing strength at the base of the column cannot exceed

$$\phi(0.85f_c'A_1)$$

except where the supporting surface is wider on all sides than the loaded area, for which case the concrete bearing strength of the supporting surface cannot exceed

$$\phi(0.85f_c'A_1)\sqrt{\frac{A_2}{A_1}}$$

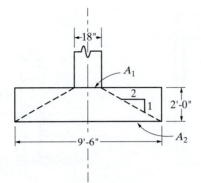

FIGURE 10-13   Determination of A_2.

As described in Section 8-4, A_2 is the lower base of the frustum of a pyramid having 1:2 sloping sides and fully contained within the support as shown in Figure 10-13. In this case, A_2 is the same as the area of the footing and

$$\sqrt{\frac{A_2}{A_1}} = \sqrt{\frac{90.3}{2.25}} = 6.3 > 2.0$$

Therefore use 2.0. Then

$$\textit{footing} \text{ bearing strength} = \phi(0.85f_c'A_1)(2.0)$$
$$= 0.65(0.85)(3.0)(18)^2(2.0)$$
$$= 1074 \text{ kips}$$

The *column*-bearing strength is computed as follows:

$$\phi(0.85)f_c'A_1 = 0.65(0.85)(4.0)(18)^2 = 716 \text{ kips}$$

The *calculated* design-bearing load is

$$P_u = 1.2(225) + 1.6(175)$$
$$= 550 \text{ kips}$$

Because $550 < 716 < 1074$, the entire column load can be transferred by concrete alone. The ACI Code, however, requires a minimum dowel area of

$$\text{required } A_s = 0.005A_g$$
$$= 0.005(18)^2 = 1.62 \text{ in.}^2$$

Use a minimum of four bars. Four No. 6 bars, $A_s = 1.76$ in.^{2}, is satisfactory. It is general practice in a situation such as this, however, to use dowels of the same diameter as the column steel. Therefore use four No. 8 dowels, and place one in each corner ($A_s = 3.16$ in.^{2}).

The development length for dowels into the column and footing must be adequate even though full load transfer can be made without dowels. The bars are in compression. The compression development length for the No. 8 dowels into the footings is ℓ_{dc} from Table A-12 and may be reduced by any applicable modification factors. We will use the modification factor for the case where the steel provided is in excess of the steel required:

$$\frac{\text{required } A_s}{\text{provided } A_s} = \frac{1.62}{3.16} = 0.51$$

Therefore,

$$\text{required } \ell_{dc} = 21.9(0.51) = 11.2 \text{ in.}$$

For bars in compression, ℓ_{dc} must not be less than 8 in. Therefore, use ℓ_{dc} of 12 in. into both the column and the footing, because f'_c for the column concrete is higher and the required ℓ_{dc} for the bars into the column would be less than that into the footing. The actual anchorage used may be observed in Figure 10-12b. The dowels should be placed adjacent to the corner longitudinal bars. Generally, these dowels are furnished with a 90° hook at their lower ends and are placed on top of the main footing reinforcement. This will tie the dowel in place and will reduce the possibility of the dowel being dislodged during construction. The hook cannot be considered effective as part of the required development length (ACI Code, Section 12.5.5).

10-6   RECTANGULAR REINFORCED CONCRETE FOOTINGS

Rectangular footings are generally used where space limitations require it. The design of these footings is very similar to that of the square column footing with the one major exception that each direction must be investigated independently. Shear is checked for two-way action in the normal way, but for one-way action, it is checked across the shorter side only. The bending moment must be considered separately for each direction. Each direction will generally require a different area of steel. The reinforcing steel running in the long direction should be placed below the short-direction steel so that it may have the larger effective depth to carry the larger bending moments in that direction.

In rectangular footings, the *distribution* of the reinforcement is different than for square footings (ACI Code, Section 15.4.4). The reinforcement in the long direction should be uniformly distributed over the shorter footing width. A part of the required reinforcement in the short direction is placed in a band equal to the length of the short side of the footing. The portion of the total required steel that should go into this band is

$$\frac{2}{\beta + 1}$$

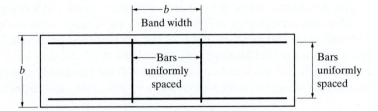

FIGURE 10-14   Rectangular footing plan.

where β is the ratio of the long side to the short side of the footing. The remainder of the reinforcement is uniformly distributed in the outer portions of the footing. This distribution is depicted in Figure 10-14. Other features of the design are similar to those for the square column footing.

Example 10-4_____

Design a reinforced concrete footing to support an 18-in.-square tied interior concrete column, as shown in Figure 10-15. One dimension of the footing is limited to a maximum of 7 ft. Assume normal-weight concrete.

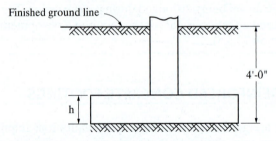

FIGURE 10-15   Sketch for Example 10-4.

Solution:

1. The design data are as follows: service dead load = 175 kips, service live load = 175 kips, allowable soil pressure = 5000 psf (5.00 ksf), f_c' for both footing and column = 3000 psi, f_y for all steel = 60,000 psi, and longitudinal column steel consists of No. 8 bars. The weight of earth = 100 lb/ft^3.

2. Assume a total footing thickness h of 24 in. subject to later check. The footing weight may then be calculated as

    $$0.150(2.0) = 0.300 \text{ ksf}$$

    Because the bottom of the footing is to be 4 ft below the finished ground line, there will be 24 in. of earth on top of the footing. This depth of earth has a weight of

    $$^{24}\!/_{12}(0.100)50.200 \text{ ksf}$$

Therefore the *net allowable* soil pressure for the superimposed loads becomes

$$5.00 - 0.300 - 0.200 = 4.50 \text{ ksf}$$

Based on service loads, the required area of footing may be calculated:

$$\text{required } A = \frac{175 + 175}{4.50} = 77.8 \text{ ft}^2$$

Use a rectangular footing 7 ft-0 in. by 11 ft-6 in. This furnishes an actual area A of 80.5 ft^2.

3. The factored soil pressure from superimposed loads to be used for the footing design may now be calculated:

$$p_u = \frac{P_u}{A} = \frac{1.2(175) + 1.6(175)}{80.5} = 6.09 \text{ ksf}$$

4. The footing thickness h was assumed to be 24 in. Therefore the effective depth, based on a 3-in. cover for bottom steel and 1-in.-diameter bars in each direction, will be

$$d = 24 - 3 - 1 = 20 \text{ in.}$$

This constitutes an *average* effective depth, which will be used for design calculations for both directions.

5. Checking the shear strength *for two-way action* (with reference to Figure 10-16a),

$$B = \text{column width} + \left(\frac{d}{2}\right)2$$

$$= 18 + 20 = 38 \text{ in.} = 3.17 \text{ ft}$$

The total factored shear acting on the critical section is

$$V_u = p_u(A - B^2)$$

$$= 6.09(80.5 - 3.17^2)$$

$$= 429 \text{ kips}$$

The shear strength of the concrete is taken as the smallest of

a. $$V_c = \left(2 + \frac{4}{\beta_c}\right)\sqrt{f_c'}\, b_o d$$

$$= \left(2 + \frac{4}{1}\right)\sqrt{3000}\,(38)(4)(20) = 999{,}000 \text{ lb}$$

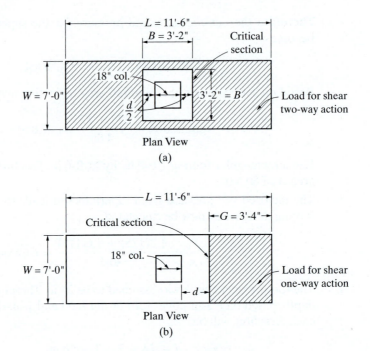

FIGURE 10-16   Footing shear analysis.

b.   With $\alpha_s = 40$ for an interior column,

$$V_c = \left(\frac{\alpha_s d}{b_0} + 2\right)\sqrt{f_c'}\, b_0 d$$

$$= \left(\frac{40(20)}{38(4)} + 2\right)\sqrt{3000}\,(38)(4)(20) = 1{,}209{,}000 \text{ lb}$$

c. $V_c = 4\sqrt{f_c'}\, b_0 d = 4\sqrt{3000}\,(38)(4)(20) = 666{,}000 \text{ lb}$

from which $V_c = 666{,}000 \text{ lb} = 666 \text{ kips}$. Thus

$$\phi V_n = \phi V_c = 0.75(666) = 500 \text{ kips}$$

Therefore

$$\phi V_n > V_u \qquad\qquad\qquad\qquad\qquad \text{(O.K.)}$$

For one-way action, consider shear across the short side only. The critical section is at a distance equal to the effective depth of the member from the face of the column (see Figure 10-16b).

The total factored shear acting on the critical section is

$$V_u = p_u W G$$

$$= 6.09(7.0)(3.33)$$

$$= 142.0 \text{ kips}$$

The nominal shear strength of the concrete is

$$V_c = 2\sqrt{f'_c}\, b_w d$$
$$= 2\sqrt{3000}\,(7.0)(12)(20)$$
$$= 184,000 \text{ lb}$$
$$= 184 \text{ kips}$$

from which

$$\phi V_n = \phi V_c = 0.75(184) = 138.0 \text{ kips}$$
$$\phi V_n < V_u$$

This is unsatisfactory. We will determine the required d based on one-way shear using $\phi V_c = V_u$.

$$\phi V_n = \phi 2\sqrt{f'_c}\, b_w d = V_u = 142.0 \text{ kips}$$

from which

$$\text{Required } d = \frac{V_u}{\phi 2\sqrt{f'_c}\, b_w} = \frac{142.0(1000 \text{ lb/kip})}{0.75(2)\sqrt{3000}\,(7.0)(12)} = 20.6 \text{ in.}$$

Therefore, the required $h = 20.6 + 3 + 1 = 24.6$ in. Use $h = 25$ in. The effect of this change on the previous calculations is small and no other revisions are considered warranted.

$$\text{new } d = 25 - 1 - 3 = 21.0 \text{ in.}$$

6. For bending moment, each direction must be considered independently, with the critical section taken at the face of the column. Using the actual factored soil pressure and assuming the footing to act as a wide cantilever beam in each direction, the design moment may be calculated. With reference to Figure 10-17b, for moment in the long direction,

$$M_u = p_u F\left(\frac{F}{2}\right)(W)$$
$$= 6.09(5)\left(\frac{5}{2}\right)(7.0)$$
$$= 533 \text{ ft-kips}$$

For moment in the short direction (Figure 10-17a),

$$M_u = p_u F\left(\frac{F}{2}\right)(L)$$
$$= 6.09(2.75)\left(\frac{2.75}{2}\right)(11.5)$$
$$= 265 \text{ ft-kips}$$

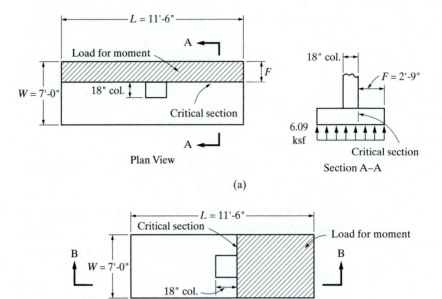

FIGURE 10-17   Footing moment analysis.

7.   Assume $\phi = 0.90$ and design the tension steel as follows. *For the long direction*, where $M_u = 533$ ft-kips,

$$\text{required } \bar{k} = \frac{M_u}{\phi b d^2} = \frac{533(12)}{(0.9)(7)(12)(21)^2}$$

$$= 0.1918 \text{ ksi}$$

From Table A-8, the required $\rho = 0.0034$, $\epsilon_t > 0.005$, and $\phi = 0.90$. Therefore

$$\text{required } A_s = \rho b d$$

$$= 0.0034(7)(12)(21)$$

$$= 6.00 \text{ in.}^2$$

Check the ACI Code minimum reinforcement requirement. From Table A-5,

$$A_{s,min} = 0.0033(7)(12)(21)$$

$$= 5.82 \text{ in.}^2$$

Of the two steel areas, the larger (6.00 in.^{2}) controls; therefore use 10 No. 7 bars ($A_s = 6.00$ in.^{2}). These bars will run in the long direction and will be distributed uniformly across the 7 ft-0 in. width. They will be placed in the bottom layer where they will have the advantage of slightly greater effective depth. The development length must be checked for these bars. Assume uncoated bars. This calculation for ℓ_d follows the eight-step procedure presented in Chapter 5, Section 5-2, and is summarized as follows:

1. $K_D = 82.2$
2. $\psi_t = 1.0, \psi_e = 1.0, \psi_s = 1.0, \lambda = 1.0$
3. $\psi_t \times \psi_e = 1.0$ (O.K.)
4. $c_b = 3.44$ in.
5. $K_{tr} = 0$
6. $(c_b + K_{tr})/d_b = 3.93$ Use 2.5
7. $K_{ER} = 1.00$
8. $\ell_d = 28.8$ in. > 12 in. (O.K.)

The development length furnished is $60 - 3 = 57$ in., which is in excess of that required (O.K.).

For the short direction, where $M_u = 265$ ft-kips,

$$\text{required } \bar{k} = \frac{M_u}{\phi b d^2} = \frac{265(12)}{0.9(11.5)(12)(21)^2}$$

$$= 0.0581 \text{ ksi}$$

From Table A-8, the required $\rho = 0.0010$, $\epsilon_t < 0.005$, and $\phi = 0.90$. Therefore

$$\text{required } A_s = \rho b d$$

$$= 0.0010 \,(11.5)(12)(21)$$

$$= 2.90 \text{ in.}^2$$

Check the ACI Code minimum reinforcement requirement. From Table A-5,

$$A_{s,min} = 0.0033(11.5)(12)(21)$$

$$= 9.56 \text{ in.}^2$$

The $A_{s,min}$ requirement need not be applied if the area of steel provided is at least one-third greater than that required (ACI Code,

Section 10.5.3), however. Therefore, the required steel area may be calculated from

$$\text{required } A_s = 1.33(2.90) = 3.86 \text{ in.}^2$$

Checking further, using the ACI Code, Section 10.5.4, as an absolute minimum, the required steel area is

$$A_s = 0.0018bh = 0.0018(11.5)(12)(25) = 6.21 \text{ in.}^2$$

Therefore 6.21 in.^{2} controls. Use 15 No. 6 bars ($A_s = 6.60$ in.^{2}). These bars will run in the short direction but will *not* be distributed uniformly across the 11 ft-6 in. side.

In rectangular footings, a portion of the total reinforcement required in the short direction is placed in a band centered on the column and having a width equal to the short side. The portion of the total required steel that goes into this band is

$$\frac{2}{\beta + 1}$$

where

$$\beta = \frac{\text{long-side dimension}}{\text{short-side dimension}} = \frac{11.5}{7.0} = 1.64$$

from which

$$\frac{2}{\beta + 1} = \frac{2}{1.64 + 1} = 0.757 = 75.7\%$$

Therefore 75.7% of 15 No. 6 bars must be placed in a band width = 7 ft-0 in. The balance of the required bars will be distributed equally in the outer portions of the footing. Thus

$$(0.757)(15) = 11.4 \text{ bars}$$

Use 12 bars in the 7 ft-0 in. band width. Because reinforcing should be symmetrical with respect to the centerline of the footing, use two bars on each side of the 7 ft-0 in. band width: Therefore, the total steel used in the short direction will be 16 No. 6 bars ($A_s = 7.04$ in.^{2}). The bar arrangement is depicted in Figure 10-18.

The development length will be checked for the No. 6 bars in the center band. Assume uncoated bars. This calculation for ℓ_d follows the eight-step procedure presented in Chapter 5, Section 5-2, and is summarized as follows:

1. $K_D = 82.2$
2. $\psi_t = 1.3, \psi_e = 1.0, \psi_s = 0.8, \lambda = 1.0$
3. $\psi_t \times \psi_e = 1.0$ (O.K.)
4. $c_b = 3.38$ in.
5. $K_{tr} = 0$

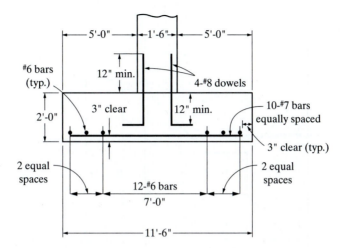

FIGURE 10-18   Design sketch for Example 10-4.

6. $(c_b + K_{tr})/d_b = 4.51$ Use 2.5

7. $K_{ER} = 0.882$

8. $\ell_d = 17.4$ in. > 12 in. (O.K.)

Use $\ell_d = 17.4$ in. (minimum). The development length furnished $= 33 - 3 = 30$ in., which is in excess of that required. (O.K.)

8.   Because the supporting surface is wider on all sides, the bearing strength for the footing may be computed as follows:

$$\text{footing bearing strength} = \phi(0.85f'_cA_1)\sqrt{\frac{A_2}{A_1}}$$

where A_2 is calculated with reference to Figure 10-19 and is seen to be a square, 7 ft-0 in. on each side.

$$\sqrt{\frac{A_2}{A_1}} = \sqrt{\frac{49.0}{2.25}} = 4.67 > 2.0$$

Therefore use 2.0. Then,

$$\text{footing bearing strength} = \phi(0.85f'_cA_1)(2.0)$$
$$= 0.65(0.85)(3.0)(18)^2(2.0)$$
$$= 1074 \text{ kips}$$

The column bearing strength is computed as follows:

$$\phi(0.85)f'_cA_1 = 0.65(0.85)(3)(18)^2 = 537 \text{ kips}$$

The factored bearing load is

$$P_u = 1.2(175) + 1.6(175)$$
$$= 490 \text{ kips}$$

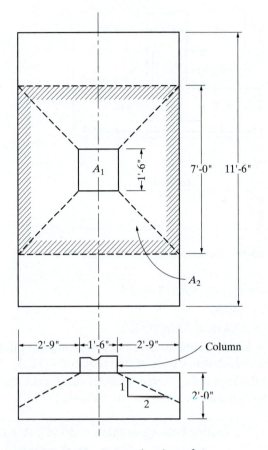

FIGURE 10-19   Determination of A_2.

Because 490 kips < 537 kips < 1074 kips, the entire column load can be transferred by concrete alone.

For proper connection between column and footing, use four No. 8 dowels (to match the column steel), one in each corner. The development length requirement for the No. 8 dowels is identical to that of Example 10-3.

10-7   ECCENTRICALLY LOADED FOOTINGS

Where footings are subject to eccentric vertical loads or to moments transmitted by the supported column, the design varies somewhat from that of the preceding sections. The soil pressure is no longer uniform across the footing width, but may be assumed to vary linearly. The resultant force should be within the middle one-third of the footing base to ensure a positive contact surface between the footing and the soil. With the soil pressure distribution known, the footing must be designed to resist all moments and shears, as were the concentrically loaded footings. The effects of load eccentricity on isolated footings may result in an undesirable large rotation of the footing, but this can be mitigated by connecting or strapping the eccentrically loaded footing to an adjacent concentrically loaded footing as discussed in Section 10-9.

10-8   COMBINED FOOTINGS

Combined footings are footings that support more than one column or wall. The two-column type of combined footing, which is relatively common, generally results from necessity. Two conditions that may lead to its use are (1) an exterior column that is immediately adjacent to a property line where it is impossible to use an individual column footing and (2) two columns that are closely spaced, causing their individual footings to be closely spaced. In these situations, a rectangular or trapezoidal combined footing would usually be used. The choice of which shape to use is based on the difference in column loads as well as on physical (dimensional) limitations. If the footing cannot be rectangular, a trapezoidal shape would then be selected.

The physical dimensions (except thickness) of the combined footing are generally established by the allowable soil pressure. In addition, the centroid of the footing area should coincide with the line of action of the resultant of the two column loads. These dimensions are usually determined using service loads in combination with an allowable soil pressure.

Example 10-5

Determine the shape and proportions of a combined footing subject to two column loads, as shown in Figure 10-20.

Solution:

1.   The design data are as follows: Service load on the footing from column A is 300 kips and from column B is 500 kips, and the allowable soil pressure is 6.00 ksf.

2.   Locate the resultant column load by a summation of moments at point Z in Figure 10-18:

$$\Sigma M_z = 300(2) + 500(18) = 800(x)$$

from which

$$x = 12 \text{ ft } 0 \text{ in. (measured from } Z)$$

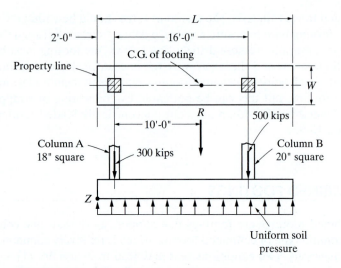

FIGURE 10-20   Sketch for Example 10-5.

3.   Assuming a rectangular shape, establish the length of footing L so that the centroid of the footing area coincides with the line of action of resultant force R:

$$\text{required } L = 12(2) = 24 \text{ ft 0 in.}$$

4.   Assume a footing thickness of 3 ft-0 in. Therefore its weight = 0.150(3) = 0.450 ksf, and the net allowable soil pressure for superimposed loads = 6.00 − 0.450 = 5.55 ksf. This neglects any soil *on* the footing.

5.   The footing area required is

$$\frac{R}{5.55} = \frac{800}{5.55} = 144.1 \text{ ft}^2$$

6.   With a length = 24 ft 0 in., the footing width W required is

$$\frac{144.1}{24} = 6 \text{ ft-0 in.}$$

7.   The actual uniform soil pressure is

$$\frac{800}{6(24)} + 0.450 = 6.00 \text{ ksf} \qquad \text{(O.K.)}$$

Example 10-6

Using the design data from Example 10-5, determine the proportions of a combined footing if the footing length is limited to 22 ft.

Solution:

1. The resultant column load is located as in Example 10-5 at a point 12 ft-0 in. from point Z.

2. Assume a footing thickness of 3 ft-0 in. Therefore its weight = 3(0.150) = 0.450 ksf, and the allowable soil pressure for superimposed loads is 6.00 − 0.450 = 5.55 ksf.

3. The footing area required is then

$$\frac{800}{5.55} = 144.1 \text{ ft}^2$$

4. Assume a trapezoidal shape. The area A of a trapezoid is (see Figure 10-21)

$$A = \frac{(b + b_1)L}{2}$$

from which

$$b + b_1 = \frac{2A}{L} = \frac{2(144.1)}{22} = 13.1 \text{ ft}$$

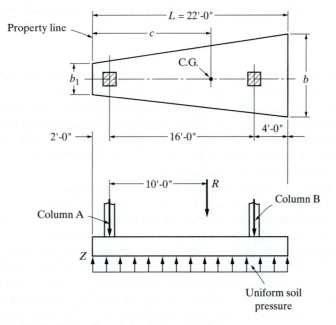

FIGURE 10-21   Sketch for Example 10-6.

5. The center of gravity of the trapezoid and the resultant force of the column loads are to coincide. The location of the center of gravity, a distance c from point Z, may be written

$$c = \frac{L(2b + b_1)}{3(b + b_1)} = \frac{L(b + b + b_1)}{3(b + b_1)} = 12 \text{ ft}$$

6. We now have two equations that contain b and b_1. The equation of step 5 may easily be solved for b by substitution of $L = 22$ ft and $(b + b_1) = 13.1$ ft:

$$\frac{22(b + 13.1)}{3(13.1)} = 12$$

from which

$$b = \frac{12(3)(13.1)}{22} - 13.1 = 8.34 \text{ ft}$$

and

$$b_1 = 13.1 - 8.34 = 4.76 \text{ ft}$$

Use $b = 8$ ft-4 in. and $b_1 = 4$ ft-9 in. Thus

$$\text{actual } A = \frac{22(4.75 + 8.33)}{2} = 143.9 \text{ ft}^2$$

The footing is very slightly undersized. No revision is warranted.

The structural design of the rectangular and trapezoidal combined footings is generally based on a uniform soil pressure, even though loading combinations will almost always introduce some eccentricity with respect to the centroid of the footing. The determination of the footing thickness and reinforcement must be based on factored loads and soil pressure to be consistent with the ACI strength method approach. The assumed footing behavior, which is briefly described, is a generally used approach to simplify the design of the footing.

In Figure 10-21, the columns are positioned relatively close to the ends of the footing. Assuming the columns as the supports and the footing subjected to an upward, uniformly distributed load caused by the uniform soil pressure, moments that create tension in the top of the footing will predominate in the longitudinal direction. Therefore the principal longitudinal reinforcement will be placed in the top of the footing equally distributed across the footing width. Somewhat smaller moments in the transverse direction will cause compression in the top of the footing. Transverse steel will be placed under each column in the bottom of the footing to distribute the column load in the transverse direction using the provisions for individual column footings. In effect, this makes the combined footing act as a wide rectangular beam in the longitudinal direction, which may then be designed using the ACI Code provisions for flexure.

Pertinent design considerations may be summarized as follows:

1. Main reinforcement (uniformly distributed) is placed in a longitudinal direction in the top of the footing, assuming the footing to be a longitudinal beam.

2. Shear should be checked considering both one-way shear at a distance d from the face of the column and two-way (punching) shear on a perimeter $d/2$ from the face of the column.

3. Stirrups or bent bars are frequently required to maintain an economical footing thickness. This assumes that the shear effect is uniform across the width of the footing.

4. Transverse reinforcement is generally uniformly placed in the bottom of the footing within a band having a width not greater than the column width plus twice the effective depth of the footing. The design treatment in the transverse direction is similar to the design treatment of the individual column footing, assuming dimensions equal to the band width as previously described and the transverse footing width.

5. Longitudinal steel is also placed in the bottom of the footing to tie together and position the stirrups and transverse steel. Although the required steel areas may be rather small, the effects of cantilever moments in the vicinity of the columns should be checked.

10-9   CANTILEVER OR STRAP FOOTINGS

A third type of combined footing is generally termed a *cantilever* or *strap footing*. This is an economical type of footing when the proximity of a property line precludes the use of other types. For instance, an isolated column footing may be too large for the area available, and the nearest column is too distant to allow a rectangular or trapezoidal combined footing to be economical. The strap footing may be regarded as two individual column footings connected by a strap beam.

In Figure 10-22, the exterior footing is placed eccentrically under the exterior column so that it does not violate the property-line limitations. This would produce a nonuniform pressure distribution under the footing, which could lead to footing rotation. To balance this rotational or overturning effect, the exterior footing is connected by a stiff beam, or strap, to the nearest interior footing, and uniform soil pressures under the footings are assumed. The strap, which may be categorized as a flexural member, is subjected to both bending moment and shear, resulting from the forces P_e and R_e acting on the exterior footing. As shown in Figure 10-22, the applied moment is counterclockwise, and the shear will be positive because $R_e > P_e$. At the interior column, there is no eccentricity between the column load P_i and the resultant soil pressure force R_i. Therefore we will assume that no moment is induced in the strap at the interior column.

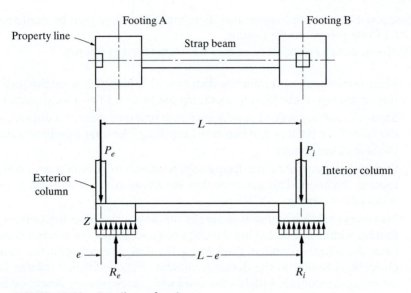

FIGURE 10-22   Cantilever footing.

We define V as the vertical shear force necessary to keep the strap in equilibrium, as shown in Figure 10-23. Then, with P_e known, V and R_e may be calculated using the principles of statics. A moment summation about R_e yields

$$P_e e = V(L - e)$$

$$V = \frac{P_e e}{L - e}$$

and from a summation of vertical forces,

$$R_e = P_e + V$$

Then, by substitution,

$$R_e = P_e + \frac{P_e e}{L - e}$$

Note that V acting downward on the strap beam also means that V is an uplift force on the interior footing. Therefore

$$R_i = P_i - V$$

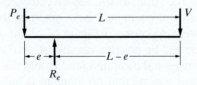

FIGURE 10-23   Strap beam.

and, by substitution,

$$R_i = P_i - \frac{P_e e}{L - e}$$

In summary, R_e becomes greater than P_e by a magnitude of V, whereas R_i becomes less than P_i by a magnitude equal to V.

The footing areas required are merely the reactions R_e and R_i based on service loads divided by the effective allowable soil pressure. These values are based on an assumed trial e and may have to be recomputed until the trial e and the actual e are the same.

The structural design of the interior footing is simply the design of an isolated column footing subject to a load R_i. The exterior footing is generally considered as under one-way *transverse* bending similar to a wall footing with longitudinal steel furnished by extending the strap steel into the footing. The selection of footing thickness and reinforcement should be based on factored loads to be consistent with the ACI strength design approach.

The strap beam is assumed to be a flexural member with no bearing on the soil underneath. Many designers make a further simplifying assumption that the beam weight is carried by the underlying soil; hence, the strap is designed as a rectangular beam subject to a constant shearing force and a linearly varying negative bending moment based on factored loads.

Example 10-7_____

Determine the size of the exterior and interior footings of a strap footing for the conditions and design data furnished in Figure 10-24.

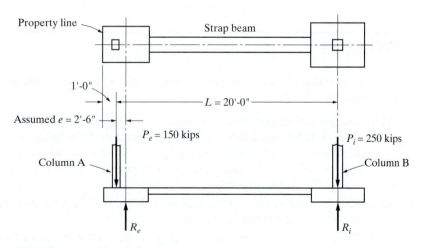

FIGURE 10-24   Sketch for Example 10-7.

Solution:

1. The design data are as follows: Service load on the footing from column A is 150 kips and from column B is 250 kips, and the allowable soil pressure is 4.00 ksf.

2. Assume that $e = $ 2 ft-6 in. and that the footing thickness is 2 ft-0 in. Therefore the footing weight $= 2(0.150) = 0.300$ ksf.

3. The net allowable soil pressure for superimposed loads $= 4.0 - 0.300 = 3.70$ ksf.

4. Determine the strap beam shear V:

$$V = \frac{P_e e}{L - e} = \frac{150(2.5)}{17.5} = 21.4 \text{ kips}$$

5. The footing reactions are

$$R_e = P_e + V = 150 + 21.4 = 171.4 \text{ kips}$$

$$R_i = P_i - V = 250 - 21.4 = 228.6 \text{ kips}$$

6. For the exterior footing, the required area is

$$\frac{171.4}{3.7} = 46.3 \text{ ft}^2$$

Use a footing 7 ft-0 in. by 7 ft-0 in. ($A = 49 \text{ ft}^2$). Note in Figure 10-24 that the actual e is then the same as the assumed e, and no revision of calculations is necessary.

For the interior footing, the required area is

$$\frac{228.6}{3.7} = 61.8 \text{ ft}^2$$

Use a footing 8 ft-0 in. by 8 ft-0 in. ($A = 64 \text{ ft}^2$).

PROBLEMS

Note: Assume that all steel is uncoated and the soil pressures given are gross allowable soil pressures.

10-1. Design a plain concrete wall footing to carry a 12-in.-thick reinforced concrete wall. Service loads are 2.3 kips/ft dead load (includes the weight of the wall) and 2.3 kips/ft live load. Use $f'_c = 3000$ psi and an allowable soil pressure of 4000 psf. Assume that the bottom of the footing is to be 4 ft below grade. The weight of earth $w_e = 100$ lb/ft³.

10-2. Redesign the footing for the wall of Problem 10-1. Service loads are 8.0 kips/ft dead load and 8.0 kips/ft live load.

10-3. Design a reinforced concrete footing for the wall of Problem 10-1 if the service loads are 6 kips/ft dead load (includes wall weight) and 15 kips/ft live load. Use $f_y = 60,000$ psi.

10-4. Design a square individual column footing (reinforced concrete) to support an 18-in.-square reinforced concrete tied interior column. Service loads are 200 kips dead load and 350 kips live load. Use $f'_c = 3000$ psi and $f_y = 60,000$ psi. The allowable soil pressure is 3500 psf, and $w_e = 100$ lb/ft³. The column, reinforced with eight No. 8 bars, has $f'_c = 5000$ psi and $f_y = 60,000$ psi. The bottom of the footing is to be 4 ft below grade.

10-5. Design a square individual column footing (reinforced concrete) to support a 16-in.-square reinforced concrete tied interior column. Service loads are 200 kips dead load and 160 kips live load. Both column and footing have $f'_c = 4000$ psi and $f_y = 60,000$ psi. The allowable soil pressure is 5000 psf, and $w_e = 100$ lb/ft³. The bottom of the footing is to be 4 ft below grade. The column is reinforced with eight No. 7 bars.

10-6. Design for load transfer from a 14-in.-square tied column to a 13-ft-0-in.-square reinforced concrete footing. Use $P_u = 650$ kips, footing $f'_c = 3000$ psi, column $f'_c = 5000$ psi, and $f_y = 60,000$ psi (all steel). The column is reinforced with eight No. 8 bars.

10-7. Redesign the footing of Problem 10-5 if there is a 7-ft-0-in. restriction on the width of the footing.

10-8. Footings are to be designed for two columns A and B spaced with $D = 16$ ft as shown. There are no dimensional limitations. The service load from column A is 100 kips and from column B is 150 kips. The allowable soil pressure is 4000 psf. Determine the appropriate footing(s) size, type, and layout. Assume thickness and disregard reinforcing.

10-9. Same as Problem 10-8, except service loads are 700 kips from column A and 900 kips from column B, and $D = 14$ ft.

10-10. For the column layout shown, there is a width restriction W of 16 ft. Here $D = 14$ ft. Determine an appropriate footing size and layout if the service load from column A is 700 kips and from column B is 800 kips. Assume thickness and disregard reinforcing. The allowable soil pressure is 4000 psf.

10-11. For the column layout shown, $D = 24$ ft. The footing for the 18-in.-square column A must be placed flush with the left side of the column. Service loads are 200 kips for column A and 300 kips for column B. The allowable soil pressure is 4500 psf. Assume a footing thickness of 2 ft-6 in. and determine an appropriate size, type, and layout for the footing.

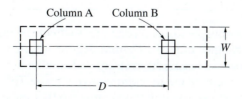

PROBLEMS 10-8 TO 10-11

Prestressed Concrete Fundamentals

11-1   INTRODUCTION

11-2   DESIGN APPROACH AND BASIC CONCEPTS

11-3   STRESS PATTERNS IN PRESTRESSED CONCRETE BEAMS

11-4   PRESTRESSED CONCRETE MATERIALS

11-5   ANALYSIS OF RECTANGULAR PRESTRESSED CONCRETE BEAMS

11-6   ALTERNATIVE METHODS OF ELASTIC ANALYSIS:
       INTERNAL COUPLE METHOD

11-7   ALTERNATIVE METHODS OF ELASTIC ANALYSIS:
       LOAD BALANCING METHOD

11-8   FLEXURAL STRENGTH ANALYSIS

11-9   NOTES ON PRESTRESSED CONCRETE DESIGN

11-1   INTRODUCTION

According to the ACI definition, prestressed concrete is a material that has had internal stresses induced to balance out, to a desired degree, tensile stresses due to externally applied loads. Because tensile stresses are undesirable in concrete members, the object of prestressing is to create compressive stresses (prestress) at the same locations as the tensile stresses within the member so that the tensile stresses will be diminished or will disappear altogether. The diminishing or elimination of

384

tensile stresses within the concrete will result in members that have fewer cracks or are crack-free at service load levels. This is one of the advantages of prestressed concrete over reinforced concrete, particularly in corrosive atmospheres. Prestressed concrete offers other advantages. Because beam cross sections are primarily in compression, diagonal tension stresses are reduced and the beams are stiffer at service loads. In addition, sections can be smaller, resulting in less dead weight.

Despite the advantages, we must consider the higher unit cost of stronger materials, the need for expensive accessories, the necessity for close inspection and quality control, and, in the case of precasting, a higher initial investment in plant.

11-2   DESIGN APPROACH AND BASIC CONCEPTS

As most of the advantages of prestressed concrete are at service load levels and as permissible stresses in the "green" concrete often control the amount of prestress force to be used, the major part of analysis and design calculations is made using service loads, permissible stresses, and basic assumptions as outlined in Sections 18.2 through 18.5 of the ACI Code. The strength requirements of the code must also be met, however. Therefore, at some point, the design must be checked using appropriate load factors and strength reduction factors.

The normal method for applying prestress force to a concrete member is through the use of steel tendons. There are two basic methods of arriving at the final prestressed member: pretensioning and post-tensioning.

Pretensioning may be defined as a method of prestressing concrete in which the tendons are tensioned before the concrete is placed. This operation, which may be performed in a casting yard, is basically a five-step process:

1. The tendons are placed in a prescribed pattern on the casting bed between two anchorages. The tendons are then tensioned to a value not to exceed 94% of the specified yield strength, but not greater than the lesser of 80% of the specified tensile strength of the tendons and the maximum value recommended by the manufacturer of the prestressing tendons or anchorages (ACI Code, Section 18.5.1). The tendons are then anchored so that the load in them is maintained.

2. If the concrete forms are not already in place, they may then be assembled around the tendons.

3. The concrete is then placed in the forms and allowed to cure. Proper quality control must be exercised, and curing may be accelerated with the use of steam or other methods. The concrete will bond to the tendons.

4. When the concrete attains a prescribed strength, normally within 24 hours or less, the tendons are cut at the anchorages. Because the tendons are now bonded to the concrete, as they are cut from their anchorages the high prestress force must be transferred to the concrete. As the high tensile force of the tendon creates a compressive force on the concrete section, the concrete will tend to shorten slightly. The stresses that exist once the tendons have been cut

PHOTO 11-1   Precast bridge girder components for post-tensioning.

are often called the stresses at *transfer*. Because there is no external load at this stage, the stresses at transfer include only those due to prestressing forces and those due to the weight of the member.

5.   The prestressed member is then removed from the forms and moved to a storage area so that the casting bed can be prepared for further use.

Pretensioned members are usually manufactured at a casting yard or plant that is somewhat removed from the job site where the members will eventually be used. In this case, they are usually delivered to the job site ready to be set in place. Where a project is of sufficient magnitude to warrant it economically, a casting yard may be built on the job site, thus decreasing transportation costs and allowing larger members to be precast without the associated transportation problems.

Figure 11-1 depicts the various stages in the manufacture of a precast, pretensioned member.

Post-tensioning may be defined as a method of prestressing concrete in which the tendons are tensioned *after* the concrete has cured. (Refer to Figure 11-2.) The operation is commonly a six-step process:

1.   Concrete forms are assembled with flexible hollow tubes (metal or plastic) placed in the forms and held at specified locations.

2.   Concrete is then placed in the forms and allowed to cure to a prescribed strength.

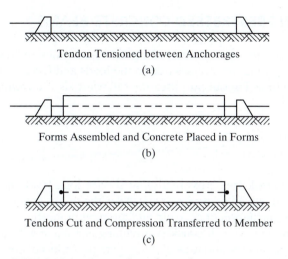

Tendon Tensioned between Anchorages

(a)

Forms Assembled and Concrete Placed in Forms

(b)

Tendons Cut and Compression Transferred to Member

(c)

FIGURE 11-1   Pretensioned member.

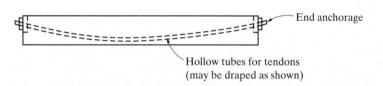

End anchorage

Hollow tubes for tendons
(may be draped as shown)

FIGURE 11-2   Post-tensioned member.

3.   Tendons are placed in the tubes. (In some systems, a complete tendon assembly is placed in the forms prior to the placing of the concrete.)

4.   The tendons are tensioned by jacking against an anchorage device or end plate that, in some cases, has been previously embedded in the end of the member. The anchorage device will incorporate some method for gripping the tendon and holding the load.

5.   If the tendons are to be bonded, the space in the tubes around the tendons may be grouted using a pumped grout. Some members use unbonded tendons.

6.   The end anchorages may be covered with a protective coating.

Although post-tensioning is sometimes performed in a plant away from the project, it is most often done at the job site, particularly for units too large to be shipped assembled or for unusual applications.

The tensioning of the tendons is normally done with hydraulic jacks. Many patented devices are available to accomplish the anchoring of the tendon ends to the concrete.

11-3   STRESS PATTERNS IN PRESTRESSED CONCRETE BEAMS

The stress pattern existing on the cross section of a prestressed concrete beam may be determined by superimposing the stresses due to the loads and forces acting on the beam at any particular time. For our purposes, the following sign convention will be adopted:

Tensile stresses are positive ($+$).
Compressive stresses are negative ($-$).

Because we will assume a crack-free cross section at service load level, the entire cross section will remain effective in carrying stress. Also, the entire concrete cross section will be used in the calculation of centroid and moment of inertia.

For purposes of explanation, we will consider a rectangular shape with tendons placed at the centroid of the section, and we will investigate the induced stresses. Although the rectangular shape is used for some applications, it is often less economical than the more complex shapes.

Example 11-1

For the section shown in Figure 11-3, determine the stresses due to prestress immediately after transfer and the stresses at midspan when the member is placed on a 20-ft simple span. Use $f'_c = 5000$ psi and assume that the concrete has attained a strength of 4000 psi at the time of transfer. Use a central prestressing force of 100 kips.

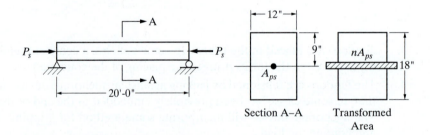

Section A–A   Transformed Area

FIGURE 11-3   Sketches for Example 11-1.

Solution:

1. Compute the stress in the concrete at the time of initial prestress. With the prestressing force P_s applied at the centroid of the section and assumed acting on the gross section A_c, the concrete stress will be uniform over the entire section. Thus

$$f = \frac{P_s}{A_c} = \frac{-100}{12(18)} = -0.463 \text{ ksi}$$

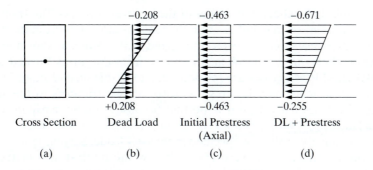

FIGURE 11-4   Midspan stresses for Example 11-1.

2.   Compute the stresses due to the beam dead load:

Weight of beam: $w_{DL} = \dfrac{12(18)}{144}(0.150) = 0.225$ kip/ft

Moment due to dead load: $M_{DL} = \dfrac{w_{DL}\ell^2}{8} = \dfrac{0.225(20)^2}{8} = 11.25$ ft-kips

Compute the moment of inertia I of the beam using the gross cross section and neglecting the transformed area of the tendons (as the steel is at the centroid, it will have no effect on the moment of inertia):

$$I_g = \frac{bh^3}{12} = \frac{12(18)^3}{12} = 5832 \text{ in.}^4$$

Dead load stresses: $f = \dfrac{Mc}{I} = \dfrac{11.25(12)(9)}{5832} = \pm 0.208$ ksi

3.   Compute the stresses due to prestress plus dead load:

$f_{\text{(initial prestress + DL)}} = -0.463 \pm 0.208$

$\qquad\qquad = -0.671$ ksi   (compression, top)

$\qquad\qquad = -0.255$ ksi   (compression, bottom)

These results are shown on the stress summation diagram of Figure 11-4.

   The tensile stresses due to the DL moment in the bottom of the beam have been completely canceled out and compression exists on the entire cross section. A limited additional positive moment may be carried by the beam without resulting in a net tensile stress in the bottom of the beam. This situation may be further improved on by *lowering the location of the tendon* to induce additional compressive stresses in the bottom of the beam.

Example 11-1 reflects two stages of the prestress process. The transfer stage occurs in a pretensioned member when the tendons are cut at the ends of the member and the prestress force has been transferred to the beam, as shown in Figure 11-1. When the beam in this problem is removed from the forms (picked up by its end points), dead load stresses are introduced, and in this second stage, both the beam weight (dead load) and the prestress force are contributors to the stress pattern within the beam. This stage is important because it occurs early in the life of the beam (sometimes within 24 hours of casting), and the concrete stresses must be held within permissible values as specified in the ACI Code, Section 18.4.

If the prestress force were placed below the neutral axis in Example 11-1, negative bending moment would occur in the member at transfer, causing the beam to curve upward and pick up its dead load. Hence, for a simple beam such as this, where the prestress force is eccentric, the stresses due to the initial prestress would never exist alone without the counteracting stresses from the dead load moment.

11-4 PRESTRESSED CONCRETE MATERIALS

The application of the prestress force both strains or "stretches" the tendons and at the same time induces tensile stresses in them. If the tendon strain is reduced for some reason, the stress will also be reduced. In prestressed concrete, this is known as a *loss of pre-stress*. This loss occurs after the prestress force has been introduced. Contributory factors to this loss are creep and shrinkage of the concrete, elastic compression of the concrete member, relaxation of the tendon stress, anchorage seating loss, and friction losses due to intended or unintended curvature in post-tensioning tendons. Estimates on the magnitudes of these losses vary. It is generally known, however, that an ordinary steel bar tensioned to its yield strength (40,000 psi) would lose its entire prestress by the time all stress losses had taken place. Therefore for prestressed concrete applications, it is necessary to use very high-strength steels, where the previously mentioned strain losses will result in a much smaller percentage of change in the original prestress force.

The most commonly used steel for pretensioned prestressed concrete is in the form of a seven-wire, uncoated, stress-relieved strand having a minimum tensile strength (f_{pu}) of 250,000 psi or 270,000 psi, depending on grade. The seven-wire strand is made up of seven cold-drawn wires. The center wire is straight, and the six outside wires are laid helically around it. All six outside wires are the same diameter, and the center wire is slightly larger. This in effect guarantees that each of the outside wires will bear on and grip the center wire.

Prestressing steel does not exhibit the definite yield point characteristic found in the normal ductile steel used in reinforcing steel (see Figure 11-5). The yield strength for prestressing wire and strand is a "specified yield strength" that is obtained from the stress–strain diagram at 1% strain, according to the American

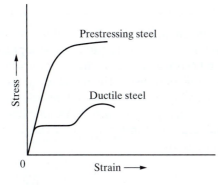

FIGURE 11-5   Comparative stress–strain curves.

Society for Testing and Materials (ASTM). Nevertheless, the specified yield point is not as important in prestressing steel as is the yield point in the ductile steels. It is a consideration when determining the ultimate strength of a beam. For more information on prestressing wire and strand, the reader is referred to ASTM Standards A416 and A421 [1].

In normal reinforced concrete members designed to ensure tension failures, the strength of the concrete is secondary in importance to the strength of the steel in the determination of the flexural strength of the member. In prestressed applications, concrete in the range from 4000 to 6000 psi is commonly used. Some of the reasons for this are as follows: (1) volumetric changes for higher-strength concrete are smaller, which will result in smaller prestress losses; (2) bearing and development stresses are higher; and (3) higher-strength concrete is more easily obtained in precast work than in cast-in-place work because of better quality control. In addition, high early strength cement (type III) is normally used to obtain as rapid a turnover time as possible for optimum use of forms.

11-5   ANALYSIS OF RECTANGULAR PRESTRESSED CONCRETE BEAMS

The analysis of flexural stresses in a prestressed member should be performed for different stages of loading—that is, the initial service load stage, which includes dead load plus prestress before losses; the final service load stage, which includes dead load plus prestress plus live load after losses; and finally the factored load stage, which involves load and strength-reduction factors. Generally, checking of prestressed members is accomplished at the service load level based on unfactored loads. The nominal strength of a member should be checked, however, using the same strength principles as for non-prestressed reinforced concrete members.

Example 11-2

For the beam of cross section shown in Figure 11-6, analyze the flexural stresses at midspan at transfer and in service. Neglect losses. Use a prestressed steel area A_{ps} of 2.0 in.^{2}, use $f_c' = 6000$ psi, and assume that the concrete has attained a strength of 5000 psi at the time of transfer. The initial prestress force = 250 kips. The service dead load = 0.25 kip/ft, which does not include the weight of the beam. The service live load = 1.0 kip/ft. Use $n = 7$. Assume that the entire cross section is effective and use the transformed area (neglecting displaced concrete) for the moment of inertia.

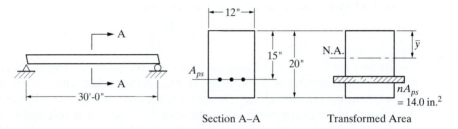

Section A–A   Transformed Area

FIGURE 11-6   Sketches for Example 11-2.

Solution:

1. The beam weight is

$$w_{\text{DL}} = \frac{20(12)}{144}(0.150) = 0.25 \text{ kip/ft}$$

The moment due to the beam weight is

$$M_{\text{DL}} = \frac{w_{\text{DL}}\ell^2}{8} = \frac{0.25(30)^2}{8} = 28.1 \text{ ft-kips}$$

The moment due to superimposed loads (DL + LL) is

$$M_{\text{DL+LL}} = \frac{w_{(\text{DL+LL})}\ell^2}{8} = \frac{1.25(30)^2}{8} = 140.6 \text{ ft-kips}$$

2. The location of the neutral axis is

$$\bar{y} = \frac{\Sigma(Ay)}{\Sigma A}$$

Using the top of the section as the reference axis,

$$\bar{y} = \frac{12(20)(10) + 14(15)}{12(20) + 14} = 10.28 \text{ in.}$$

The eccentricity e of the strands from the neutral axis is

$$15 - 10.28 = 4.72 \text{ in.}$$

The moment of inertia about the neutral axis is

$$I = \frac{12(20)^3}{12} + 12(20)\left(10.28 - \frac{20}{2}\right)^2 + 14(4.72)^2$$

$$= 8331 \text{ in.}^4$$

3. The stresses may now be calculated. These are summarized in Figure 11-7.

   a. Initial prestress: As a result of the eccentric prestressing force, the induced stress at the initial prestress stage will *not* be uniform but may be computed from

   $$f = -\frac{P_s}{A_c} \pm \frac{Mc}{I}$$

   $$= -\frac{P_s}{A_c} \pm \frac{P_s(e)c}{I}$$

   where P_s/A_c represents the axial effect of the prestress force and $\dfrac{P_s(e)c}{I}$ is the eccentric, or moment, effect. Then

   $$-\frac{P_s}{A_c} = -\frac{250}{12(20)} = -1.04 \text{ ksi} \qquad \text{(compression top and bottom)}$$

   $$+\frac{P_s(e)c}{I} = \frac{250(4.72)(10.28)}{8331} = +1.46 \text{ ksi} \qquad \text{(tension in top)}$$

   $$-\frac{P_s(e)c}{I} = -\frac{250(4.72)(9.72)}{8331} = -1.38 \text{ ksi} \qquad \text{(compression in bottom)}$$

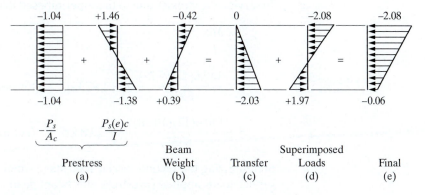

FIGURE 11-7   Midspan stress summary for Example 11-2.

b.   The stresses due to the beam weight are

$$f = \pm \frac{Mc}{I}$$

$$= +\frac{28.1(12)(9.72)}{8331} = +0.39 \text{ ksi} \qquad \text{(tension in bottom)}$$

$$= -\frac{28.1(12)(10.28)}{8331} = -0.42 \text{ ksi} \qquad \text{(compression in top)}$$

Summarizing the initial service load stage at the time of transfer [prestress plus beam weight (DL)]:

Top of beam: $-1.04 + 1.46 - 0.42 = 0$

Bottom of beam: $-1.04 - 1.38 + 0.39 = -2.03 \text{ ksi}$ (compression)

The permissible stresses (ACI Code, Section 18.4) immediately after prestress transfer (before losses) are in terms of f'_{ci}, which is the specified compressive strength (psi) of concrete at the time of initial prestress:

$$\text{compression} = 0.60 \, f'_{ci}$$

$$= 0.60(5000) = 3000 \text{ psi} = 3.0 \text{ ksi}$$

$$\text{tension} - 3\sqrt{f'_{ci}}$$

$$= 3\sqrt{5000} = 212 \text{ psi} = 0.212 \text{ ksi}$$

Because 2.03 ksi < 3.0 ksi and 0 < 0.212 ksi, the beam is satisfactory at this stage. Should the tensile stress exceed $3\sqrt{f'_{ci}}$, additional bonded reinforcement shall be provided.

Note in Figure 11-7c that no tensile stress exists at transfer.

c.   In service, the stresses due to the superimposed loads (DL + LL) are

$$f = \pm \frac{Mc}{I}$$

$$= +\frac{140.6(12)(9.72)}{8331} = +1.97 \text{ ksi} \qquad \text{(tension in bottom)}$$

$$= -\frac{140.6(12)(10.28)}{8331} = -2.08 \text{ ksi} \qquad \text{(compression in top)}$$

Summarizing the second service load stage when the beam has had service loads applied (prestress plus beam dead load plus superimposed loads) gives us

Top of beam: $0 - 2.08 = -2.08$ ksi   (compression)

Bottom of beam: $-2.03 + 1.97 = -0.06$ ksi   (compression)

The permissible compressive stress in the concrete at the service load level (ACI Code, Section 18.4.2) due to prestress plus total load is $0.60 f'_c$:

$$0.60(6000) = 3600 \text{ psi} = 3.60 \text{ ksi}$$

The permissible stress just given assumes that losses have been accounted for. Our analysis has neglected losses. Aside from this, the stresses in the beam would be satisfactory, because 2.08 ksi < 3.60 ksi, and no tensile stress exists.

The final stress distribution (Figure 11-7e) in the beam of Example 11-2 shows that the entire beam cross section is under compression. This is the stress pattern that exists at *midspan* (as the applied moments calculated were midspan moments). It is evident that moments will decrease toward the supports of a simply supported beam and that the stress pattern will change drastically. For example, if the eccentricity of the tendon were constant in the beam in question, the net prestress (Figure 11-7a) would exist at the beam ends, where moment due to beam weight and applied loads is zero. Because the tensile stress is undesirable, the location of the tendon is changed in the area of the end of the beam, so that eccentricity is decreased. This results in a *curved* or *draped* tendon within the member, as shown in Figure 11-8. Post-tensioned members may have curved tendons accurately placed to satisfy design requirements, but in pretensioned members, because of the nature of the fabrication process, only approximate curves are formed by forcing the tendon up or down at a few points.

Although shear was not included in the beam analysis, the reader should be aware that the draping of the tendons, in addition to affecting the flexural stresses at the beam ends, produces a force acting vertically upward, which has the effect of reducing the shear force due to dead and live loads. Prestressed concrete flexural members are available in numerous shapes suitable for various applications. A few of the more common ones are shown in Figure 11-9.

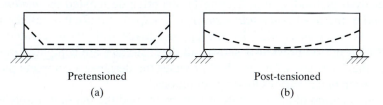

Pretensioned

(a)

Post-tensioned

(b)

FIGURE 11-8 Draped tendons.

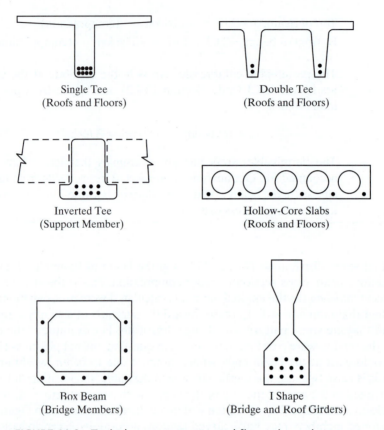

FIGURE 11-9   Typical precast, prestressed flexural members.

11-6   ALTERNATIVE METHODS OF ELASTIC ANALYSIS: INTERNAL COUPLE METHOD

The ACI Code requires that prestressed beams be analyzed elastically to establish whether the stresses developed by the combined action of service loads and prestressing force are within specified allowable stresses. In effect, this limit on the magnitude of stress, specifically at transfer or with service loads in place, controls cracking and subsequently prevents crushing of the concrete.

The analysis method used in Section 11-5 is often designated the *method of superposition* or the *combined loading concept*. To explain briefly, it is based on superimposing the stresses created by the applied loads with those created by the prestressing force. As previously shown, all these stresses are computed independently and then combined. This method provides the designer a complete picture of stress variation under various loading conditions.

A second method of elastic analysis may be termed the *internal couple method*. In a sense, this is analogous to the internal couple of reinforced concrete beams,

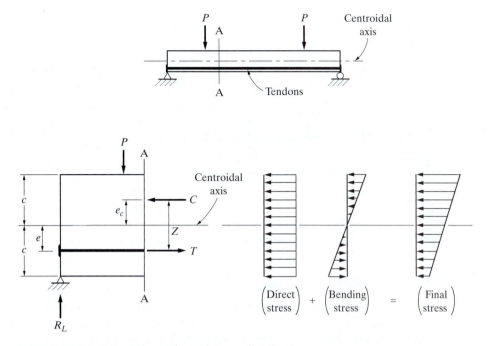

FIGURE 11-10 Internal couple and stress distributions.

discussed in Chapter 2, where steel takes the tension and concrete takes the compression, as shown in Figure 11-10. The two materials form an internal couple resisting the moment due to the external loads. For a prestressed beam, the couple consists of the tendon prestress force P_s, which may be designated the tensile force T (therefore, $P_s = T$), and the resultant of the longitudinal compressive stresses in the concrete, C.

Note, however, that the internal resisting couple (or moment) in a prestressed concrete beam differs somewhat from that in a reinforced concrete beam. Under service load conditions, the forces C and T in a prestressed concrete beam remain virtually constant, and the lever arm Z increases with increasing moment (see Figure 11-10). In a reinforced concrete beam, the lever arm Z of the internal couple is considered essentially constant, whereas the forces C and T increase with increasing moment. If the moment due to the applied loads is M and the prestress force P_s (which is also the tension T) is essentially constant, it follows that

$$Z = \frac{M}{P_s} = \frac{M}{T}$$

This, in effect, locates the *center of compression* (the location at which C acts). Once the position of C has been found, the stress distribution can be established by combining the direct stress created by C with the bending stress created by C acting at an eccentricity e_c from the centroidal axis of the section, as shown in Figure 11-10.

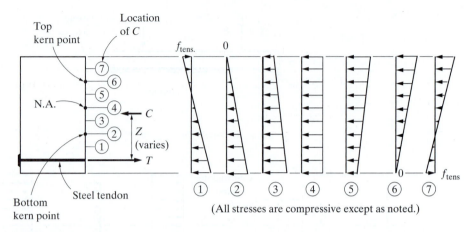

FIGURE 11-11   Stress distribution in concrete for various locations of C (elastic analysis).

Because the prestress force is essentially constant as long as the concrete remains uncracked, the position of the resultant compression C on the cross section with respect to the position of the tendon varies directly with the magnitude of the moment. If the moment is zero (which occurs at transfer), the arm Z between C and T will be zero and the resultant of the compressive stresses C will be located at the level of the tendon. As the location of C varies due to an increasing moment, the stress distribution shape will vary and could be any one of those shown in Figure 11-11.

Numerically, the stress at any point on a cross section may be obtained using

$$f = \frac{C}{A_g} \pm \frac{Ce_c y}{I_g}$$

where

e_c = eccentricity of C with respect to the centroidal axis

y = distance from the centroidal axis to any point on the cross section (for the outer fibers, $y = c$)

A_g = cross-sectional area of the *gross* concrete section

I_g = moment of inertia with respect to the centroidal axis of the *gross* section

C = total compression force acting on the cross section

It is common practice in the analysis and design of rectangular-shaped prestressed concrete beams to base the stress calculations in the elastic range on the properties of the gross concrete section. This, in effect, neglects the transformed area of the steel tendons as well as any displaced concrete due to the presence of the tendons or ducts. This is an acceptable approach, because the differences in the resulting stresses are quite small.

As the moment due to the applied loads changes and the internal couple lever arm Z changes, small tensile stresses may develop on a given cross section. If no tensile stresses are to be permitted either at transfer or under full service loads, C cannot be located outside the *kern* of the cross section. When a resultant force, acting by itself, is located within the kern area, no tensile stress will be produced. When the resultant force acts at the boundary of the kern area, it will produce a triangular stress distribution [2]. The boundary of the kern area is referred to as the *kern point*. As shown in Figure 11-11, when C is at the top or bottom kern point, the triangular stress distribution results.

11-7 ALTERNATIVE METHODS OF ELASTIC ANALYSIS: LOAD BALANCING METHOD

A third method of elastic analysis is called the *load balancing method*. According to the Post-Tensioning Institute, this method is by far the most widely used method for analysis and design of post-tensioned structures. It is a technique of balancing the external load by selecting a prestressing force and tendon profile that creates a transverse load acting opposite to the external load. This transverse load may be equal to either the full external applied load or only part of it.

For example, to balance a uniformly distributed load (w) acting on a simply supported beam, a parabolic tendon profile with zero end eccentricities would be selected. The prestressing force P_s needed would be a function of the load to be balanced as well as the acceptable sag of the tendon. If the transverse load created by the tendons exactly balanced the external load, a uniform compressive stress distribution P_s/A_g will develop over the beam cross section, and the beam will remain essentially level with no deflection or camber. To balance the load, the end eccentricities should be zero; otherwise, an end moment will be developed that disturbs the uniform stress distribution.

If only a portion of the external load is balanced, a net moment in the beam at any point will develop from that portion of the load that is *not* balanced by the prestressing. It is only this net moment that must be considered in computing the bending stress. Therefore the stress acting at any point on a cross section may be expressed as

$$f = \frac{P_s}{A_g} \pm \frac{M_{net}y}{I_g}$$

where

P_s = prestressing force applied to the member (this is also designated T)

M_{net} = net unbalanced moment on the section

and A_g, y, and I_g are as previously defined. The computed stress due to the uniform compression and the unbalanced moment must then (as in the other methods) be compared with code-allowable stresses.

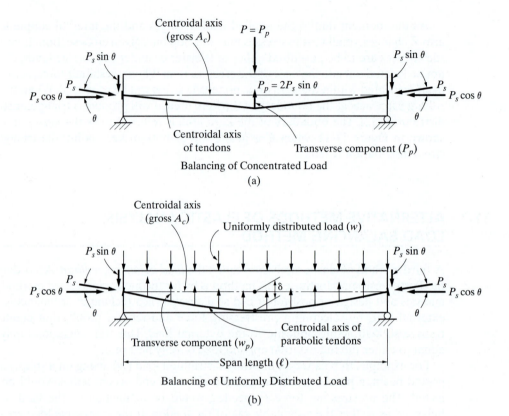

FIGURE 11-12   Load balancing: (a) balancing of concentrated load; (b) balancing of uniformly distributed load.

The externally applied load that will be balanced need not necessarily be uniformly distributed. It can also be a concentrated load or be a combination of both types. For a uniformly distributed load, a shallow parabolic tendon profile is generally selected, whereas a linear profile with a sharp directional change is used for a concentrated load. These are illustrated in Figure 11-12.

With respect to the balancing of a concentrated load, note the sharp directional change of the tendon under the external load at midspan. This creates an upward component

$$P_p = 2P_s \sin \theta$$

In the case shown, P_p exactly balances the applied load P. Therefore, if we neglect its own weight, the beam is not subject to net *transverse* load. At the ends of the beam, the horizontal components of P_s, which are shown as $P_s \cos \theta$ (and which are collinear with the centroidal axis of the beam), create a uniform compressive stress along the entire length of the beam. Therefore the stress in the beam at any section may be expressed as

$$f = \frac{P_s \cos \theta}{A_g}$$

and for small values of θ,

$$f = \frac{P_s}{A_g}$$

If any additional external load is applied, the beam will act as an elastic, homogeneous concrete beam (up to the point of cracking), and a bending stress will develop, which can be evaluated by

$$f = \frac{M_{net}c}{I_g}$$

where M_{net} is the moment developed by any load applied in addition to P.

With respect to the balancing of the uniformly distributed load, note that the tendon profile is that of a parabolic curve whose upward transverse component w_p in lb/ft is given by

$$w_p = \frac{8P_s\delta}{\ell^2} \tag{11-1}$$

Assuming that the externally applied load w, including the weight of the beam, is exactly balanced by the transverse component w_p, there is no bending in the beam, and the beam is subject to a uniform compressive stress calculated from

$$f = \frac{P_s}{A_g}$$

If the transverse component is different from the applied external load, the bending moment developed will induce a bending stress, which can be evaluated by

$$f = \frac{M_{net}c}{I_g}$$

where M_{net} is the moment developed by a load applied over and above w.

It is questionable as to what portion of the external load should be balanced by the prestress. If too much of the applied load, such as DL plus ½LL is to be balanced, excessive prestress may be required. The designer must exercise judgment in determining the proper amount of loading to be balanced by prestressing.

Example 11-3 _____

The rectangular prestressed beam shown in Figure 11-13 carries uniformly distributed service loads of 1.0 kip/ft LL and 1.0 kip/ft DL (which includes the weight of the beam). A parabolic tendon, configured as shown, will be used. The tendon is to furnish a uniformly distributed upward balancing load of 1.5 kips/ft (DL + ½LL). Calculate the required prestressing force and determine the net moment at midspan.

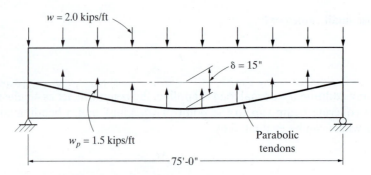

FIGURE 11-13   Sketch for Example 11-3.

Solution:

Solving Equation (11-1) for required P_s:

$$\text{required } P_s = \frac{w_p \ell^2}{8\delta} = \frac{1.5(75)^2}{8(15/12)} = 844 \text{ kips}$$

The net moment at midspan is calculated from

$$M = \frac{w\ell^2}{8} = \frac{(2.0 - 1.5)(75)^2}{8} = 352 \text{ ft-kips}$$

Example 11-4 illustrates the three methods of elastic analysis. In this example, we will neglect the transformed area of the steel as well as any displaced concrete, and we will use the gross properties of the cross section to compute stresses. Note that this varies from the solution of Example 11-2.

Example 11-4 _____

A rectangular prestressed concrete beam shown in Figure 11-14 is simply supported and has a span length of 25 ft-0 in. The beam carries a superimposed

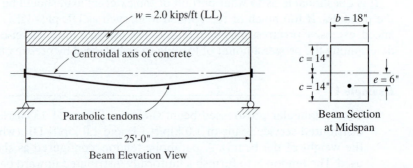

FIGURE 11-14   Sketches for Example 11-4.

service LL of 2.0 kips/ft. The only service DL is the weight of the beam. The parabolic prestressing tendon is located as shown and is subjected to an effective prestress force of 300 kips (neglect any prestress losses). Determine the outer fiber flexural stresses at midspan at transfer (prestress + beam weight) and when the member is under full service load conditions (prestress + beam weight + service LL).

Solution:

a. Using the method of superposition
1. Calculate moments after determining the beam weight:

$$w_{DL} = \frac{28(18)}{144}(0.150) = 0.525 \text{ kip/ft}$$

The moment due to beam weight is

$$M_{DL} = \frac{w_{DL}\ell^2}{8} = \frac{0.525(25)^2}{8} = 41.0 \text{ ft-kips}$$

The moment due to superimposed LL is

$$M_{LL} = \frac{w_{LL}\ell^2}{8} = \frac{2.0(25)^2}{8} = 156.3 \text{ ft-kips}$$

2. Using the gross concrete section and neglecting the transformed steel area, the neutral axis coincides with the centroidal axis. Therefore, the distance from the neutral axis to the outer fiber is 14 in. The eccentricity e of the tendon is given as 6 in. The moment of inertia about the neutral axis is found from

$$I_g = \frac{18(28)^3}{12} = 32{,}928 \text{ in.}^4$$

3. Calculate the stresses.
a. At transfer, the prestressing stress is determined from

$$f = -\frac{P_s}{A_g} \pm \frac{P_s ec}{I_g}$$

$$= -\frac{300}{18(28)} \pm \frac{300(6)(14)}{32{,}928}$$

$$= -0.595 \pm 0.765$$

from which

$$\text{stress at top} = -0.595 + 0.765 = 0.170 \text{ ksi} \qquad \text{(tension)}$$

$$\text{stress at bottom} = -0.595 - 0.765 = -1.36 \text{ ksi} \qquad \text{(compression)}$$

The stress due to the service DL (beam weight) is calculated from

$$f = \pm \frac{Mc}{I} = \pm \frac{41.0(12)(14)}{32,928} = \pm 0.209 \text{ ksi}$$

Combining the preceding stresses for the transfer condition gives us

$$\text{stress at top} = +0.170 - 0.209 = -0.039 \text{ ksi} \qquad \text{(compression)}$$

$$\text{stress at bottom} = -1.36 + 0.209 = -1.15 \text{ ksi} \qquad \text{(compression)}$$

     b.   Under full service load conditions, stress due to superimposed LL is calculated from

$$f = \pm \frac{Mc}{I} = \pm \frac{156.3(12)(14)}{32,928} = \pm 0.797 \text{ ksi}$$

Combining the preceding stresses for the full service load condition gives us

$$\text{stress at top} = -0.039 - 0.797 = -0.836 \text{ ksi} \qquad \text{(compression)}$$

$$\text{stress at bottom} = -1.15 + 0.797 = -0.353 \text{ ksi} \qquad \text{(compression)}$$

   b.   Using the internal couple method

     1.   Given the prestress force of 300 kips and noting that $\Sigma H_F = 0$, we see that

$$P_s = T = C = 300 \text{ kips}$$

     a.   At transfer (with reference to Figure 11-15),

$$\text{lever arm } Z = \frac{M}{T} = \frac{41.0(12)}{300} = 1.64 \text{ in}$$

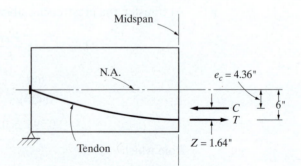

FIGURE 11-15   Location of C.

The eccentricity of C to the neutral axis is calculated from

$$e_c = 6 - 1.64 = 4.36 \text{ in.} \qquad \text{(below middepth)}$$

$$f = -\frac{C}{A_g} \pm \frac{Ce_c c}{I_g}$$

$$= -\frac{300}{18(28)} \pm \frac{300(4.36)(14)}{32,928}$$

$$= -0.595 \pm 0.556$$

$$\text{stress at top} = -0.595 + 0.556 = -0.039 \text{ ksi} \qquad \text{(compression)}$$

$$\text{stress at bottom} = -0.595 - 0.556 = -1.15 \text{ ksi} \qquad \text{(compression)}$$

b. Under full service load (with reference to Figure 11-16),

$$\text{lever arm } Z = \frac{M_{DL} + M_{LL}}{T} = \frac{(156.3 + 41.0)(12)}{300} = 7.89 \text{ in.}$$

The eccentricity of C to the neutral axis is calculated from

$$e_c = 7.89 - 6 = 1.89 \text{ in.} \qquad \text{(above middepth)}$$

$$f = -\frac{C}{A_g} \pm \frac{Ce_c c}{I_g}$$

$$= -\frac{300}{18(28)} \pm \frac{300(1.89)(14)}{32,928}$$

$$= -0.595 \pm 0.241$$

$$\text{stress at top} = -0.595 - 0.241 = -0.836 \text{ ksi} \qquad \text{(compression)}$$

$$\text{stress at bottom} = -0.595 + 0.241 = -0.354 \text{ ksi} \qquad \text{(compression)}$$

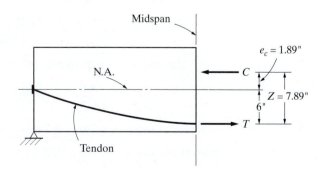

FIGURE 11-16 Location of C.

c.   Using the load balancing method

    1.   Let $e = \delta = 6$ in. The parabolic tendons produce an upward balancing load of

$$w_p = \frac{8P_s\delta}{\ell^2} = \frac{8(300)(0.5)}{25^2} = 1.92 \text{ kips/ft}$$

a.   At transfer, the unbalanced load on the beam is

$$1.92 - 0.525 = 1.395 \text{ kips/ft} \qquad \text{(upward)}$$

The net moment at midspan is

$$M_{net} = \frac{1.395(25)^2}{8} = 109.0 \text{ ft-kips} \qquad \text{(tension on top)}$$

The bending stress due to the net moment is

$$f = \pm\frac{M_{net}c}{I_g} = \pm\frac{109(12)(14)}{32,928} = \pm0.556 \text{ ksi}$$

The uniform compressive stress is

$$-\frac{P_s}{A_g} = -\frac{300}{18(28)} = -0.595 \text{ ksi}$$

$$\text{stress at top} = -0.595 + 0.556 = -0.039 \text{ ksi} \qquad \text{(compression)}$$

$$\text{stress at bottom} = -0.595 - 0.556 = -1.15 \text{ ksi} \qquad \text{(compression)}$$

b.   Under full service load, the unbalanced load on the beam is

$$2.0 + 0.525 - 1.92 = 0.605 \text{ kip/ft} \qquad \text{(downward)}$$

The net moment at midspan is

$$M_{net} = \frac{0.605(25)^2}{8} = 47.3 \text{ ft-kips} \qquad \text{(tension on bottom)}$$

The bending stress due to the net moment is

$$f = \pm\frac{M_{net}c}{I_g} = \pm\frac{47.3(12)(14)}{32,928} = \pm0.241 \text{ ksi}$$

The uniform compressive stress is

$$-\frac{P_s}{A_g} = -\frac{300}{18(28)} = -0.595 \text{ ksi}$$

$$\text{stress at top} = -0.595 - 0.241 = -0.836 \text{ ksi} \qquad \text{(compression)}$$

$$\text{stress at bottom} = -0.595 + 0.241 = -0.354 \text{ ksi} \qquad \text{(compression)}$$

Note that the results agree (with slight round-off error) using the three elastic analysis methods.

The load balancing method offers advantages of simplicity and clarity when with continuous beams and slabs. It is the recommended method for those situations, particularly for preliminary designs. For simple spans, none of the three methods has any particular advantage over the others; all are equally applicable.

11-8   FLEXURAL STRENGTH ANALYSIS

As part of the design and analysis procedure of a prestressed concrete beam, the ACI Code requires that the moment due to factored service loads, M_u, not exceed the flexural design strength ϕM_n of the member. The design strength of prestressed beams may be computed using strength equations similar to those for reinforced concrete members, discussed in Chapter 2. The checking of the flexural (nominal) strength of a prestressed beam ensures that the beam is designed with an adequate factor of safety against failure.

The expression for the nominal bending strength of a rectangular-shaped prestressed beam is developed from the internal couple of an underreinforced beam at failure, as shown in Figure 11-17.

Assuming failure is initiated by the steel yielding, the magnitude of the internal couple is

$$M_n = A_{ps}f_{ps}\left(d_p - \frac{a}{2}\right)$$

where

A_{ps} = area of prestressed reinforcement in the tension zone

d_p = distance from extreme compression fiber to the centroid of the prestressed reinforcement

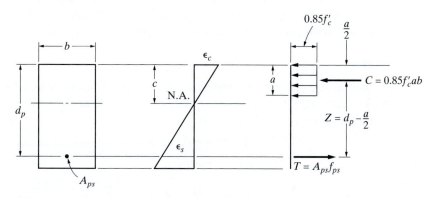

FIGURE 11-17   Equivalent stress block for strength analysis.

a = depth of stress block, determined by equating T and C and calculated from

$$a = \frac{A_{ps}f_{ps}}{0.85f_c'b}$$

f_{ps} = stress in the prestressed reinforcement at nominal strength

For members with bonded prestressing tendons and no non-prestressed tension or compression reinforcement, the value of f_{ps} can be obtained from

$$f_{ps} = f_{pu}\left(1 - \frac{\gamma_p}{\beta_1}\rho_p\frac{f_{pu}}{f_c'}\right) \qquad \text{[Mod. ACI Eq. (18-3)]}$$

where

f_{pu} = ultimate tensile strength of the prestressing steel

γ_p = factor based on the type of prestressing steel: 0.55 when f_{py}/f_{pu} is not less than 0.80, 0.40 when f_{py}/f_{pu} is not less than 0.85, and 0.28 when f_{py}/f_{pu} is not less than 0.90, where f_{py} is the specified yield stress of the prestressing tendons (psi)

ρ_p = reinforcement ratio A_{ps}/bd_p

and β_1 is as defined in Chapter 2, Section 2-6. The expression for f_{ps} is valid when $f_{se} > f_{pu}/2$, where f_{se} is the effective stress in prestressing steel after losses. Here f_{ps} represents the average stress in the prestressing steel at failure. It is analogous to f_y, but as the high-strength prestressing steels do not have a well-defined yield point, it can be predicted using ACI Equation (18-3).

Example 11-5 _____

Calculate the flexural strength ϕM_n of the prestressed beam of Example 11-4 and compare with M_u. Use f_c' = 5000 psi (normal-weight concrete) and seven-wire-strand grade 270 with f_{pu} = 270,000 psi (ordinary strand with f_{py} = 0.85f_{pu}). Use bonded prestressing tendons and neglect all prestress losses. Therefore, for the purpose of this problem, $f_{se} = f_{pu}$. Use A_{ps} = 1.224 in.^{2}. The beam has no non-prestressed tension or compression steel.

Solution:

Because prestress losses are neglected, $f_{se} > f_{pu}/2$. Therefore ACI Equation (18-3) is applicable. For the given conditions, γ_p = 0.40 and β_1 = 0.80. The reinforcement ratio is calculated from

$$\rho_p = \frac{A_{ps}}{bd_p} = \frac{1.224}{18(20)} = 0.00340$$

$$f_{ps} = f_{pu}\left(1 - \frac{\gamma_p}{\beta_1}\rho_p\frac{f_{pu}}{f_c'}\right)$$

$$= 270\left[1 - \frac{0.40}{0.80}(0.00340)\left(\frac{270}{5}\right)\right]$$

$$= 245 \text{ ksi}$$

Prestressed concrete sections are subject to the same conditions of being tension-controlled, compression-controlled, or transition sections as regular reinforced concrete sections, as discussed in Section 2-8. The appropriate values of ϕ, as previously discussed, apply. It is convenient to check the net tensile strain limit of 0.005 (for tension-controlled sections) through the use of a *reinforcement index* ω_p, where

$$\omega_p = \frac{\rho_p f_{ps}}{f_c'}$$

A reinforcement index of $0.32\beta_1$ corresponds to a net tensile stain of 0.005, the lower limit for a tension-controlled section. Therefore, checking ω_p:

$$\omega_p = \frac{\rho_p f_{ps}}{f_c'} = \frac{0.00340(245,000)}{5000} = 0.167$$

$$0.32\beta_1 = 0.32(0.80) = 0.256$$

Because $\omega_p < 0.256$, this is a tension-controlled section and $\phi = 0.90$.
    Compute the nominal moment strength:

$$a = \frac{A_{ps}f_{ps}}{0.85f_c'b} = \frac{1.224(245)}{0.85(5)(18)} = 3.92 \text{ in.}$$

$$M_n = A_{ps}f_{ps}\left(d_p - \frac{a}{2}\right)$$

$$= 1.224(245)\left(20 - \frac{3.92}{2}\right)$$

$$= 5410 \text{ in.-kips}$$

$$= 451 \text{ ft-kips}$$

$$\phi M_n = 0.90(451) = 406 \text{ ft-kips}$$

Compute the factored service load moment:

$$M_u = 1.2M_{\text{DL}} + 1.6M_{\text{LL}}$$

$$= 1.2(41.0) + 1.6(156.3) = 299 \text{ ft-kips}$$

Thus $M_u < \phi M_n$ (299 ft-kips < 406 ft-kips). Therefore O.K.

If a prestressed beam is satisfactorily designed based on service loads and then, when checked by the strength equations, is found to have insufficient strength to resist the factored loads, non-prestressed reinforcement may be added to increase the factor of safety. In addition, the ACI Code, Section 18.8.2, requires that the total amount of reinforcement (prestressed and non-prestressed) be sufficient to develop a factored load equal to at least 1.2 times the cracking load calculated from the modulus of rupture of the concrete. It is permissible to waive this requirement when a flexural member has shear and flexural strength at least twice that required by Code Section 9.2.

11-9   NOTES ON PRESTRESSED CONCRETE DESIGN

Because the shape and dimensions of a prestressed concrete member may be established by a trial procedure or even assumed based on physical limitations, the design may be reduced to finding the prestress force, tendon profile, and the amount of the prestress steel area. The design problem then is further reduced to an analysis-type problem whereby service load stresses are checked at various stages and the flexural strength of the member is checked against the moment due to the factored service loads.

The intent of this chapter, however, is to furnish a conceptual approach to prestressed concrete members. Therefore, many significant topics normally considered in an analysis or design problem have been omitted. These include total and design procedures, along with shear design, deflection, prestress loss, and block stresses. These topics are beyond the scope of our text. References [3] through [6] are among many other texts and publications available for further reading.

REFERENCES

[1]   "ASTM Standards." American Society for Testing and Materials, 100 Barr Harbor Drive, West Conshohocken, PA 19428.

[2]   George F. Limbrunner and Leonard Spiegel. *Applied Statics and Strength of Materials*, 5th ed. Upper Saddle River, NJ: Prentice Hall, 2009, pp. 470–473.

[3]   T. Y. Lin and Ned H. Burns. *Design of Prestressed Concrete Structures*, 3rd ed. New York: John Wiley & Sons, Inc., 1981.

[4]   *Post-Tensioning Manual*, 6th ed. Phoenix, AZ: Post-Tensioning Institute, 2006.

[5]   Edward G. Nawy. *Prestressed Concrete: A Fundamental Approach*. Upper Saddle River, NJ: Prentice Hall, 2005.

[6]   Arthur H. Nilson. *Design of Prestressed Concrete*, 2nd ed. New York: John Wiley & Sons, Inc., 1987.

PROBLEMS

11-1. The *plain concrete* beam shown having a rectangular cross section 10-in. wide and 18-in. deep is simply supported on a single span of 20 ft. Assuming no loads other than the dead load of the beam itself, find the bending stresses that exist at midspan.

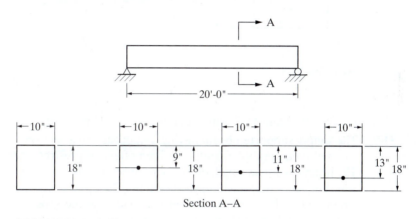

Section A–A

PROBLEMS 11-1 TO 11-4

11-2. A beam identical to that of Problem 11-1 is prestressed with a force P_s of 185 kips. The tendons are located at the center of the beam. Use $A_{ps} = 1.23$ in.^{2}. Find the stresses in the beam at midspan at transfer. There is no load other than the weight of the beam. Draw complete stress diagrams.

11-3. The tendons in the beam of Problem 11-2 are placed 11 in. below the top of the beam at midspan. Assuming that no tension is allowed, determine the maximum service uniform load that the beam can carry in addition to its own weight as governed by the stresses at midspan. Assume that $f'_{ci} = 3500$ psi and $f'_c = 5000$ psi.

11-4. Same as Problem 11-3, except that the tendons are 13 in. below the top of the beam at midspan and permissible stresses are as defined in the ACI Code, Section 18.4.

11-5. For the beam shown, calculate the outer fiber stresses at midspan and at the end supports at the following stages: (a) transfer (prestress + beam weight) and (b) under full service load. The beam is simply supported with a span length of 20 ft-0 in. and carries a superimposed live load of 2.7 kips/ft. The tendon is straight and located 9 in. below the neutral axis of the beam. The prestress force is 250 kips. Neglect prestress losses and use the gross concrete section properties. Use the method of superposition for the solution.

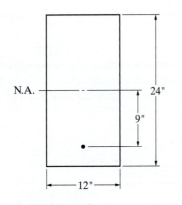

PROBLEM 11-5

11-6. Rework Problem 11-5 using the internal couple method.

11-7. The simply supported beam having a midspan cross section as shown is pre-stressed by a parabolic tendon with an effective prestress force of 233 kips. The beam has a span length of 30 ft and carries a total service load of 1.20 kips/ft, which includes the weight of the beam of 0.21 kip/ft. Calculate the outer fiber stresses at midspan under full service load. Use the gross properties of the cross section. Use the load balancing method.

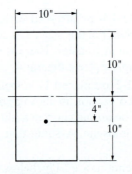

PROBLEM 11-7

11-8. For the beam of Problem 11-5, calculate the outer fiber stresses at midspan for the following stages: (a) transfer and (b) under full service load. Assume the prestress tendons are parabolically draped with a midspan eccentricity of 9 in. as shown and zero eccentricity at the end supports. Neglect prestress losses and use the gross concrete section properties. The prestress force is still 250 kips. Use the load balancing method.

11-9. Calculate the flexural strength ϕM_n of the prestressed beam shown and compare it with the factored design moment M_u. The service load moments are 125 ft-kips for dead load and 383 ft-kips for live load. Use $f'_c = 5000$ psi (normal-weight concrete) and grade 270 seven-wire strand with $f_{pu} = 270{,}000$ psi ($f_{py} = 0.85 f_{pu}$). The tendons are bonded. Neglect all prestress losses.

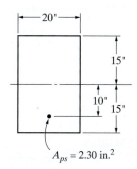

PROBLEM 11-9

Chapter 12

Concrete Formwork

12-1 INTRODUCTION

Forms are temporary structures whose purpose is to provide containment for the fresh concrete and to support it until it can support itself. Forms usually must be *engineered* structures as they are required to support loads composed of the fresh concrete, construction materials, equipment, workers, impact of various kinds, and sometimes wind. The forms must support all the applied loads without collapse or excessive deflection. In addition, the forms must provide for the molding of the concrete to the desired dimensions, shape, and surface finish.

The cost of formwork is significant, generally amounting to anywhere from 40% to 60% of the cost of a concrete structure. Economy in formwork design depends partly on the ingenuity and experience of the form designer, whether a contractor or an engineer. Judgment with respect to the development of a forming system could

414

both expedite a project and reduce costs. Although forms may be job built (design examples are presented later in this chapter), many proprietary forming systems are available. Specialty companies will design and construct forms using patented systems. In multiple-reuse situations, steel formwork has replaced wood formwork to some degree, although the use of wood is still substantial because of its availability and ease of fabrication.

Whatever the eventual type of forming may be, economy in formwork begins when the architect/engineer, during the design of the structure, examines each member and considers ease of forming and reuse of forms.

12-2   FORMWORK REQUIREMENTS

The essential requirements of good formwork may be categorized as quality, safety, and economy.

Quality in formwork requires that the forms be

1. *Accurate:* The size, shape, position, and alignment of structural elements will depend on the accuracy with which the forms are built. The designer (engineer or architect) should specify the allowable tolerances in form dimensions for form design and construction, but not make these tolerances finer than necessary, as this could add unnecessarily to formwork cost as well as delay the project.

2. *Rigid:* Forms must be sufficiently rigid to prevent movement, bulging, or sagging during the placing of concrete. Formwork must therefore be adequately propped, braced, and tied. Special consideration may have to be given to such items as corner details and the effect of any uplift pressures.

3. *Tight jointed:* Joints that are insufficiently tight will leak cement paste. The surface of the concrete will thus be disfigured by fins of the cement paste, and honeycombing may result adjacent to the leaking joint.

4. *Properly finished:* The formwork in contact with the concrete should be so arranged and jointed as to produce a concrete surface of good appearance. Wires, nails, screws, and form surface flaws must not be allowed to disfigure the concrete surface. In some cases a provision of special form lining may be necessary to achieve the desired surface finish.

Safety in formwork requires that the forms be

1. *Strong:* To ensure the safety of the structure and the protection of the workers, it is essential that formwork be designed to carry the full load and side pressures from freshly placed concrete, together with construction traffic and any necessary equipment. On large jobs the design of formwork is usually left to a specialist, but on smaller or routine jobs it may be left to the carpenter supervisor using rule-of-thumb methods.

2. *Sound:* The materials used to construct the forms must be of the correct size and quantity, of good quality, and sufficiently durable for the job.

Economy in formwork requires that the forms be

1. *Simple:* For formwork construction to be economical, it must be designed to be simple to erect and dismantle. Modular dimensions should be used.
2. *Easily handled:* The sizes of form panels or units should be such that they are not too heavy to handle.
3. *Standardized:* Comparative ease of assembly and the possibility of reuse will lower formwork costs when sizes are standardized.
4. *Reusable:* Formwork intended for reuse should be designed for easy removal. If formwork panels have to be ripped down, the concrete may be damaged, materials will be wasted, time will be lost, and expense will be incurred in repairing and replacing damaged panels.

12-3   FORMWORK MATERIALS AND ACCESSORIES

Materials used for forms for concrete structures include lumber, plywood, hardboard, fiberglass, plastic and rubber liners, steel, paper and cardboard, aluminum, fiber forms, and plaster of Paris. Additional materials include nails, bolts, screws, form ties, form clamps, anchors, various types of inserts, and various types of form oils and compounds as well as their accessories. Forms frequently involve the use of two or more materials, such as plywood facing attached to steel frames for wall panels.

Form *lumber* generally consists of the softwoods with the species used being the type available in the local area. Various species are usually grouped together for grading and marketing purposes. Some of these groups and their included species commonly used for formwork are

1. Douglas fir–larch (Douglas fir, western larch)
2. Douglas fir–south (only Douglas fir from southern growth areas)
3. Hem–fir (Western hemlock, California red fir, grand fir, noble fir, white fir, Pacific silver fir)
4. Spruce–pine–fir (alpine fir, balsam fir, black spruce, Englemann spruce, jack pine, lodge-pole pine, red spruce, white spruce)
5. Southern pine (loblolly pine, longleaf pine, shortleaf pine, slash pine)

Form lumber is made up of standard sizes, either rough or dressed, with the grading quality of *construction grade* or *No. 2 grade* usually specified. Shoring or falsework requiring greater capacity should be of *select structural grade* lumber. Partially seasoned wood should be used because it has been found to be the most stable. Green lumber will warp and crack, whereas kiln-dried lumber will swell excessively when it becomes wet. Boards of various thicknesses may be used for sheathing when the imprint of the boards on the concrete surface is desired for architectural reasons.

Plywood is available as flat panels (4 ft × 8 ft sheets) made of thin sheets of wood (called plies) with a total thickness ranging from ¼ in. to 1⅛ in. It is commonly used for sheathing or lining forms because it gives smooth concrete surfaces that require a minimum of hand finishing and because the relatively large sheets are economical and easy to use. U.S. plywood is built up of an odd number of plies with the grain of adjacent layers perpendicular. The plies are dried and joined under pressure with glues that make the joints as strong or stronger than the wood itself. The alternating direction of the grains of adjoining layers equalizes strains and thus minimizes shrinkage and warping of the plywood panels. Generally, the grain direction of the outer layers is parallel to the long dimension of the panel. For maximum strength, the direction of the grain in the face layers should be placed parallel to the span.

The plywood should be the exterior grade because of its waterproof glue. It should also be factory treated with a form oil or parting compound. With proper care, plywood can be reused many times.

Coated plywood is occasionally used. Called *overlaid* or *plastic coated*, it is ordinary exterior plywood with a resin-impregnated fiber facing material fused to one or both sides of the plywood sheet. The overlay covers the grain of the wood, resulting in a smoother and more durable forming surface. The resins used in overlay production are hard and resist water, chemicals, and abrasion. Two grades, high density and medium density, are available, and their difference is in the density of the surfacing material. Coated plywood is generally reused considerably more than the plain plywood form.

Nearly any exterior type of plywood can be used for concrete forming. A special panel called *plyform* is manufactured specifically for that purpose, however. Plyform is exterior-type plywood limited to certain wood species and veneer grades to ensure high performance and is manufactured in two classes, class I and class II. Class I is the strongest, stiffest, and most widely available type. Both classes have smooth, solid surfaces and can have many reuses. They are mill-oiled unless otherwise specified. Plyform is manufactured in thicknesses ranging from ¹⁵⁄₃₂ in. to 1⅛ in., and the most commonly used thicknesses are ⅝ in. and ¾ in.

Plyform can also be manufactured with a *high-density overlaid (HDO) surface* (on one side or both sides). HDO plyform has an exceptionally smooth, hard surface for the smoothest possible concrete finishes and the maximum number of reuses. The HDO surface is a hard, semiopaque resin-fiber overlay heat-fused to panel faces. It can have as many as 200 reuses, and light oiling between pours is recommended.

Structural I plyform is stronger and stiffer than plyform class I or II and is sometimes used when the face grain must be parallel to supports and there is a heavy loading against the forms. It is also available with HDO faces.

Hardboard (fiberboard) is made up of wood particles that have been impregnated, pressed, and baked. When used as a form liner, which is generally the case, hardboard must be backed with sheathing of some kind because it does not have the strength that plywood has. Tempered hardboard, which is preferred for formwork, has been impregnated with drying oil or other material that makes the board less absorbent and improves its strength. Hardboard has limited reuse capability.

Fiberglass is a glass-fiber-reinforced plastic. It is especially suitable for repetitive production of complicated shapes, particularly precast elements. The initial cost is high, but the extensive reuse of one mold without evidence of wear contributes to economy. Suitable reinforcement and backing must be provided so that the mold will not be deformed when it is handled. Molds can be used with or without form oil. Wear of the mold surface will be slightly less if an oil is used.

Plastic and rubber form liners generally come in sheets, either flexible or rigid, and are attached to the solid form sheathing. The term *form liners* includes any sheet, plate, or layer of material attached directly to the inside face of forms to improve or alter the surface texture and quality of the finished concrete. Substances that are applied by brushing, dipping, mopping, or spraying to preserve the form material and to make form stripping easier are referred to as *form coatings*. Some of these coatings are so effective as to approximate the form liner in function. Repetitive use, a great variety of textures and patterns, and ease of stripping are among the advantages of these form liners.

Various types of rigid insulation board are used as form liners and attached to the form sheathing. The boards are generally left in place, either bonded to the concrete or held in place by form plank clips.

Steel forms have been widely used as special-purpose forms. Steel panel systems have been fabricated and used, and steel framing and bracing are important in the construction of many wood and plywood panel systems. Patented steel pans and dome components form the underside of waffle slabs and pan joists. Corrugated steel sheets serve as permanent bottom forms for decks—that is, they remain in place after the concrete has been placed. Fabricated steel form systems are extensively used in the precast and prestressed concrete industries due to their suitability for being reused extensively.

Earth as a form is used in subsurface construction where the soil is stable enough to retain the desired shape of the concrete structure. With the resulting rough surface finish, the use of earth as a form is generally limited to footings and foundations.

Fiber forms are composed of multiple layers of heavy paper bonded together and impregnated with waxes and resins to create a water-repellent cardboard. These are generally single-use, cylindrical-shape molds for columns or other applications where preformed shapes are desirable. "Sonotube" fiber forms are patented light-weight units that can be adapted to various applications.

Form accessories consist of a multitude of items necessary for form planning and construction. Types, availability, and detailed information can usually be obtained from manufacturers' catalogs as well as local concrete construction specialties distributors.

Some of the more common types of accessories are fasteners, spreaders, ties, anchors, hangers, clamps, and inserts. Typical examples of some of the many types available are shown in Figure 12-1. The most common *fastener* used is the double-headed nail, which may be withdrawn with a minimum of effort and little damage to the forms. *Spreaders* are braces that are inserted in forms to keep the faces a proper distance apart until the concrete is placed. Preferably, they are removed during or after the placement of the concrete so that they are not cast in the concrete. They may be wood blocks or proprietary metal pieces. *Ties* are tensile units adapted to holding concrete forms secure against the lateral pressure of the fresh concrete, with

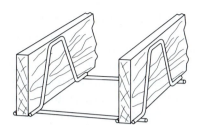

Plank Spreaders for Footing Forms

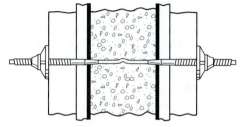

Spreader

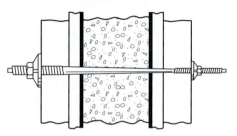

Taper Tie

She-Bolt Tie

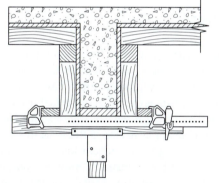

Beam Form Clamp

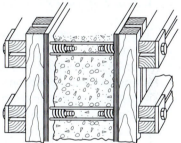

Coil Tie System

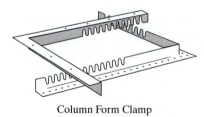

Column Form Clamp

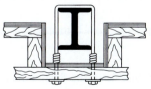

Coil Tie Hanger

Wedge Insert

Hanger Insert

FIGURE 12-1   Typical form accessories. (Courtesy of A. H. Harris and Sons, Inc.)

or without provision for spacing the forms a definite distance apart. They pass through the concrete and are fastened on each side. Various types of ties are available, many of which are patented items. *Anchors* are devices used to secure formwork to previously placed concrete of adequate strength and are normally embedded in concrete during placement. There are two basic parts, the embedded anchoring device and the external fastener, which is removed after use. *Hangers* are devices used for suspending one object from another, such as the hardware attached to a building frame to support forms. *Inserts* are of many designs and are attached to the forms in such a way that they remain in the concrete when the forms are stripped. They provide for anchorage of brick or stone veneers, pipe hangers, suspended ceilings, duct work, and any building hardware or components that must be firmly attached to the concrete. *Clamps* consist of many varied devices but serve a similar purpose to that of the tie. Beam form clamps firmly hold the beam sides and bottom together with a minimum of nailing required. Column clamps or yokes encircle column forms and hold them together securely, withstanding the lateral pressure of the freshly placed concrete.

12-4   LOADS AND PRESSURES ON FORMS

The concrete form is a structure. Like all structures, it must be designed and constructed with due regard to the effects of the imposed loads. A well-designed form must have adequate strength to resist failure; in addition, it must have sufficient stiffness so that deflection will not be excessive. Efficient and economical use of the material of which the forms are built is also an integral part of the design process.

The first consideration in the design of a form for concrete is the load to be supported. The loads may be considered in two categories: vertical loads and lateral loads. The lateral loads may be further categorized as externally applied load (such as wind) and internally applied pressure due to the contained fresh concrete. These loads must be carried to the ground by the formwork system or resisted by in-place construction that has adequate strength for the purpose.

Vertical loads consist of dead load and live load. The dead load consists of the weight of the formwork and the freshly placed concrete. Live load consists of the weight of workers and equipment, stored materials, and impact due to moving loads. ACI 347-04. *Guide to Formwork for Concrete* [1], recommends that the formwork be designed for the following:

1.   Minimum live load of 50 psf of horizontal projection (75 psf if motorized carts are used for delivery of the concrete).
2.   Minimum combined dead and live load of 100 psf (125 psf if motorized carts are used for delivery of the concrete).

Note that these minimum design loads would be applicable to the design of structural floor and roof slabs (not to the design of slabs on ground).

Lateral pressure is exerted on all nonhorizontal surfaces that contain the fresh concrete. The pressures are imposed on wall and column forms and in essence are hydraulic loadings. The amount of lateral pressure is influenced by the weight, chemistry, and temperature of the concrete, the vertical rate of placement, size and shape of the form, height of the form, and method of consolidation (hand spaded or mechanically vibrated). The amount and location of reinforcing also affect lateral pressure, although their effect is small and is usually neglected. ACI 347-04 provides formulas that can be used to reasonably predict concrete pressure on wall forms and column forms. In the formulas, the following notation is used:

p = lateral pressure (psf)

R = rate of placement (ft/h)

T = temperature of concrete in the forms (°F)

h = height of fluid or plastic concrete above point considered (ft)

w = unit weight of fresh concrete (pcf)

C_C = chemistry coefficient

C_W = unit weight coefficient

A liquid head formula is the basis for lateral pressure calculations:

$$p = wh \qquad \text{[ACI 347-04 Eq. (2.1)]}$$

This formula will apply where a form is filled rapidly, before any hardening of the concrete occurs, and in situations where the conditions of Equation (2.2), Equation (2.3) and Equation (2.4) are not met. The limits stated for Equation (2.2), Equation (2.3) and Equation (2.4) do not apply to Equation (2.1).

For the special conditions of

a.   concrete having a slump of 7 in. or less and

b.   placed with normal internal vibration to a depth of 4 ft or less,

formwork can be designed for a lateral pressure as follows:

1.   For columns (note that a column, for purposes of applying the pressure formulas, is defined as a vertical element with no plan dimension exceeding 6.5 ft),

$$p_{max} = C_W C_C \left[150 + 9000 \frac{R}{T} \right] \qquad \text{[ACI 347-04 Eq. (2.2)]}$$

with a minimum of $600\, C_W$ lb/ft^2, but in no case greater than wh.

2.   For walls, with a rate of placement of less than 7 ft/hr and a placement height not exceeding 14 ft,

$$p_{max} = C_W C_C \left[150 + \frac{9000R}{T} \right] \qquad \text{[ACI 347-04 Eq. (2.3)]}$$

with a minimum of $600 C_W$ lb/ft^2, but in no case greater than wh.

3.   For walls with a placement rate of less than 7 ft/h, where placement height exceeds 14 ft, and for all walls with a placement rate of 7 to 15 ft/h,

$$p_{max} = C_W C_C \left[150 + \frac{43{,}400}{T} + \frac{2800R}{T} \right] \qquad \text{[ACI 347-04 Eq. (2.4)]}$$

with a minimum of 600 C_W lb/ft^2, but in no case greater than wh.

The unit weight coefficient C_W has the value of 1.0 for concretes having a unit weight in the range of 140 to 150 lb/ft^3. For concrete unit weights outside this range, see Table 2.1 in ACI 347-04. The chemistry coefficient C_C has the value of 1.0 for Types I, II, and III cements without retarders. For other types of cements or blends containing slag or fly ash, see Table 2.2 in ACI 347-04. For the illustrations in this text, we will assume both C_W and C_C equal to 1.0.

Formwork should also be designed to resist all foreseeable *lateral loads*, such as seismic forces, wind, cable tensions, inclined supports, dumping of concrete, and impact due to equipment. ACI 347-04 recommends for slabs a minimum horizontal design load of 100 lb per lineal foot of floor edge or 2% of the total dead load of the floor, whichever is greater. Wall forms exposed to the elements should be designed for a minimum wind load of 15 psf, and bracing for wall forms should be designed for a lateral load of at least 100 lb per lineal foot of wall applied at the top.

In addition, formwork should be designed for any special loads likely to occur, such as unsymmetrical placement of concrete and uplift. Form designers must be alert to provide for any and all special loading conditions.

12-5   THE DESIGN APPROACH

The design of job-built forms may be considered largely as beam and post design. The bending members usually span several supports and are therefore indeterminate. Assumptions and approximations are made that simplify the calculations and facilitate the design process.

After establishing the appropriate design loads, the sheathing and supporting members are analyzed or designed in sequence. Bending members (sheathing, joists, studs, stringers, or wales) are considered uniformly loaded and supported on (1) a single span, (2) two spans, or (3) three or more spans. The uniform load assumption is common practice unless the spacing of point loads exceeds one-third to one-half of the span between supports, in which case the worst loading condition is investigated.

Each bending member should be analyzed or designed for bending moment, shear, and deflection. In addition, vertical supports (shoring) and lateral bracing (if applicable) must be analyzed or designed for either compressive or tensile loads. Bearing stresses at supports must be investigated (except for the sheathing).

In the analysis or design of the component parts, the traditional stress equations are used (see any strength of materials text). These expressions for stress are as follows. For bending stress:

$$f_b = \frac{M}{S}$$

For shear stress:

$$f_v = \frac{1.5V}{A} \qquad \text{(for rectangular wood members)}$$

$$f_v = \frac{VQ}{Ib} \quad \text{or} \quad \frac{V}{Ib/Q} \qquad \text{(for plywood)}$$

For compression stress (both parallel and perpendicular to the grain):

$$f_{comp} = \frac{P}{A}$$

For tension stress:

$$f_{tens} = \frac{P}{A}$$

where

f_b = calculated unit stress in bending (psi)

f_v = calculated unit stress in shear (psi)

f_{comp} = calculated unit stress in compression (either parallel or perpendicular to the grain) (psi)

f_{tens} = calculated unit stress in tension (psi)

M = maximum bending moment (in.-lb)

S = section modulus (in.^{3})

P = concentrated load (lb)

V = maximum shear (lb)

A = cross-sectional area (in.^{2})

Ib/Q = rolling shear constant (in.^{2}/ft)

If we equate allowable unit stresses to the maximum unit stresses developed in a beam subjected to a uniformly distributed load w (lb/ft), expressions can be derived for the maximum allowable span length. Therefore, knowing the design loads and member section properties, a maximum allowable span length can be computed.

Table 12-1 contains expressions for maximum allowable span length as governed by moment, shear, and deflection. The reader may wish to derive these expressions from basic principles. The moment expressions are based on the maximum positive or negative moment. The maximum shear considered is the shear that exists at d distance from the support, where d is the depth of the member. Shear considerations for plywood vary from those accorded sawn lumber because of the cross-directional way in which the plies of the plywood are assembled. When plywood is loaded in flexure, the plies with the grain perpendicular to the direction of the span are the weakest aspect of the plywood with respect to shear. The wood fibers in these plies roll at stresses below the fiber shear strength parallel to the grain, hence the name *rolling shear*. For a discussion of the

TABLE 12-1   Concrete Form Design Equations (to Determine Allowable Span Length)

	One span	Two spans	Three or more spans
Bending moment	$\ell = 9.8\sqrt{\dfrac{F_b S}{w}}$	$\ell = 9.8\sqrt{\dfrac{F_h S}{w}}$	$\ell = 10.95\sqrt{\dfrac{F_b S}{w}}$
Shear	$\ell = \dfrac{16 F_v A}{w} + 2d$ $\ell^a = \dfrac{24 F_v (Ib/Q)}{w} + 2d$	$\ell = \dfrac{12.8 F_v A}{w} + 2d$ $\ell^a = \dfrac{19.2 F_v (Ib/Q)}{w} + 2d$	$\ell = \dfrac{13.3 F_v A}{w} + 2d$ $\ell^a = \dfrac{20 F_v (Ib/Q)}{w} + 2d$
Deflection $\Delta_{\text{all.}} = \dfrac{\ell}{240}$	$\ell = 1.57\sqrt[3]{\dfrac{EI}{w}}$	$\ell = 2.10\sqrt[3]{\dfrac{EI}{w}}$	$\ell = 1.94\sqrt[3]{\dfrac{EI}{w}}$
Deflection $\Delta_{\text{all.}} = \dfrac{\ell}{360}$	$\ell = 1.37\sqrt[3]{\dfrac{EI}{w}}$	$\ell = 1.83\sqrt[3]{\dfrac{EI}{w}}$	$\ell = 1.69\sqrt[3]{\dfrac{EI}{w}}$
When deflection Δ (in.) is specified	$\ell = 5.51\sqrt[4]{\dfrac{\Delta EI}{w}}$	$\ell = 6.86\sqrt[4]{\dfrac{\Delta EI}{w}}$	$\ell = 6.46\sqrt[4]{\dfrac{\Delta EI}{w}}$

Notes: ℓ, span length (center to center of supports) (in.); F_b, allowable bending stress (psi); S, section modulus (in.^{3}); w, uniform load (lb/ft); F_v, allowable shear stress (psi); A, cross-sectional area (in.^{2}); d, depth of member (in.); E, modulus of elasticity (psi); I, moment of inertia (in.^{4}); Ib/Q, rolling shear constant for plywood (in.^{2}/ft); Δ, deflection (in.).

^{a}For plywood only.

properties of plywood with respect to shear, see [2]. Deflection of forms must be limited to minimize unsightly bulges in the resulting concrete surface. The deflection limit may be specified as a fraction of span (that is, $\ell/240$), as a limit ($\frac{1}{8}$ in.), or as the smaller of the two. The architect/engineer must decide on the deflection limits for the formwork based on bulging and sagging that can be tolerated in the surfaces of the finished structure. The casting of test panels may be warranted in some cases. Limiting deflection to $\frac{1}{360}$ of the span is acceptable in many cases where surfaces are coarse-textured and there is little reflection of light. Three design equations for deflection are given in Table 12-1. Alternatively, the designer may wish to compute the required size of the members when the design loads and span lengths are known. The same basic principles apply.

Section properties for selected thicknesses of plyform are given in Table 12-2. These section properties reflect that various species of wood used in manufacturing plywood have different stiffness and strength properties. Those species with similar properties are assigned to a species group. To simplify plywood design, the effects of using different species groups in a given panel as well as the effects of the cross-banded construction of the panel have been taken into consideration in establishing the section properties.

TABLE 12-2   Section Properties and Design Values for Plyform

Thickness t (in.)	Approx. weight (psf)	Properties for stress applied parallel with face grain			Properties for stress applied perpendicular to face grain		
		Moment of inertia I (in.^{4}/ft)	Effective section modulus KS (in.^{3}/ft)	Rolling shear const. Ib/Q (in.^{2}/ft)	Moment of inertia I (in.^{4}/ft)	Effective section modulus KS (in.^{3}/ft)	Rolling shear const. Ib/Q (in.^{2}/ft)
Class I							
1/2	1.5	0.077	0.268	5.153	0.024	0.130	2.739
5/8	1.8	0.130	0.358	5.717	0.038	0.175	3.094
3/4	2.2	0.199	0.455	7.187	0.092	0.306	4.063
7/8	2.6	0.296	0.584	8.555	0.151	0.422	6.028
1	3.0	0.427	0.737	9.374	0.270	0.634	7.014
Class II							
1/2	1.5	0.075	0.267	4.891	0.020	0.167	2.727
5/8	1.8	0.130	0.357	5.593	0.032	0.225	3.074
3/4	2.2	0.198	0.454	6.631	0.075	0.392	4.049
7/8	2.6	0.300	0.591	7.990	0.123	0.542	5.997
1	3.0	0.421	0.754	8.614	0.220	0.812	6.987
Structural I							
1/2	1.5	0.078	0.271	4.908	0.029	0.178	2.725
5/8	1.8	0.131	0.361	5.258	0.045	0.238	3.073
3/4	2.2	0.202	0.464	6.189	0.108	0.418	4.047
7/8	2.6	0.317	0.626	7.539	0.179	0.579	5.991
1	3.0	0.479	0.827	7.978	0.321	0.870	6.981

	Design values		
	Plyform class I	Plyform class II	Structural I plyform
Modulus of elasticity (psi)	1,650,000	1,430,000	1,650,000
Bending stress (psi)	1930	1330	1930
Rolling shear stress (psi)	72	72	102

Source: APA—The Engineered Wood Association [3].

Note: All properties adjusted to account for reduced effectiveness of plies with grain perpendicular to applied stress.

In calculating these section properties, all plies were transformed to properties of the face ply. As a result, the designer need not be concerned with the actual panel layup but only with the allowable stresses for the face ply and the given section properties of Table 12-2. The section properties of Table 12-2 are generally the minimums that can be expected. Hence the actual panel obtained in the marketplace will usually have a section property greater than that represented in the table.

The plyform design values presented in Table 12-2 are based on wet strength and 7 day load duration, so no further adjustment in these values is required except for the modulus of elasticity. The modulus shown is an adjusted value based on the assumption that shear deflection is computed separately from bending deflection. These values should be used for bending deflection calculations (which is the usual case). To calculate shear deflection, the modulus should be reduced to 1,500,000 psi for class I and structural plyform and to 1,300,000 for class II plyform.

Section properties for selected members of standard dressed (S4S) sawn lumber are given in Table 12-3. Typical base design values for visually graded dimension lumber are furnished in Table 12-4. This table is simplified and brief and is primarily intended as a resource to accompany the examples and problems of this text. Those

TABLE 12-3   Properties of Structural Lumber

Nominal size (in.)	Standard dressed size (S4S) (in.)	Area of section, A (in.^{2})	Moment of inertia,[a] I (in.^{4})	Section Modulus,[a] S (in.^{3})	Weight[b] (lb/ft)
2 × 4	1½ × 3½	5.25	5.36	3.06	1.28
2 × 6	1½ × 5½	8.25	20.80	7.56	2.01
2 × 8	1½ × 7¼	10.88	47.63	13.14	2.64
2 × 10	1½ × 9¼	13.88	98.93	21.39	3.37
2 × 12	1½ × 11¼	16.88	178.0	31.64	4.10
3 × 6	2½ × 5½	13.75	34.66	12.60	3.34
3 × 8	2½ × 7¼	18.13	79.39	21.90	4.41
3 × 10	2½ × 9¼	23.13	164.9	35.65	5.62
3 × 12	2½ × 11¼	28.13	296.6	52.73	6.84
4 × 4	3½ × 3½	12.25	12.51	7.15	2.98
4 × 6	3½ × 5½	19.25	48.53	17.65	4.68
4 × 8	3½ × 7¼	25.38	111.1	30.66	6.17
4 × 10	3½ × 9¼	32.38	230.8	49.91	7.87
4 × 12	3½ × 11¼	39.38	415.3	73.83	9.57
6 × 6	5½ × 5½	30.25	76.26	27.73	7.35
6 × 8	5½ × 7½	41.25	193.4	51.56	10.03
6 × 10	5½ × 9½	52.25	393.0	82.73	12.70
6 × 12	5½ × 11½	63.25	697.1	121.2	15.37

[a]I and S are about the strong axis.
[b]Weight in lb/ft when the unit weight of wood is 35 lb/ft^3.

TABLE 12-4   Base Design Values for Visually Graded Dimension Lumber (Normal Load Duration and Dry Service Condition)

#2 Grade (2"–4" thick, 2" and wider)

Species	Design values (psi)					
	F_b	F_t	F_v	$F_{c\perp}$	F_c	E
Douglas fir–larch	900	575	180	625	1350	1,600,000
Douglas fir–south	850	525	180	520	1350	1,200,000
Hem–fir	850	525	150	405	1300	1,300,000
Spruce–Pine–fir	875	450	135	425	1150	1,400,000

Construction grade (2"–4" thick, 2"–4" wide)

Species	Design values (psi)					
	F_b	F_t	F_v	$F_{c\perp}$	F_c	E
Douglas fir–larch	1000	650	180	625	1650	1,500,000
Douglas fir–South	975	600	180	520	1650	1,200,000
Hem–fir	975	600	150	405	1550	1,300,000
Spruce–pine–fir	1000	500	135	425	1400	1,300,000

Southern pine (2"–4" thick)

#2 Grade	Design values (psi)					
	F_b	F_t	F_v	$F_{c\perp}$	F_c	E
2"–4" wide	1500	825	175	565	1650	1,600,000
5"–6" wide	1250	725	175	565	1600	1,600,000
8" wide	1200	650	175	565	1550	1,600,000
10" wide	1050	575	175	565	1500	1,600,000
12" wide	975	550	175	565	1450	1,600,000
Construction grade 4" wide	1100	625	175	565	1800	1,500,000

Adjustment factors	F_b	F_t	F_v	$F_{c\perp}$	F_c	E
Load duration C_D*	1.25	1.25	1.25	—	1.25	—
Wet service C_M	0.85**	1.0	0.97	0.67	0.80***	0.90

Source: Courtesy of American Forest & Paper Association, Washington, D.C. [4].

*Values shown are typical for forming applications.

**When $F_b(C_F) \le 1150$ psi, $C_M = 1.0$.

***When $F_c(C_F) \le 750$ psi, $C_M = 1.0$; for southern pine, when $F_c \le 750$ psi, $C_M = 1.0$

who require more detailed information with respect to section properties and design values should obtain Reference [4].

The base design values in Table 12-4 must be modified by applicable adjustment factors that are appropriate for the conditions under which the wood is used. Among the adjustment factors itemized in [4] are

$$C_D = \text{load duration factor}$$

$$C_M = \text{wet service factor}$$

$$C_t = \text{temperature factor}$$

$$C_F = \text{size factor (not applicable to southern pine)}$$

$$C_{fu} = \text{flat use factor}$$

$$C_r = \text{repetitive member factor}$$

$$C_L = \text{beam stability factor}$$

$$C_P = \text{column stability factor}$$

$$C_i = \text{incising factor}$$

$$C_b = \text{bearing area factor}$$

The product of the base design value (F_b for moment or F_v for shear) and any applicable adjustment factors provides an allowable unit stress for use in the design and analysis of wood members. The allowable unit stress is denoted as a primed value (e.g., F_b'), which indicates that it is an adjusted design value calculated for specific conditions. For instance, for allowable shear stress,

$$F_v' = F_v C_D C_M C_t C_i$$

If no adjustment factors apply, then the allowable unit stress will be equal to the tabulated base design value (e.g., $F_v' = F_v$). One required value that is not an allowable stress is modulus of elasticity E. It will be determined in a similar fashion as the product of a base design value and applicable adjustment factors.

Determination of Allowable Bending Stress F_b'

Applicable adjustment factors for allowable bending stress are load duration, wet service, temperature, size, flat use, repetitive member, and beam stability.

The load duration factor, C_D, for 7 days or less duration of load (common for form design) is given in Table 12-4.

The wet service factor, C_M, is applicable when the moisture content of the wood is more than 19% (which would occur in a wet service situation such as when the wood is in contact with fresh concrete) and the wood is at ordinary temperature. Table 12-4 shows C_M factors.

The temperature factor, C_t, is generally not applicable in form design and may be disregarded.

TABLE 12-5   Size and Flat Use Factors for Bending Stress F_b

Width of lumber	Size factor $C_F{}^a$		Flat use factor C_{fu}^b	
	2"–3" thick	4" thick	2"–3" thick	4" thick
2"–3"	1.5	1.5	1.0	—
4"	1.5	1.5	1.1	1.0
5"	1.4	1.4	1.1	1.05
6"	1.3	1.3	1.15	1.05
8"	1.2	1.3	1.15	1.05
10"	1.1	1.2	1.2	1.1
12"	1.0	1.1	1.2	1.1

Source: Courtesy of American Forest & Paper Association, Washington, D.C. [4].
[a] Applicable to No. 2 grade.
[b] Applicable to No. 2 grade and construction grade.

The size factor, C_F, is based on member size and is shown in Table 12-5. Construction-grade-material allowable stresses are not adjusted for the size of the member. Southern pine base design values shown in Table 12-4 have size adjustments already included. The size factor C_F applies only to visually graded sawn lumber members.

The flat use factor, C_{fu}, is also shown in Table 12-5 and is applicable when dimension lumber 2 in. to 4 in. thick is loaded on the wide face.

The repetitive member factor, C_r, applies only to F_b and to members 2 in. to 4 in. thick. ACI Committee 347, however, recommends against application of C_r for cases where base stresses have already been increased the 25% permitted for short-duration loads [5].

The beam stability factor, C_L, is essentially a reduction factor for F_b where insufficient lateral restraint is furnished for the bending member. Because formwork assemblies are usually designed to meet lateral support criteria as prescribed by the *National Design Specification for Wood Construction* [4], the use of C_L is generally not applicable. For the purpose of this text, all form bending-members are assumed to have adequate lateral support.

Determination of Allowable Shear Stress F_v'

Applicable adjustment factors for allowable shear stress are wet service, load duration, temperature, and incising factor.

The wet service factor C_M and the load duration factor C_D are furnished in Table 12-4. The temperature factor is generally not applicable in form design.

The incising adjustment factor C_i is a measure of the incisions that have to be made in a wood member to enhance its ability to receive pressure treatment. Thus, incising is used to increase the depth of penetration of wood preservatives in a wood member. The C_i factor is taken as 1.0 for non-incised wood, such as in the examples and problems of this chapter.

Determination of Allowable Stress for Compression Perpendicular to the Grain $F'_{c\perp}$

Applicable adjustment factors for allowable stress perpendicular to the grain are wet service, temperature, and bearing area. Note that the stress increase for short-term loading does not apply to $F_{c\perp}$.

The wet service factor C_M is furnished in Table 12-4. Again, the temperature factor is generally not applicable in form design.

The bearing area factor, C_b, is applicable where the bearing length is less than 6 in. long and at least 3 in. from the end of the member. It may be calculated from

$$C_b = \frac{\ell_b + 0.375}{\ell_b}$$

where ℓ_b is the length of the bearing (in.) measured parallel to the grain. For bearing at the end of a member and lengths of bearing equal to 6 in. or more, use $C_b = 1.0$. Note that C_b will never be less than 1.0. Therefore, it is conservative to omit it.

Determination of Modulus of Elasticity E'

The only applicable adjustment factor for modulus of elasticity in form design is the wet service C_M. It is shown in Table 12-4.

12-6   DESIGN OF FORMWORK FOR SLABS

Figure 12-2 shows a typical structural system for job-built forms for elevated slabs. The sequence of design is first to consider a strip of sheathing of the specified thickness and 12 in. in width. The maximum allowable span may then be determined based on the allowable values of bending stress, shear stress, and deflection for the sheathing. The lower of the computed values will determine the maximum spacing of the joists. This span value, usually rounded down to some lower modular value, becomes the spacing of the joists.

Based on the joist spacing used, the joist itself is analyzed to determine its maximum allowable span. Each joist must support the load from the sheathing halfway over to the adjacent joist on either side. Therefore, the *width* of the load area carried by the joist is equal to the spacing of the joists. The joist span selected becomes the spacing of the stringers. Again, a modular value is selected for stringer spacing.

Based on the selected stringer spacing, the process is repeated to determine the maximum stringer span (distance between vertical supports or shores). Notice in the design of the stringers that the joist loads are actually applied to the stringer as a series of concentrated loads at the points where the joists rest on the stringer. It is simpler and sufficiently accurate to treat the load on the stringer as a uniformly distributed load, however. Again, the width of the uniform design load applied to the stringer is equal to the stringer spacing. The calculated stringer span must next be checked against the capacity of the shores used to support the stringers. The load on each shore is equal to the shore spacing multiplied by the load per foot of stringer.

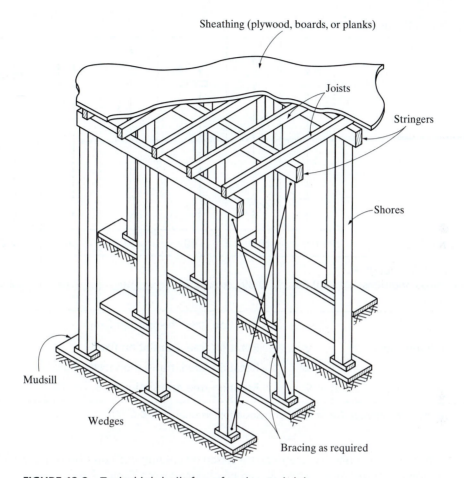

FIGURE 12-2   Typical job-built form for elevated slab.

Thus the maximum shore spacing (or stringer span) is limited to the lower span length as governed by stringer strength or shore strength. In addition, it is necessary to check the bearing at the point where each joist rests on the stringer. This is done by dividing the load at this point by the bearing area and comparing the resulting stress with the allowable unit stress in compression *perpendicular* to the grain. A similar procedure is applied at the point where each stringer rests on a vertical support.

The stringers shown in Figure 12-2 are supported by solid, rectangular wood shores that are columns (which we will assume are axially loaded). As with all axially loaded columns, the allowable load is a function of the slenderness ratio ℓ/d, the ratio of the unbraced length of the member to its least lateral dimension (not to exceed 50). The slenderness ratio ℓ/d is further modified and expressed as ℓ_e/d, where ℓ_e represents an effective unbraced column length and is defined as

$$\ell_e = K_e \ell$$

TABLE 12-6   Effective Length Factor K_e

Buckling modes						
Theoretical K_e value	0.5	0.7	1.0	1.0	2.0	2.0
Recommended design K_e when ideal conditions approximated	0.65	0.80	1.2	1.0	2.10	2.4
End condition code		Rotation fixed, translation fixed				
		Rotation free, translation fixed				
		Rotation fixed, translation free				
		Rotation free, translation free				

Source: Courtesy of American Forest & Paper Association, Washington, D.C. [4].

where K_e is an effective length factor based on column end conditions that affect rotation and translation. Factor K_e can be obtained from Table 12-6.

The following design approach applies for the determination of the allowable stress for compression parallel to the grain for simple, solid-sawn lumber columns, as recommended in [4].

Determination of Allowable Stress for Compression Parallel to the Grain F_c'

Applicable adjustment factors for allowable stress for compression parallel to the grain are load duration, wet service, temperature, size, and column stability. This may be expressed as

$$F_c' = F_c C_D C_M C_t C_F C_P$$

where

F_c = base design value for compression parallel to the grain (psi)

C_p = column stability factor

and all other quantities have been previously defined. This allowable stress is applicable to all values of the slenderness ratio (≤ 50) and replaces the previously used short-, intermediate-, and long-column equations of earlier design specifications. Note that although the slenderness ratio ℓ_e/d for solid columns shall not exceed 50, during construction this limit is increased to 75.

Because wood shores (columns) are generally reused repeatedly, ACI Committee 347 does not recommend the use of any adjustment factor that provides increased stresses for short load duration [5]. In addition, temperature and wet service adjustments for wood shores are generally not required or considered. Therefore only the column size factor and the column stability factor are normally considered in shore design.

The size factor, C_F, for compression parallel to the grain for the two grades indicated can be taken from Table 12-7.

The column stability factor, C_p, is a function of the effective slenderness ratio ℓ_e/d of the shore, the adjusted modulus of elasticity E', and the adjusted base value of compression parallel to the grain before C_p is applied.

The unadjusted modulus of elasticity is normally used for wood shore design as shores are rarely in the wet service condition. Therefore $E' = E$.

The column stability factor C_P for solid-sawn lumber can be calculated from Equation (3.7-1) in [4]. For convenience, and to use with the limited properties of Table 12-4, this equation is modified and written as

$$C_P = \frac{1+\alpha}{1.6} - \sqrt{\left(\frac{1+\alpha}{1.6}\right)^2 - \frac{\alpha}{0.8}}$$

where

$$\alpha = \frac{0.3\, E'}{\left(\dfrac{\ell_e}{d}\right)^2 F_c^*}$$

and F_c^* is the adjusted base value of compression stress parallel to the grain before application of C_P.

TABLE 12-7   Size Factor C_F for Compression Parallel to the Grain

Width of lumber	Size factor C_F	
	No. 2 grade	Const. grade
2″–4″	1.15	1.0
5″	1.1	—
6″	1.1	—
8″	1.05	—
10″	1.0	—
12″	1.0	—

Source: Courtesy of American Forest & Paper Association, Washington, D.C. [4].

Various tables are available to simplify the design approach and to expedite the selection of the formwork structural members [5]. These should be used with caution because of job-specific conditions and possible tabular limitations.

Example 12-1

Design the formwork for a 6-in. structural concrete floor slab. The floor system is of the type shown in Figure 12-2. Use ¾-in. class I plyform for the sheathing and No. 2 Douglas fir–larch for the rest of the lumber. The maximum deflection for the sheathing may be taken as $\frac{1}{240}$ of the span. The maximum deflection for other bending members is taken as $\frac{1}{360}$ of the span. Based on end conditions, the effective unbraced length of the shores may be taken as 10 ft. The following conditions for design have been established:

1. Joists and stringers will be designed based on adequate lateral support.
2. Joists, stringers, and shores are used under dry service conditions.

Solution:

1. *Sheathing design* (find the joist spacing): Consider a 12-in.-wide strip of sheathing perpendicular to the supporting joists. The sheathing acts as a beam and is continuous over three or more supports. Determine the maximum allowable span for the sheathing. This becomes the maximum spacing for the joists (which support the sheathing).

   a. The design values for ¾-in. class I plyform (see Table 12-2) are

   $$E = 1,650,000 \text{ psi} \qquad \text{(modulus of elasticity)}$$

   $$F_b = 1930 \text{ psi} \qquad \text{(bending stress)}$$

   $$F_v = 72 \text{ psi} \qquad \text{(rolling shear stress)}$$

   The plyform properties with face grain parallel to span (perpendicular to the joists) are

   $$I = 0.199 \text{ in.}^4/\text{ft}$$

   $$S = 0.455 \text{ in.}^3/\text{ft}$$

   $$Ib/Q = 7.187 \text{ in.}^2/\text{ft} \qquad \text{(rolling shear constant)}$$

   $$w_s = 2.2 \text{ psf} \qquad \begin{array}{l}\text{(weight of sheathing: for a 12 in.}\\ \text{strip, this will be lb/ft)}\end{array}$$

   b. The loading on the sheathing is

DL (slab): (6/12)(150)	=	75 psf
LL (min.):	=	50 psf
Neglect sheathing weight		125 psf $= w$

c. The maximum joist spacing based on the bending moment formula (from Table 12-1) is

$$\ell = 10.95\sqrt{\frac{F_b S}{w}}$$

$$= 10.95\sqrt{\frac{1930(0.455)}{125}} = 29 \text{ in.}$$

d. The maximum joist spacing based on shear is

$$\ell = \frac{20F_v(Ib/Q)}{w} + 2d$$

$$= \frac{20(72)(7.187)}{125} + 2(0.75) = 84.3 \text{ in.}$$

e. The maximum joist spacing based on deflection (maximum deflection is $\frac{1}{240}$ of the span) is

$$\ell = 1.94\sqrt[3]{\frac{EI}{w}}$$

$$= 1.94\sqrt[3]{\frac{1,650,000(0.199)}{125}}$$

$$= 26.8 \text{ in.}$$

Deflection controls. The maximum spacing of supporting joists should not exceed 26.8 in. Use joists at 24 in. o.c.

2. *Joist design* (find the stringer spacing): Assume 2 × 8 (S4S) joists and a 7-day maximum duration of load. Consider the joists as uniformly loaded beams continuous over three or more spans (the supports are the stringers).

a. Obtain allowable stresses using base design values from Table 12-4 and appropriate adjustment factors.

1. Bending: $F_b = 900$ psi. Adjustment factors:

a. Load duration factor from Table 12-4: $C_D = 1.25$.

b. Size factor from Table 12-5: $C_F = 1.2$.

Therefore

$$F_b' = F_b C_D C_F$$

$$= 900(1.25)(1.2) = 1350 \text{ psi}$$

2. Shear: $F_v = 180$ psi. Adjustment factor:
Load duration factor from Table 12-4: $C_D = 1.25$.
Therefore

$$F_v' = F_v C_D = 180(1.25) = 225 \text{ psi}$$

3. Modulus of elasticity: $E = 1,600,000$ psi. No adjustment factors apply. Thus

$$E' = E = 1,600,000 \text{ psi}$$

Properties for the 2×8 (S4S) from Table 12-3 are

$$S = 13.14 \text{ in.}^3$$

$$I = 47.63 \text{ in.}^4$$

$$A = 10.88 \text{ in.}^2$$

b. Loading: Because joists support a 2-ft width of sheathing, the loading on the joist is

$$w = 125 \text{ psf } (2) = 250 \text{ lb/ft}$$

Assume the weight of the sheathing and joists to be 5 psf. Then

$$w = 250 + 5(2) = 260 \text{ lb/ft}$$

c. The maximum stringer spacing based on bending moment is

$$\ell = 10.95\sqrt{\frac{F_b'S}{w}}$$

$$= 10.95\sqrt{\frac{1350(13.14)}{260}} = 90.4 \text{ in.}$$

d. The maximum stringer spacing based on shear is

$$\ell = \frac{13.3F_v'A}{w} + 2d$$

$$= \frac{13.3(225)(10.88)}{260} + 2(7.25)$$

$$= 139.7 \text{ in.}$$

e. The maximum stringer spacing based on deflection (maximum deflection is $\frac{1}{360}$ of the span) is

$$\ell = 1.69\sqrt[3]{\frac{E'I}{w}}$$

$$= 1.69\sqrt[3]{\frac{1,600,000(47.63)}{260}}$$

$$= 112.3 \text{ in.}$$

Bending governs. Therefore the maximum spacing of supporting stringers cannot exceed 90.4 in. Use a stringer spacing of 7 ft 0 in. (84 in.).

3. *Stringer design* (find the shore spacing): Use 4 in. × 8 in. (S4S) stringers and a 7-day maximum duration of load. Consider the stringers to be uniformly loaded beams continuous over three or more supports. The supports are the shores. The loads from the joists are concentrated loads, but for simplicity, we will assume a uniformly distributed load.

   a. Obtain allowable stresses using base design values from Table 12-4 and appropriate adjustment factors.

      1. Bending: F_b = 900 psi. Adjustment factors:

         a. Load duration factor from Table 12-4: C_D = 1.25.

         b. Size factor from Table 12-5: C_F = 1.3.

         Therefore

         $$F_b' = F_b C_D C_F$$
         $$= 900(1.25)(1.3) = 1463 \text{ psi}$$

      2. Shear: F_v = 180 psi. Adjustment factor:
         Load duration factor from Table 12-4: C_D = 1.25.
         Therefore

         $$F_v' = F_v C_D = 180(1.25) = 225 \text{ psi}$$

      3. Modulus of elasticity: E = 1,600,000 psi. No adjustment factors apply. Thus

         $$E' = E = 1,600,000 \text{ psi}$$

      Properties for the 4 × 8 (S4S) from Table 12-3 are

      $$S = 30.66 \text{ in.}^3$$
      $$I = 111.1 \text{ in.}^4$$
      $$A = 25.38 \text{ in.}^2$$

   b. Loading: Each stringer supports a 7 ft-0 in.-wide strip of design load. Assuming a formwork weight of 5 psf, the uniformly distributed load on a stringer is calculated from

      $$w = 7.0(125 + 5) = 910 \text{ lb/ft}$$

   c. The maximum shore spacing based on bending (of the stringers) is

      $$\ell = 10.95\sqrt{\frac{F_b'S}{w}}$$
      $$= 10.95\sqrt{\frac{1463(30.66)}{910}} = 76.9 \text{ in.}$$

   d. The maximum shore spacing based on shear is

      $$\ell = \frac{13.3F_v'A}{w} + 2d$$
      $$= \frac{13.3(225)(25.38)}{910} + 2(7.25) = 98.0 \text{ in.}$$

e.   The maximum shore spacing based on deflection is

$$\ell = 1.69 \sqrt[3]{\frac{EI}{w}}$$

$$= 1.69 \sqrt[3]{\frac{1,600,000(111.1)}{910}} = 98.1 \text{ in.}$$

Bending governs. Therefore the maximum spacing of supporting shores cannot exceed 76.9 in. Use a shore spacing of 6 ft-0 in. (72.0 in.).

4.   *Design of shores:* The stringers are spaced at 6 ft-6 in. on center and are supported by shores at 6 ft-0 in. on center. Therefore each shore must support a floor area of

$$7.0(6.0) = 42 \text{ ft}^2$$

Again, assuming formwork weight of 5 psf, the load per shore is calculated as

$$42(125 + 5) = 5460 \text{ lb}$$

Although commercial shores are usually readily available to support this load, we will design 4 × 4 wood shores. The effective unbraced length of the shore, ℓ_e, is 10 ft-0 in. in each direction. The capacity of the 4 × 4 (S4S) shore is calculated using the recommendations of Reference [4], as discussed previously.

a.   The base design value for compression parallel to the grain from Table 12-4 is $F_c = 1350$ psi.

b.   Adjustment factors:

  1.   Size factor from Table 12-7: $C_F = 1.15$.

  2.   For the column stability factor C_P, initially the following items must be established:

   a.   For modulus of elasticity, there is no adjustment factor:

$$E' = E = 1,600,000 \text{ psi}$$

   b.   Find F_c^*:

$$F_c^* = F_c C_F = 1350(1.15) = 1553 \text{ psi}$$

   c. $$\frac{\ell_e}{d} = \frac{10(12)}{3.5} = 34.3 < 50 \qquad\qquad \text{(O.K.)}$$

   d.   Solve for α:

$$\alpha = \frac{0.3E'}{\left(\dfrac{\ell_e}{d}\right)^2 F_c^*}$$

$$= \frac{0.3(1,600,000)}{34.3^2(1553)}$$

$$= 0.263$$

Solve for C_P:

$$C_P = \frac{1 + \alpha}{1.6} - \sqrt{\left(\frac{1 + \alpha}{1.6}\right)^2 - \frac{\alpha}{0.8}}$$

$$= \frac{1 + 0.263}{1.6} - \sqrt{\left(\frac{1 + 0.263}{1.6}\right)^2 - \frac{0.263}{0.8}}$$

$$= 0.247$$

c.   Compute the allowable stress F_c':

$$F_c' = F_c C_F C_P = 1350(1.15)(0.247) = 383 \text{ psi}$$

Therefore the allowable load is

$$P = F_c'A = 383(3.5)^2 = 4690 \text{ lb} \qquad \text{(N.G.)}$$

$$4690 \text{ lb} < 5460 \text{ lb}$$

There are several possible solutions. The shore size could be increased (try a 4×6 [S4S]), use lateral bracing (horizontal lacing) at midheight to reduce the effective length of the shore, or reduce the shore spacing. The latter choice is the simplest approach.

$$\text{new required spacing} = \frac{4690}{5460}(72 \text{ in.}) = 61.8 \text{ in.}$$

If a shore spacing of 5 ft-0 in. is used,

$$\text{load per shore} = 7.0(5.0)(125 + 5) = 4550 \text{ lb}$$

$$4550 \text{ lb} < 4690 \text{ lb} \qquad \text{(O.K.)}$$

Use 4×4(S4S) shores at 5 ft-0 in. on center.

5.   *Bearing stresses:*

a.   Where stringers bear on shores (4×8 stringers on 4×4 shores),

$$\text{contact area} = 3.5(3.5) = 12.25 \text{ in.}^2$$

$$\text{total load on shore} = 4550 \text{ lb}$$

The actual bearing stress perpendicular to the grain of the stringer is

$$\frac{4550}{12.25} = 371 \text{ psi}$$

Determine the allowable bearing stress perpendicular to the grain of the stringer. From Table 12-4, $F_{c\perp} = 625$ psi. The adjustment factor applicable in this case is C_b, since the length of bearing ℓ_b is 3.5 in.:

$$C_b = \frac{\ell_b + 0.375}{\ell_b} = \frac{3.88}{3.5} = 1.107$$

Therefore

$$F_{c\perp}' = 625(1.107) = 692 \text{ psi} \qquad \text{(O.K.)}$$

$$371 \text{ psi} < 692 \text{ psi}$$

b.   Where joists bear on stringers (2 × 8 joists on 4 × 8 stringers),

$$\text{contact area} = 1.5(3.5) = 5.25 \text{ in.}^2$$

Recalling from the design of the joists that the loading on each joist was 260 lb/ft, the load on the stringer from each joist is

$$260(7.0) = 1820 \text{ lb}$$

The actual bearing stress perpendicular to the grain is

$$\frac{1820}{5.25} = 347 \text{ psi}$$

Determine the allowable bearing stress perpendicular to the grain. Because the calculated bearing stress is low relative to the 625 psi base design value for $F_{c\perp}'$, the increase in $F_{c\perp}'$ due to C_b may be disregarded. Thus

$$347 \text{ psi} < 625 \text{ psi} \qquad \text{(O.K.)}$$

6.   *Lateral bracing:* For floor systems, the minimum load to be used in designing lateral bracing is the greater of 100 lb per lineal foot of floor edge or 2% of the total dead load of the floor. We will assume the slab to be 80 ft × 100 ft and placed in one operation. Guy wire bracing capable of carrying a load of 4000 lb each will be used on all four sides of the slab area attached at slab elevation and making a 45° angle with the ground. Guy wires can resist only tensile forces.

Calculating lateral load H as 2% of the dead load of the floor, again assuming the formwork to be 5 psf, yields

$$H = 0.02(75 + 5)(80)(100) = 12,800 \text{ lb}$$

Distributing this load along the long side yields

$$\frac{12,800}{100} = 128 \text{ lb/ft} > 100 \text{ lb/ft}$$

and along the short side yields

$$\frac{12,800}{80} = 160 \text{ lb/ft} > 100 \text{ lb/ft}$$

These results are shown in Figure 12-3a. For determination of the guy wire spacing, the 160 lb/ft lateral load will be used. From Figure 12-3b, the tension in the guy wire T is calculated as

$$\frac{T}{1.414} = \frac{160}{1}$$

$$T = 226.2 \text{ lb per ft of slab being braced}$$

The maximum spacing for the guy wires is

$$\frac{4000}{226.2} = 17.7 \text{ ft}$$

Use guy wires spaced at 15 ft (max.) on center on all sides.

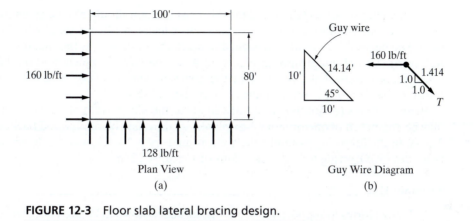

FIGURE 12-3   Floor slab lateral bracing design.

12-7   DESIGN OF FORMWORK FOR BEAMS

Figure 12-4 shows one of several common types of beam forms. The usual design procedure involves consideration of the vertical loads, with the following components to be designed: the beam bottom, the ledger that supports joists, and the supporting shores. Bearing stresses must also be checked.

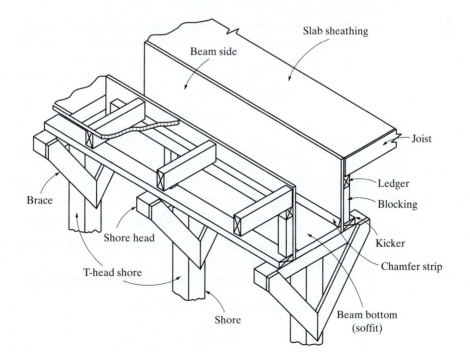

FIGURE 12-4   Typical beam formwork.

For the deeper beams (24 in. and more), consideration should also be given to the lateral pressure produced by the fresh concrete, which must be resisted by the beam sides. The beam sides would be designed in much the same way as the sheathing in a wall form. Also of importance in Figure 12-4 are the *kickers*, which hold the beam sides in place against the pressure of the concrete, and *blocking*, which serves to transmit the slab load from the ledgers to the T-head shores.

Beam bottoms (or soffits) are usually made to the exact width of the beam. They may be composed of one or more 2-in. planks, or they may be of plywood backed by 2 × 4s. In the following example, the soffit is made of a 2 × 12, which is finished on two sides (S2S), giving it final dimensions of 1½ in. × 12 in.

Example 12-2 _____

Design forms to support the 12 in. × 20 in. beam shown in Figure 12-5. The beam is to support a 4-in.-thick reinforced concrete slab. Use Douglas fir–larch No. 2 grade. The maximum allowable deflection is to be $\frac{1}{360}$ of the span for bending members. The unsupported shore height will be based on an assumed floor-to-floor height of 10 ft, from which the depth of the beam will be subtracted. All bending members are to be designed based on adequate lateral support.

Solution:

1. *Beam bottom design* (compute the maximum spacing between shores): Assume a 7-day maximum duration of load. Assume the plank to be continuous over three or more supports.

   a. Obtain allowable stresses using base design values from Table 12-4 and appropriate adjustment factors from Tables 12-4 and 12-5.

      1. Bending: $F_b = 900$ psi. Adjustment factors: $C_D = 1.25$, $C_F = 1.0$, $C_{fu} = 1.2$. Because $F_b C_F < 1150$ psi, $C_M = 1.0$ (see footnote, Table 12-4). Therefore

      $$F_b' = F_b C_D C_F C_M C_{fu}$$
      $$= 900(1.25)(1.0)(1.0)(1.2)$$
      $$= 1350 \text{ psi}$$

      2. Shear: $F_v = 180$ psi. Adjustment factors: $C_D = 1.25$ and $C_M = 0.97$. Therefore

      $$F_v' = F_v C_D C_M = 180(1.25)(0.97) = 218.3 \text{ psi}$$

      3. Modulus of elasticity: $E = 1,600,000$ psi. Adjustment factor: $C_M = 0.9$. Therefore

      $$E' = E C_M = 1,600,000(0.9) = 1,440,000 \text{ psi}$$

      Properties for the 2 × 12 (S2S) are

      $$S = \frac{bh^2}{6} = \frac{12(1.5)^2}{6} = 4.5 \text{ in.}^3$$

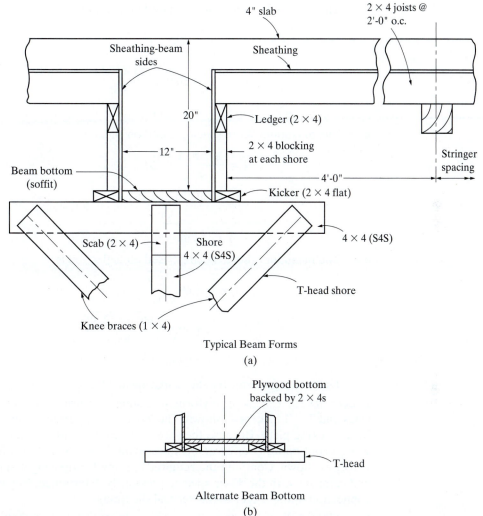

FIGURE 12-5 Beam form design.

$$I = \frac{bh^3}{12} = \frac{12(1.5)^3}{12} = 3.38 \text{ in.}^4$$

b.   The loading on the beam soffit is calculated as

$$\text{DL (reinforced concrete beam): } \frac{12(20)}{144}(150) = 250 \text{ lb/ft}$$

$$\text{LL (use 50 psf): } \frac{12}{12}(50) = 50 \text{ lb/ft}$$

$$\text{total load} = 250 + 50 = 300 \text{ lb/ft}$$

c.   The maximum shore spacing based on bending is

$$\ell = 10.95\sqrt{\frac{F_b'S}{w}}$$

$$= 10.95\sqrt{\frac{1350(4.5)}{300}}$$

$$= 49.3 \text{ in.}$$

d.   The maximum shore spacing based on shear is

$$\ell = \frac{13.3F_v'A}{w} + 2d$$

$$= \frac{13.3(218.3)(12)(1.5)}{300} + 2(1.5)$$

$$= 177.2 \text{ in.}$$

e.   The maximum shore spacing based on deflection is

$$\ell = 1.69\sqrt[3]{\frac{E'I}{w}}$$

$$= 1.69\sqrt[3]{\frac{1,440,000(3.38)}{300}}$$

$$= 42.8 \text{ in.}$$

Deflection governs. Try shore spacing at 42 in. o.c.

2.   *Ledger design:* Use ¾-in. plyform sheathing (vertically) for the beam sides and 2 × 4 kickers as shown. The ledger is supported at each shore by a blocking piece. Because the shores are to be 42 in. o.c., the ledger will be continuous over three or more spans. Use 2 × 4s (S4S) for the ledger as shown. Compute the required spacing for the ledger supports and compare with the 42-in. spacing previously determined. Neglect the connection of the ledger to the vertical sheathing.

a.   Obtain allowable stresses using base design values from Table 12-4 and appropriate adjustment factors from Tables 12-4 and 12-5.

1.   Bending: $F_b = 900$ psi. Adjustment factors: $C_D = 1.25$ and $C_F = 1.5$. Therefore

$$F_b' = F_bC_DC_F$$

$$= 900(1.25)(1.5) = 1688 \text{ psi}$$

2.   Shear: $F_v = 180$ psi. Adjustment factor: $C_D = 1.25$. Therefore

$$F_v' = F_vC_D = 180(1.25) = 225 \text{ psi}$$

3.   Modulus of elasticity: $E = 1,600,000$ psi. No adjustment factors apply. Thus

$$E' = E = 1,600,000 \text{ psi}$$

Properties for the 2 × 4 (S4S) from Table 12-3 are

$$A = 5.25 \text{ in.}^2$$
$$I = 5.36 \text{ in.}^4$$
$$S = 3.06 \text{ in.}^3$$

b. Loading: The 2 × 4 joists are supported by the ledger, as shown in Figure 12-5. Although the ledger is loaded with point loads, a uniform load will be assumed for simplicity. The loading on the slab sheathing is

$$\text{DL (slab): } \left(\frac{4}{12}\right)(150) = \quad 50 \text{ psf}$$

$$\text{LL (min.)} = \quad 50 \text{ psf}$$

$$\text{assume sheathing weight} = \quad \underline{\quad 5 \text{ psf}}$$
$$105 \text{ psf}$$

Because the span of the joists from the ledger to the adjacent stringer is 4 ft-0 in. (see Figure 12-5), the load to the ledger is calculated as

$$w = \tfrac{1}{2}(4)(105) = 210 \text{ lb/ft}$$

c. The maximum blocking spacing based on bending is

$$\ell = 10.95\sqrt{\frac{F_b' S}{w}}$$

$$= 10.95\sqrt{\frac{1688(3.06)}{210}}$$

$$= 54.3 \text{ in.}$$

d. The maximum blocking spacing based on shear is

$$\ell = \frac{13.3 F_v' A}{w} + 2d$$

$$= \frac{13.3(225)(5.25)}{210} + 2(3.5)$$

$$= 81.8 \text{ in.}$$

e. The maximum blocking spacing based on deflection is

$$\ell = 1.69\sqrt[3]{\frac{E' I}{w}}$$

$$= 1.69\sqrt[3]{\frac{1,600,000(5.36)}{210}}$$

$$= 58.2 \text{ in.}$$

Because all the three foregoing spacings *exceed* 42 in., the 2 × 4 ledgers supported by blocking at 42 in. on center are satisfactory.

3. *Design of the shores:* The shores are spaced 42 in. (or 3.5 ft) on center, and each must support a loading of

$$\text{from beam bottom: } 300(3.5) = 1050 \text{ lb}$$

$$\text{from slab forms (two sides): } 210(3.5)(2) = 1470 \text{ lb}$$

$$\text{total load per shore} = 1050 + 1470 = 2520 \text{ lb}$$

Assume 4×4 (S4S) wood shores. The unsupported shore height will be based on an assumed floor-to-floor height of 10 ft-0 in., from which the depth of the beam will be subtracted. The unsupported height is

$$\ell = 10(12) - 20 = 100 \text{ in.} = 8.33 \text{ ft}$$

We will assume the shores are pin-connected. From Table 12-6, $K_e = 1.0$. Therefore, the effective unbraced length is

$$\ell_e = K_e \ell = 1.0(8.33) = 8.33 \text{ ft}$$

a. The base design value for compression parallel to the grain from Table 12-4 is $F_c = 1350$ psi.

b. Adjustment factors:
   1. Size factor from Table 12-7: $C_F = 1.15$.
   2. For the column stability factor C_P, initially the following items must be established:

   a. For modulus of elasticity, there is no adjustment factor:

   $$E' = E = 1,600,000 \text{ psi}$$

   b. Find F_c^*:

   $$F_c^* = F_c C_F = 1350(1.15) = 1553 \text{ psi}$$

   c.
   $$\frac{\ell_e}{d} = \frac{8.33(12)}{3.5} = 28.6 < 50 \qquad \text{(O.K.)}$$

   d. Solve for α:

   $$\alpha = \frac{0.3E'}{\left(\dfrac{\ell_e}{d}\right)^2 F_c^*}$$

   $$= \frac{0.3(1,600,000)}{28.6^2(1553)}$$

   $$= 0.378$$

Solve for C_P:

$$C_p = \frac{1 + \alpha}{1.6} - \sqrt{\left(\frac{1 + \alpha}{1.6}\right)^2 - \frac{\alpha}{0.8}}$$

$$= \frac{1 + 0.378}{1.6} - \sqrt{\left(\frac{1 + 0.378}{1.6}\right)^2 - \frac{0.378}{0.8}}$$

$$= 0.342$$

c.   Compute the allowable stress F_c':

$$F_c' = F_c C_F C_P = 1350(1.15)(0.342) = 531 \text{ psi}$$

Therefore the allowable load is

$$P = F_c'A = 531(3.5)^2 = 6500 \text{ lb} \qquad \text{(O.K.)}$$

$$6500 \text{ lb} > 2520 \text{ lb}$$

Use 4×4 (S4S) shores spaced 42 in. on center.

4.   *Bearing stresses:*

a.   Assume 4×4 (S4S) T-heads on the 4×4 (S4S) shores. Actual bearing stress perpendicular to the grain of the T-head is calculated as

$$\frac{\text{shore load}}{\text{contact area}} = \frac{2520}{3.5^2} = 206 \text{ psi}$$

From Table 12-4, the base design value for $F_{c\perp}$ is 625 psi. We will neglect the adjustment factor for bearing area (due to $C_b > 1.0$ because $3\frac{1}{2}$ in. < 6 in.). Therefore, use $F_{c\perp}' = 625$ psi. Thus

$$206 \text{ psi} < 625 \text{ psi} \qquad \text{(O.K.)}$$

b.   Check bearing stress between the 2×4 ledger and the 2×4 blocking. The load from the ledger to the blocking is $210 (3.5) = 735$ lb. The actual bearing stress perpendicular to the grain of the ledger is

$$\frac{\text{load}}{\text{contact area}} = \frac{735}{1.5(3.5)} = 140 \text{ psi}$$

The allowable bearing stress, from part a, is 625 psi. Thus,

$$140 \text{ psi} < 625 \text{ psi} \qquad \text{(O.K.)}$$

c.   Check bearing stress between the 2×4 joists and the 2×4 ledger. The joist loading is $(105 \text{ psf})(2 \text{ ft}) = 210$ lb/ft, and the span of the joist is 4 ft. Therefore, the load from the joist to the ledger is $(210 \text{ lb/ft})(2 \text{ ft}) = 420$ lb. The actual bearing stress perpendicular to the grain of the ledger (and the joists) is

$$\frac{\text{load}}{\text{contact area}} = \frac{420}{1.5(1.5)} = 186.7 \text{ psi}$$

The allowable bearing stress is determined as in part a (the bearing area adjustment factor C_b is not applicable because this is end bearing):

$$186.7 \text{ psi} < 625 \text{ psi} \qquad \text{(O.K.)}$$

d.   Check the bearing of the beam soffit on the 4 × 4 (S4S) T-heads. The load from the beam soffit is (300 lb/ft)(3.5 ft) = 1050 lb. The actual bearing stress perpendicular to the grain of the T-head is

$$\frac{\text{load}}{\text{contact area}} = \frac{1050}{12(3.5)} = 25 \text{ psi}$$

For the allowable bearing stress, we will neglect the bearing area adjustment factor. The contact surface may be subjected to a wet condition, so this adjustment factor (C_M = 0.67 from Table 12-4) will be used:

$$F'_{c\perp} = F_{c\perp} C_M = 625(0.67) = 419 \text{ psi}$$

$$25 \text{ psi} < 419 \text{ psi} \qquad\qquad\qquad\qquad (\text{O.K.})$$

12-8   WALL FORM DESIGN

The design procedure for wall forms is similar to that used for slab forms, substituting studs for joists, wales for stringers, and ties for shores. See Figure 12-6 for locations of these members.

The maximum lateral pressure against the sheathing must be determined first. We will assume conditions such that C_C and C_W are both 1.0. With the sheathing thickness specified, the maximum allowable span for the sheathing is computed based on bending, shear, and deflection. This will be the maximum stud spacing. (An alternative approach would be to establish the stud spacing and then calculate the required thickness of the sheathing.)

Next, the maximum allowable stud span is calculated based on stud size and loading, considering bending, shear, and deflection. This will be the maximum wale spacing. (An alternative approach would be to establish the wale spacing and then calculate the required size of the studs.)

The next step is to determine the maximum allowable spacing of wale supports (tie spacing). This is calculated based on wale size and loading. (An alternative approach would be to preselect the tie spacing and then calculate the wale size.) Double wales are commonly used to avoid the necessity of drilling wales for tie insertion.

The load supported by each tie must be computed and compared with the tie capacity. The load on each tie is calculated as the design load (psf) multiplied by the tie spacing (ft) and wale spacing (ft). If the load exceeds the tie strength, a stronger tie must be used or the tie spacing must be reduced.

Bearing stresses must also be checked where the studs bear on the wales and where the tie ends bear on the wales. Maximum bearing stress must not exceed the allowable compression stress perpendicular to the grain or crushing will result.

Finally, lateral bracing must be designed to resist any expected lateral loads, such as wind loads.

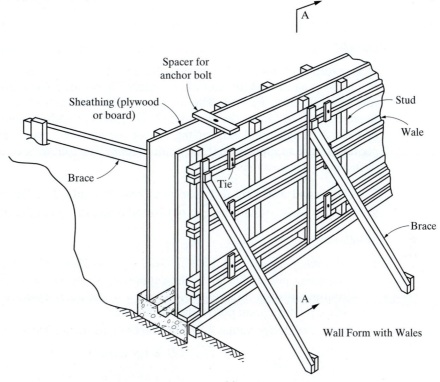

Wall Form with Wales

(a)

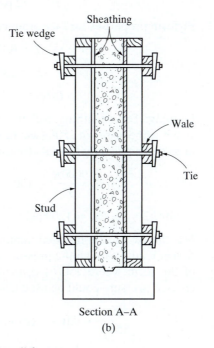

Section A–A

(b)

FIGURE 12-6 Typical wall forms.

Example 12-3

Design formwork for an 8-ft-high wall. Refer to Figure 12-6. The concrete is to be placed at a rate of 4 ft/h and will be internally vibrated. Concrete temperature is expected to be 90°F. The maximum allowable deflection of bending members is to be $1/360$ of the span. Use $3/4$-in. class I plyform for the sheathing and No. 2 Douglas fir–larch for the rest of the lumber. Assume all bending members to be supported on three or more spans. The following conditions for design have been established:

1. Studs and wales are to be designed based on adequate lateral support.
2. Studs, wales, and bracing are to be used under dry service conditions.

Solution:

1. *Sheathing design* (find the stud spacing): Consider a 12-in.-wide strip of sheathing perpendicular to the supporting studs and acting as a beam continuous over three or more spans. The sheathing spans horizontally. Place the face grain perpendicular to studs.

    a. The design values for $3/4$-in. class I plyform (see Table 12-2) are

    $$E = 1,650,000 \text{ psi}$$

    $$F_b = 1930 \text{ psi}$$

    $$F_v = 72 \text{ psi}$$

    Plyform properties for face grain parallel to the span are

    $$I = 0.199 \text{ in.}^4$$

    $$S = 0.455 \text{ in.}^3$$

    $$Ib/Q = 7.187 \text{ in.}^2/\text{ft}$$

    b. Loading: The sheathing will be designed for concrete pressure, which is the lesser of 150h (where h will be taken as 8 ft), 2000 psf, or as determined by formula (noting that $R < 7$ ft/h and $h < 14$ ft):

    $$p = 150 + \frac{9000R}{T} = 150 + \frac{9000(4)}{90} = 550 \text{ psf} < 600 \text{ psf}$$

    $$150(h) = 150(8) = 1200 \text{ psf}$$

    Use the ACI-recommended minimum of 600 psf for the sheathing design load, although the pressure could be decreased near the top of the form where 150(h) controls. The length in which the decreased pressure could be used may be calculated as

    $$\frac{600}{150} = 4.0 \text{ ft} \qquad \text{(from the top of the form)}$$

    It is conservative to design the full height for 600 psf, however.

c. The maximum stud spacing based on bending is

$$\ell = 10.95\sqrt{\frac{F_b S}{w}}$$

$$= 10.95\sqrt{\frac{1930(0.455)}{600}} = 13.25 \text{ in.}$$

d. The maximum stud spacing based on shear is

$$\ell = \frac{20 F_v (Ib/Q)}{w} + 2d$$

$$= \frac{20(72)(7.187)}{600} + 2(0.75) = 18.75 \text{ in.}$$

e. The maximum stud spacing based on deflection is

$$\ell = 1.69 \sqrt[3]{\frac{EI}{w}}$$

$$= 1.69 \sqrt[3]{\frac{1,650,000(0.199)}{600}} = 13.82 \text{ in.}$$

Bending is critical. Use a stud spacing of 12 in. o.c.

2. *Stud design* (compute the wale spacing): Assume 2 × 4 (S4S) studs and a 7-day maximum duration of load.

    a. Obtain allowable stresses using base design values from Table 12-4 and appropriate adjustment factors from Tables 12-4 and 12-5.

        1. Bending: $F_b = 900$ psi. Adjustment factors: $C_D = 1.25$ and $C_F = 1.5$. Therefore

$$F_b' = F_b C_D C_F$$

$$= 900(1.25)(1.5) = 1688 \text{ psi}$$

        2. Shear: $F_v = 180$ psi. Adjustment factor: $C_D = 1.25$. Therefore

$$F_v' = F_v C_D = 180(1.25) = 225 \text{ psi}$$

        3. Modulus of elasticity: $E = 1,600,000$ psi. No adjustment factors apply. Thus

$$E' = E = 1,600,000 \text{ psi}$$

For the 2 × 4 lumber, from Table 12-3,

$$A = 5.25 \text{ in.}^2$$

$$I = 5.36 \text{ in.}^4$$

$$S = 3.06 \text{ in.}^3$$

    b. Loading: Since the stud spacing is 12 in. o.c., the load w will be 600 lb/ft (see step 1, part [b]).

c.   The maximum wale spacing based on bending is

$$\ell = 10.95\sqrt{\frac{F_b'S}{w}}$$

$$= 10.95\sqrt{\frac{1688(3.06)}{600}} = 32.1 \text{ in.}$$

d.   The maximum wale spacing based on shear is

$$\ell = \frac{13.3F_v'A}{w} + 2d$$

$$= \frac{13.3(225)(5.25)}{600} + 2(3.5) = 33.2 \text{ in.}$$

e.   The maximum wale spacing based on deflection is

$$\ell = 1.69\sqrt[3]{\frac{E'I}{w}}$$

$$= 1.69\sqrt[3]{\frac{1,600,000(5.36)}{600}} = 41.0 \text{ in.}$$

Therefore, bending governs. Use a wale spacing of 24 in. o.c. (maximum).

3.   *Wale design* (compute the tie spacing): Double 2 × 4 (S4S) wales will be assumed, and the allowable stresses will be adjusted for a 7-day maximum duration of load.

a.   Design values: Allowable stresses and E will be the same as for the studs. Properties for the wales will be twice those for the studs because the wales are doubled. Thus

$$A = 10.5 \text{ in.}^2$$

$$I = 10.72 \text{ in.}^4$$

$$S = 6.12 \text{ in.}^3$$

b.   Loading: Each wale will support a strip of wall form that has a height equal to the spacing of the wales:

$$w = \frac{24}{12}(600) = 1200 \text{ lb/ft}$$

c.   The maximum tie spacing based on bending is

$$\ell = 10.95\sqrt{\frac{F_b'S}{w}}$$

$$= 10.95\sqrt{\frac{1688(6.12)}{1200}} = 32.1 \text{ in.}$$

PHOTO 12-1   Wall forms and bracing.

d.   The maximum tie spacing based on shear is

$$\ell = \frac{13.3 F_v' A}{w} + 2d$$

$$= \frac{13.3(225)(10.5)}{1200} + 2(3.5) = 33.2 \text{ in.}$$

e.   The maximum tie spacing based on deflection is

$$\ell = 1.69 \sqrt[3]{\frac{E'I}{w}}$$

$$= 1.69 \sqrt[3]{\frac{1,600,000(10.72)}{1200}} = 41.0 \text{ in.}$$

Bending governs. A modular spacing of 24 in. would be desirable. Use tie spacing of 24 in. o.c.

4.   Check the load on the ties (P_{tie}) with the capacity of the ties. Assume the tie capacity to be 3000 lb (ties of various capacities are widely available). Also assume that the ties have 1½-in. wedges bearing on the wales. Then

$$P_{\text{tie}} = (\text{wale spacing}) \times (\text{tie spacing}) \times (\text{pressure})$$

$$= \frac{24}{12}\left(\frac{24}{12}\right)(600) = 2400 \text{ lb}$$

2400 lb < 3000 lb

Therefore, the capacity of the tie is satisfactory.

5.   Check bearing stresses.

a.   Where tie wedges bear on wales (wedges are $1\frac{1}{2}$ in. wide),

$$P_{\text{tie}} = 2400 \text{ lb}$$

$$\text{bearing contact area} = (2)(1.5)(1.5) = 4.5 \text{ in.}^2$$

$$\text{bearing stress (actual)} = \frac{2400}{4.5} = 533 \text{ psi}$$

The allowable compressive stress perpendicular to the grain is $F'_{c\perp} = 625$ psi (neglect bearing area adjustment factor). Thus

$$533 \text{ psi} < 625 \text{ psi} \qquad\qquad \text{(O.K.)}$$

b.   Where studs bear on wales (double wales),

$$\text{bearing contact area} = (2)(1.5)(1.5) = 4.5 \text{ in.}^2$$

The load on the wale from the stud is

$$P = (\text{load/ft on stud}) \times (\text{wale spacing})$$

$$= 600\left(\frac{24}{12}\right) = 1200 \text{ lb}$$

$$\text{actual bearing stress} = \frac{1200}{4.5} = 267 \text{ psi}$$

As in part (a), $F'_{c\perp} = 625$ psi. Thus

$$267 \text{ psi} < 625 \text{ psi} \qquad\qquad \text{(O.K.)}$$

6.   Lateral bracing should be designed for wall forms based on the greater of wind load (using 15 psf as a minimum) or 100 lb/ft applied at the top of the wall. Calculate wind load on wall forms using the minimum 15 psf:

$$\text{wind load} = (15 \text{ psf})(1 \text{ ft})(8 \text{ ft}) = 120 \text{ lb per ft of wall}$$

This load would be considered to act at midheight of the wall, 4 ft above the base, and would create an overturning moment about the base of

$$M_{OT} = 120 \text{ lb/ft}(4 \text{ ft}) = 480 \text{ ft-lb} \qquad (\text{per ft of wall})$$

The equivalent force, acting at the top of the wall, that would create the same overturning moment is

$$\text{force} = \frac{480 \text{ ft-lb}}{8 \text{ ft}} = 60 \text{ lb} \qquad (\text{per ft of wall})$$

$$60 \text{ lb/ft} < 100 \text{ lb/ft}$$

Therefore use 100 lb/ft. This load, assumed to act at the top of the wall, can act in either direction. If *guy wires* are used, they must be placed on *both* sides of the wall. If *wooden strut bracing* is used, it can resist tension or compression and therefore *single-side* bracing may be used.

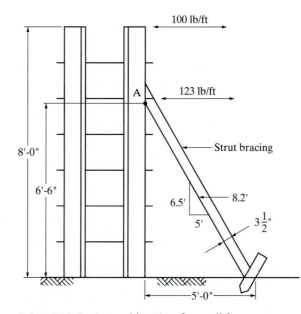

100 lb/ft

A

123 lb/ft

Strut bracing

8'-0"

6'-6"

6.5'

8.2'

5'

$3\frac{1}{2}$"

5'-0"

FIGURE 12-7   Lateral bracing for wall form.

In this problem use single-side strut bracing, as shown in Figure 12-7, and design for compression. The horizontal load H on the strut at point A is calculated by considering moment taken at the base of the wall:

$$H(6.5) = 100(8)$$

$$H = 123 \text{ lb}$$

The force F in the strut, using the slope triangle shown, is determined from

$$\frac{F}{8.2} = \frac{123}{5}$$

Therefore $F = 202$ lb (per foot of wall).

Use double 2 × 4 (S4S) lumber for the strut and compute the capacity as a compression member. (This will be adequate for tension also.) For No. 2 grade Douglas fir–larch:

a.   The base design value for compression parallel to the grain from Table 12-4 is $F_c = 1350$ psi.

b.   Adjustment factors:

  1.   Size factor from Table 12-7: $C_F = 1.15$.

  2.   For the column stability factor C_P, initially the following items must be established:

    a.   For modulus of elasticity, there is no adjustment factor:

$$E' = E = 1,600,000 \text{ psi}$$

b.   Find F_c^*:

$$F_c^* = F_cC_F = 1350(1.15) = 1553 \text{ psi}$$

c.   Assume that the ends are pin connected. Therefore, $K_e = 1.0$ and $\ell_e = 8.2$ ft. Then

$$\frac{\ell_e}{d} = \frac{8.2(12)}{3.0} = 32.8 < 50 \qquad \text{(O.K.)}$$

d.   Solve for α:

$$\alpha = \frac{0.3E'}{\left(\dfrac{\ell_e}{d}\right)^2 F_c^*}$$

$$= \frac{0.3(1,600,000)}{32.8^2(1553)}$$

$$= 0.287$$

Solve for C_P:

$$C_P = \frac{1 + \alpha}{1.6} - \sqrt{\left(\frac{1 + \alpha}{1.6}\right)^2 - \frac{\alpha}{0.8}}$$

$$= \frac{1 + 0.287}{1.6} - \sqrt{\left(\frac{1 + 0.287}{1.6}\right)^2 - \frac{0.287}{0.8}}$$

$$- 0.267$$

c.   Compute the allowable stress F_c':

$$F_c' = F_cC_FC_P = 1350(1.15)(0.267) = 415 \text{ psi}$$

Therefore the allowable load is

$$P = F_c'A = 415(2)(5.25) = 4360 \text{ lb}$$

The maximum allowable strut spacing is calculated from

$$\frac{4360 \text{ lb}}{202 \text{ lb/ft}} = 21.6 \text{ ft}$$

Use struts at 21 ft-0 in. on center.

Single 2 × 4 struts could have been used if an intermediate brace were used to reduce the unbraced length.

12-9   FORMS FOR COLUMNS

Concrete columns are usually one of five shapes: square, rectangular, L-shaped, octagonal, or round. Forms for the first four shapes are generally made of sheathing, consisting of vertical planks or plywood, with wood yokes and steel bolts, patented steel clamps, or steel bands used to resist the concrete pressure acting on the sheathing. Forms for round columns may be wood, steel plate, or patented fiber tubes.

Because forms for columns are usually filled rapidly, frequently in less than 60 min, the pressure on the sheathing will be high, especially for tall columns. The ACI recommendations for lateral concrete pressure in column forms were discussed in Section 12-4. AC1 347-04 (Equation [2.2]) is applicable for concrete having a slump of 7 in. or less and placed with normal internal vibration to a depth of 4 ft or less. However, assuming normal weight concrete (150 pcf), the pressure should not be taken as greater than 150h (psf), where h is the depth in feet below the upper surface of the freshly placed concrete. Thus the *maximum* pressure at the bottom of a form 10-ft high should be taken as 150(10) = 1500 psf regardless of the rate of filling the form or concrete temperature. It is suggested that the pressure be conservatively calculated using the equation

$$p = 150h$$

Figure 12-8a illustrates typical construction of a column form using plywood sheathing backed by vertical stiffening members and clamped with adjustable metal column clamps. The sheathing must be selected to span between the stiffening members using the concrete pressure that exists at the bottom of the column form. The vertical stiffening members must span between the column clamps, the spacing of which can be increased as the pressure decreases toward the top of the form.

Figure 12-8b illustrates a method suitable for forming smaller columns where no vertical stiffening members are required and the plywood sheathing is backed

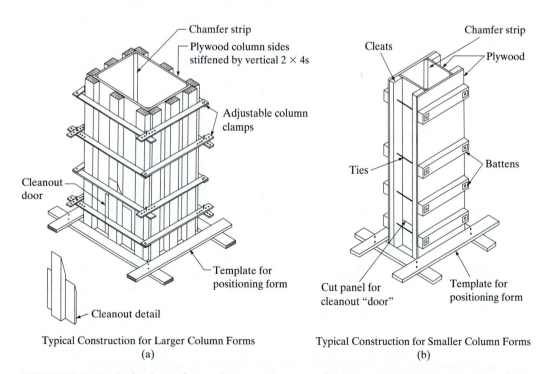

Typical Construction for Larger Column Forms
(a)

Typical Construction for Smaller Column Forms
(b)

FIGURE 12-8   Typical column forms. (*Source:* Courtesy of the American Concrete Institute [5].)

directly by battens that are part of a wood and bolt column yoke. Column clamps can be used in this situation as well. If the thickness of the sheathing is selected, the design consists of determining the maximum safe spacing of the column clamps considering the pressure from the concrete as well as the permissible deflection, allowable bending stress, and allowable shearing stress.

The sheathing span length ℓ may be calculated for moment, shear, and deflection, with the shortest of these span lengths being the controlling value.

Because the pressure against the forms varies with height, however, the determination of the optimum clamp spacing becomes laborious. As a result, tables have been developed that aid in quick determination of support (clamp) spacing. Table 12.8 is an

TABLE 12-8   Safe Span in Inches for Class I Plyform, Continuous over Four or More Supports

Pressure (psf)	Stress parallel to gain				Stress perpendicular to gain			
	1/2 in.	5/8 in.	3/4 in.	1 in.	1/2 in.	5/8 in.	3/4 in.	1 in.
75	20	24	26	32	14	16	21	28
100	18	22	24	30	12	14	19	26
125	17	20	23	28	12	13	18	25
150	16	19	22	27	11	13	17	24
175	15	18	21	26	10	12	16	23
200	15	17	20	25	10	11	15	22
300	13	15	17	22	9	10	13	19
400	12	14	16	20	8	9	12	18
500	11	13	15	18	7	8	11	16
600	10	12	13	17	7	8	11	15
700	9	11	12	16	6	8	10	14
800	9	10	11	15	6	7	9	14
900	8	10	11	14	5	6	8	13
1000	8	9	10	13	5	6	7	12
1100	8	9	10	12	5	5	7	11
1200	7	8	9	12	4	5	6	10
1300	7	8	9	11	4	5	6	10
1400	6	7	9	11	4	4	6	9
1500	6	7	8	11	4	4	5	9
1600	6	6	8	10	—	4	5	8
1700	5	6	8	10	—	4	5	8
1800	5	6	7	9	—	4	5	8
1900	5	6	7	9	—	4	5	7
2000	5	5	7	9	—	—	4	7
2200	4	5	6	8	—	—	4	7
2400	4	5	6	8	—	—	4	6
2600	4	4	5	7	—	—	4	6
2800	4	4	5	7	—	—	4	6
3000	—	4	5	6	—	—	—	5

Notes: F_b = 1930 psi; F_v = 72 psi; E = 1,650,000 psi; allowable deflection = span/360 but not greater than 1/16 in. Safe spans less than 4 in. are not shown. Tabulated spans are rounded to the nearest inch.

example of one such table specifically set up for plywood sheathing. The tabular values are based on the assumption that the lateral pressure is uniform between clamps and of an intensity equal to that at the lower clamp.

Clamps must also be investigated to determine if they can resist the applied loads. The manufacturer usually has recommended capacities for steel column clamps. Wood-yoke-type clamps with tie rods must be designed.

Generally, the type of forming system is a function of column size and height. As column size increases, either the thickness of the sheathing must be increased or vertical stiffeners must be added to prevent sheathing deflection. If vertical supports or stiffeners are used (see Figure 12-8) in combination with a plywood sheathing, the sheathing should span between the vertical supports and the plywood face grain should be horizontal (in the direction of the span) for maximum strength. The clamp spacing is then a function of the vertical support member strength. If plywood sheathing spans between clamps (without vertical supports), the face grain should be vertical (in the direction of the span) for maximum strength.

Example 12-4

Determine a clamp spacing pattern for column form sheathing made up of $\frac{3}{4}$-in.-thick plywood. The column height is to be 12 ft-0 in. Assume the sheathing continuous over four or more supports and its face grain parallel to the span (vertical). Use class I plyform design values of

$$F_b = 1930 \text{ psi}$$

$$F_v = 72 \text{ psi}$$

$$E = 1{,}650{,}000 \text{ psi}$$

with allowable deflection of span/360 but not greater than $\frac{1}{16}$ in.

Solution:

1. Table 12-8 is used to determine the maximum span of the plywood between clamps. This depends on the pressure on the form, which is determined from

$$p = wh$$

   where

   w = unit weight of concrete (pcf)

   h = depth of fresh concrete (ft)

2. Denoting the vertical distance from the bottom of the form as y (ft), the pressure is determined from

$$p = wh = 150(12 - y)$$

   For some arbitrary values of y, the calculated pressures and the maximum spans (clamp spacings) from Table 12-8 are shown in Table 12-9.

TABLE 12-9   Clamp Spacing Determination

y^a(ft)	Pressure (psf)	Maximum clamp spacing (in.)[b]
0	1800	7
3	1350	9
6	900	11
9	450	15
11.5	75	26

[a]Quantity y is measured upward from the bottom of the form.
[b]From Table 12-8.

3.   A plot of maximum clamp spacing as a function of distance above top-of-footing is shown in Figure 12-9. The final clamp layout, also shown in Figure 12-9, is determined by trial and error. This procedure is similar to stirrup design (Chapter 4). One should attempt to minimize the number of clamps without having too many different-size spacings.

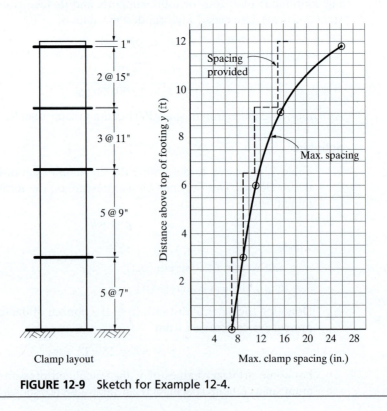

Clamp layout   Max. clamp spacing (in.)

FIGURE 12-9   Sketch for Example 12-4.

REFERENCES

[1] *Guide to Formwork for Concrete* (ACI 347-04). American Concrete Institute, 38800 Country Club Drive, Farmington Hills, MI 48331.

[2] *Plywood Design Specification*. APA—The Engineered Wood Association, P.O. Box 11700, Tacoma, WA 98411-0700, January 1997.

[3] *Concrete Forming*. APA—The Engineered Wood Association, P.O. Box 11700, Tacoma, WA 98411-0700, 2004.

[4] *National Design Specification for Wood Construction* (ANSI/AF&PA NDS-2005), with supplements and commentary. American Forest & Paper Association/American Wood Council, 1111 19th Street, N.W., Suite 800, Washington, D.C. 20036, 2005.

[5] M. K. Hurd, *Formwork for Concrete*, SP-4, 6th ed. American Concrete Institute, 38800 Country Club Drive, Farmington Hills, MI 48331, 1995.

PROBLEMS

Where applicable in these problems, and unless otherwise noted, *assume:*

1. The vertical live load is to be 75 psf (motorized carts).
2. Bending members are continuous over three or more spans and are assumed to have adequate lateral support.
3. There is a 7-day duration of load.
4. Lumber is No. 2 grade hem-fir.
5. The forming weight is 5 psf; neglect sheathing weight.
6. The maximum deflection is $\frac{1}{360}$ of the span.
7. All lumber is (S4S).
8. Plywood is placed with the face grain perpendicular to the supports.
9. No available information exists with respect to wood splits, checks, and shakes. Assume $C_i = 1.0$, $C_C = 1.0$, and $C_W = 1.0$.

12-1. Using basic principles, derive the equations of Table 12-1 for allowable span length ℓ as governed by moment, shear, and deflection. Handbooks (such as the AISC *Manual*) may be helpful for moment, shear, and deflection equations. Show load diagrams.

    (a) One span

    (b) Two spans

    (c) Three spans

12-2. A slab form is to be built for an 8-in.-thick concrete slab. The plywood sheathing face grain is perpendicular to the joists. The maximum allowed deflection is to be the smaller of $1/360$ of the span or $1/8$ in. Compare the maximum allowable joist spacing for class II plyform $1/2$-in. thick and 1-in. thick. Draw a sectional view showing slab, sheathing, and joists.

12-3. A 6-in.-thick concrete slab is to be formed using plywood supported on 2×8 (S4S) joists that are spaced 1 ft-6 in. on center. Find the maximum allowable stringer spacing if maximum allowable deflection is 1/240 of the span. Draw a sketch of the formwork.

12-4. For the slab of Problem 12-3, assume that double 2×10 stringers spaced 5 ft-0 in. o.c. will support the joists. Determine the maximum shore spacing. Draw a sketch of the formwork. (Neglect shore capacity.)

12-5. In Problem 12-4, if the shores were to be spaced 6 ft-0 in. o.c., select a new stringer size. The stringer may be either doubled 2-in.-thick planks or a single wood beam.

12-6. Compare the capacities of 6 ft-0 in. and 12 ft-0 in. 4×4 (S4S) wooden shores. What are the capacities if the shores are full nominal size (4 in. $\times$ 4 in.)?

12-7. Design a soffit (beam bottom) for an $11\frac{1}{4}$ in. $\times$ 24 in. reinforced concrete beam form. Use a 2×12. Determine the maximum shore spacing. Draw a sectional view through the beam.

12-8. Design a plywood beam bottom form shown as an alternative in Figure 12-5. The beam will be 14 in. $\times$ 24 in. First design the class I plyform to span the clear span between 2×4s (single span, maximum deflection of $1/16$-in.). Then determine maximum shore spacing assuming that the weight of the beam is supported by the 2×4s, which back the beam bottom.

12-9. In Problem 12-8, replace the 2×4s with 2×6s and determine the new maximum shore spacing.

12-10. A 12-ft-high concrete wall is to be placed at a rate of 6 ft/h, and the temperature is expected to be 90°F. What is the *maximum* lateral pressure due to fresh concrete for which the wall forms must be designed? Draw a diagram showing lateral pressure versus distance from the bottom of the wall for the full 12-ft height of wall.

12-11. Design the formwork for an 8-in.-thick reinforced concrete floor slab. Use $3/4$-in. class II plyform and No. 2 grade Douglas fir–larch for the other lumber. Use 2×8 joists and double 2×8 stringers. Use 4×4 shores with an unsupported height of 10 ft. Assume that the slab will be 120 ft $\times$ 120 ft in plan. Use guy wire bracing (at 45° to the 10-ft height) that has a tensile capacity of 4400 lb.

12-12. Design forms to support reinforced concrete beams as shown in Figure 12-5a with the following changes. The beam will be $11\frac{1}{4}$ in. $\times$ 22 in., joists are 1 ft-8 in. o.c., and stringers are spaced 7 ft-0 in. o.c. Shores will be full nominal size 4×4s with an unsupported height of 10 ft. All lumber will be No. 2 grade Southern Pine.

12-13. Design the formwork for a 12-ft-high reinforced concrete wall. Concrete will be placed at a rate not to exceed 5 ft/h and will be internally vibrated. Temperature is expected to be 80°F. Use $3/4$-in. class I plyform for the sheathing and No. 2 grade Douglas fir–larch for the rest of the lumber. Use 2 × 4s for studs and doubled 2 × 6s for wales. Ties will have a capacity of 5000 lb (2-in.-wide wedges). For lateral bracing design, assume the wind to be 15 psf and use a guy wire (at 45° to the 10-ft height) that has a tensile capacity of 4400 lb.

12-14. Determine a clamp spacing pattern for a column form that has sheathing of 1-in.-thick plywood. The column height is to be 14 ft-0 in. Use Table 12-8. Assume that $F_b = 1930$ psi, $F_v = 72$ psi, and $E = 1,650,000$ and that the face grain is parallel to the span between clamps.

Chapter 13

Detailing Reinforced Concrete Structures

13-1 INTRODUCTION

The contract documents package for a typical building as developed by an architect/engineer's office commonly includes both drawings and specifications. The drawings typically concern the following areas: site, architectural, structural, mechanical, and electrical. The specifications supplement and amplify the drawings. The contract documents package is the product that results from what may be categorized as the planning and design phase of a project.

The next sequential phase may be categorized as the construction phase. It includes many subcategories for reinforced concrete structures, two of which are

PHOTO 13-1   Reactor containment foundation mat. Seabrook Station, New Hampshire.

detailing and fabricating of the reinforcing steel. As described in the *ACI Detailing Manual-2004* [SP-66(04)] [1], detailing consists of the preparation of placing drawings, reinforcing bar details, and bar lists that are used for the fabrication and placement of the reinforcement in a structure. Fabricating consists of the actual shopwork required for the reinforcing steel, such as cutting, bending, bundling, and tagging.

Most bar fabricators not only supply the reinforcing steel but also prepare the placing drawings and bar lists, fabricate the bars, and deliver to the project site. In some cases, the bar fabricator may also act as the placing subcontractor.

It is general practice in the United States for all reinforced concrete used in building projects to be designed, detailed, and fabricated in accordance with the latest ACI Code. In addition, the Concrete Reinforcing Steel Institute regularly publishes its *Manual of Standard Practice* [2], which contains the latest recommendations of the reinforcing steel industry for standardization of materials and practices.

Techniques have also been developed that make use of electronic computers and other data-processing equipment to facilitate the generation of bar lists and other components of the detailing process. This not only aids in standardization and accuracy of the documents produced but also can be readily incorporated into the stock control system and the shopwork planning of the reinforcing steel fabricator.

13-2   PLACING DRAWINGS

The placing drawing (commonly called a shop drawing) consists of a plan view with sufficient sections to clarify and define bar placement. As such, it is the guide that the ironworkers will use as they place the reinforcing steel on the job. In addition, the placing drawing will contain typical views of beams, girders, joists, columns, and other members as necessary. Frequently, tabulations called "schedules" are used to list similar members, which vary in size, shape, and reinforcement details. A bar list, bending details, or both may or may not be shown on the placing drawing, as some fabricators not only have their own preferred format but prefer the list to be prepared as a separate entity.

The preparation of the placing drawing is based on the complete set of contract documents and generally contains only the information necessary for bar fabrication and placing. Building dimensions are not shown unless they are necessary to locate the steel properly.

The structural drawing (Figure 13-1), which is a part of the structural plans of the contract documents, is the drawing on which the placing drawing is based. Figure 13-2 shows a placing drawing of the same system shown in Figure 13-1. This building is an example of a framing system that uses girders between columns to support beams, which in turn support one-way slabs. The girders support only two-thirds of the beams; the remainder frame directly into the columns.

The structural drawing, as may be observed, shows the floor plan view locating and identifying the structural elements, along with views of typical beams, girders, and slabs and their accompanying schedules. The placing drawing supplements the structural drawing by furnishing all the information necessary for bar fabrication and placement. The precise size, shape, dimensions, and location of each bar are furnished, using a marking system discussed later in this chapter. Information relative to bar supports may also be included. Figure 13-2 illustrates a placing drawing that includes placing data for bar supports, as well as a bending details schedule.

Placing drawings, in addition to controlling the placement of the steel in the forms, serve as the basis for ordering the steel. Therefore, a proper interpretation of the contract documents by the fabricator is absolutely essential. Generally, all placing drawings are submitted to the architect or engineer for checking and approval before shop fabrication begins.

13-3   MARKING SYSTEMS AND BAR MARKS

With respect to buildings, two identification systems are required. The first involves the identification of the various structural members, and the second involves the identification of the individual bars within the members. The marking system for the structural members may consist of an alphabetical-numerical identification for each beam, girder, and slab, with the columns designated numerically as in Figures 13-1 and 13-2. Also used is a system of alphabetical and numerical coordinates in which the centerlines of columns are numbered consecutively in one direction and lettered consecutively in the other. A coordinate system may be observed in the foundation-engineering drawing of Figure 13-3, where a column may have a coordinate designation such as B2 or C3. The system is generally established on the architectural and

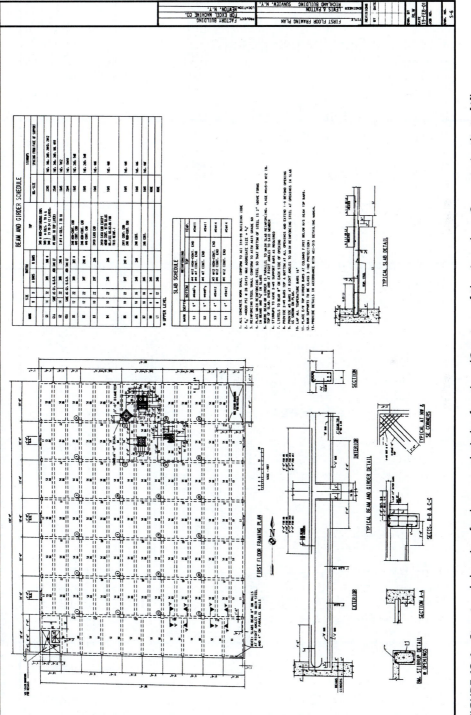

FIGURE 13-1 Structural drawing for beam and girder framing. (Courtesy of the American Concrete Institute [1])

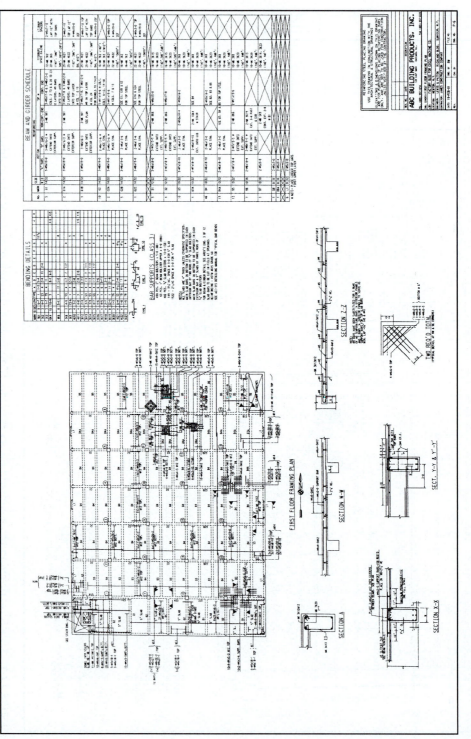

FIGURE 13-2 Placing drawing for beam and girder framing. (Courtesy of the American Concrete Institute [1])

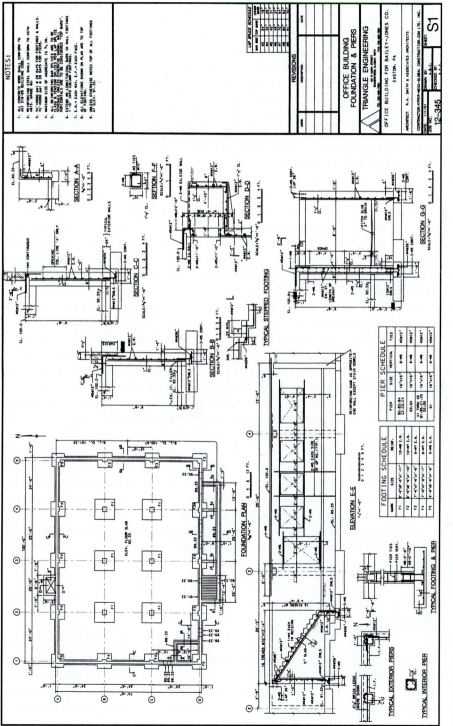

FIGURE 13-3 Structural engineering drawing for foundations. (Courtesy of the American Concrete Institute [1])

structural drawings and adopted by the detailer, unless the detailer requires a more precise identification system.

Footings, as may be observed in Figure 13-3, are generally designated with an F prefix followed by a number, such as F1 and F3, without regard to a coordinate system. Footing piers or pedestals may be identified using the coordinate system, such as B2 or D4, or may be designated with a P prefix followed by a number, such as P1 or P3. Beams, joists, girders, lintels, slabs, and walls are generally given designations that indicate the specific floor in the building, the type of member, and an identifying number. For example, 1G2 indicates a first-floor girder numbered 2, and RB4 indicates a roof beam numbered 4.

In some situations, the floor designation is omitted, as shown in Figures 13-1 and 13-2. The beams and girders are then designated with a prefix B or G, respectively, followed by a number (for example, B4 or G2). In some cases, suffixes have been added, such as G2A, indicating that there is a difference in the member.

Along with a marking system established for the structural members, a system of identifying and marking the reinforcing bars must be established. In buildings, only bent bars are furnished with a mark number or designation. The straight bar has its own identification by virtue of its size and length. Numerous systems are in use throughout the industry, with the system choice generally a function of the building type, size, and complexity as well as the standards of each fabricator.

One common system is the use throughout the project of an arbitrary letter such as K followed by consecutive numbers, without regard to bar location or shape. This system may be observed in the placing drawing of Figure 13-2. The letter is prefixed with the size of the bar. For example, 8K19 represents a No. 8 bar whose shape and dimensions may be observed in the bending details schedule and whose location, in this case, may be established by the beam and girder schedule as being the top reinforcing at the noncontinuous end of the B6 members.

An alternative system is to use many letters rather than one arbitrary letter. Column bars may be designated with a C and footing bars with an F. For example, a 7F5 would be a No. 7 footing bar, whose shape and dimensions would be established in a bending details schedule and whose location would be observed in a footing schedule, typical footing details, or the foundation plan.

Other acceptable systems are currently being used. Of primary importance in any system that is chosen is that it should be simple, logical, easy to understand, and not lead to ambiguity or confusion.

13-4   SCHEDULES

Schedules generally appear both on engineering drawings and placing drawings. Typical schedules may be observed in Figures 13-1, 13-2, and 13-3. On the engineering drawings (Figure 13-1), it is a tabular form indicating a member mark number, concrete dimensions, and the member reinforcing steel size and location. Specific design details pertinent to the member reinforcing may also be furnished in the schedule. On the engineering drawing, the schedule must be correlated with a typical section and plan view to be meaningful. Schedules are generally used for the typical members, among which are slabs, beams, girders, joists, columns, footings, and piers. Typical footing and pier schedules may be observed in Figure 13-3, and typical beam, girder, and slab schedules may be observed in Figure 13-1.

Similar schedules are used on the placing drawings, but additional information is furnished. The placing drawing schedule is more detailed and generally indicates the number of bars, member mark number, and physical dimensions of member reinforcing steel. In addition, it indicates size and length of straight bars, mark numbers (if bent bars), location of bars, and spacing, along with all specific notes and comments relative to the reinforcing bars. This schedule must also be worked together with typical sections and plan views. The schedules are generally accompanied by a bar list, indicating bending details that may or may not be presented on the placing drawing. There is no standard format for either engineering drawing or placing drawing schedules. They are merely a convenient technique of presenting information for a group of similar items, such as groups of beams, girders, columns, and footings. The schedule format will vary somewhat to conform to the requirements of a particular job.

A schedule for all structural components may not be necessary on a placing drawing. In Figure 13-2 the slab reinforcement is shown directly on the plan, so there is no need for a slab schedule.

13-5   FABRICATING STANDARDS

The fabrication process consists of cutting, bending, bundling, and tagging the re-inforcing steel. Our discussion will primarily be limited to the cutting and bending because of their effect on a member's structural capacity. Bending, which includes the making of standard hooks, is generally accomplished in accordance with the requirements of the ACI Code, the provisions of which have been discussed in Chapter 5.

In the fabricating shop, bars to be bent are first cut to length as stipulated and then sent to a special bending department, where they are bent as designated in the bending details schedule or bar lists. The common types of bent bars have been standardized throughout the industry, and applicable configurations are generally incorporated in the placing drawing or bar lists in conjunction with the bending details (see Figure 13-2). Each configuration, sometimes called a bar type, has a designation such as 7, 8, or 9 and S1 or T1, with each dimension designated by a letter. Typical bar bends are shown in Figure 13-4. In addition, standards have been established with respect to the details of the hooks and bends. The ACI Code (Sections 7.1, 7.2, and 7.3) establishes minimum requirements and is graphically portrayed in Figure 13-5. This table also shows the extra length of bar needed for the hook (A or G), which must be added to the sum of all other detailed dimensions to arrive at the total length of bar. It is common practice to show all bar dimensions as out to out

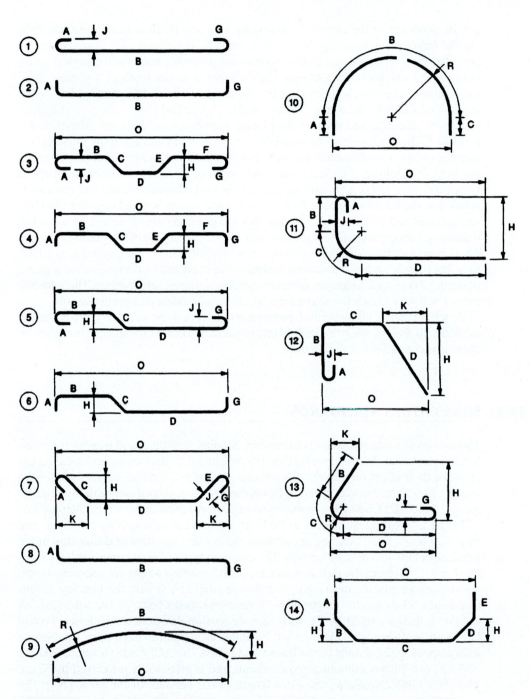

FIGURE 13-4   Typical bar bends. (Courtesy of the American Concrete Institute [1])

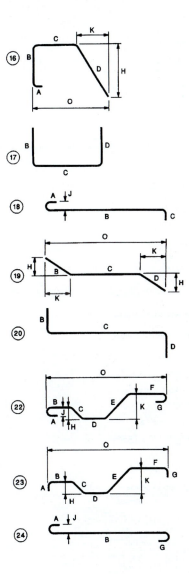

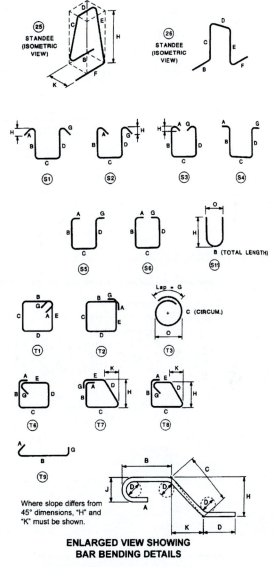

Where slope differs from
45° dimensions, "H" and
"K" must be shown.

**ENLARGED VIEW SHOWING
BAR BENDING DETAILS**

Notes:
1. All dimensions are out-to-out of bar except "A" and "G" on standard 180 and 135 degree hooks.
2. "J" dimensions on 180 degree hooks to be shown only where necessary to restrict hook size, otherwise ACI standard hooks are to be used.
3. Where "J" is not shown, "J" will be kept equal or less than "H" on Types 3, 5, and 22. Where "J" can exceed "H," it should be shown.
4. "H" dimension stirrups to be shown where necessary to fit within concrete.
5. Where bars are to be bent more accurately than standard fabricating toler-

ances, bending dimensions that require closer fabrication should have limits indicated.
6. Figures in circles show types.
7. For recommended diameter "D" of bends and hooks, see Section 3.7.1; for recommended hook dimensions, see Table 1.
8. Type S1 through S6, S11, T1 through T3, T6 through T9: apply to bar sizes No. 3 through 8 (No. 10 through 25).
9. Unless otherwise noted, diameter "D" is the same for all bends and hooks on a bar (except for Types 11 and 13).

FIGURE 13-4 (*Continued*)

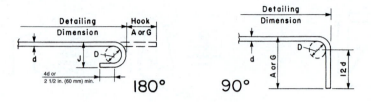

RECOMMENDED END HOOKS
All grades
D = Finished bend diameters

Bar size, No.	D, in (mm)	180 degree hook		90 degree hook
		A or G, ft-in (mm)	J, ft-in (mm)	A or G, ft-in (mm)
3 (10)	2 1/4 (60)	5 (125)	3 (80)	6 (155)
4 (13)	3 (80)	6 (155)	4 (105)	8 (200)
5 (16)	3 3/4 (95)	7 (180)	5 (130)	10 (250)
6 (19)	4 1/2 (115)	8 (205)	6 (155)	1-0 (300)
7 (22)	5 1/4 (135)	10 (250)	7 (175)	1-2 (375)
8 (25)	6 (155)	11 (275)	8 (205)	1-4 (425)
9 (29)	9 1/2 (240)	1-3 (375)	11 3/4 (300)	1-7 (475)
10 (32)	10 3/4 (275)	1-5 (425)	1-1 1/4 (335)	1-10 (550)
11 (36)	12 (305)	1-7 (475)	1-2 3/4 (375)	2-0 (600)
14 (43)	18 1/4 (465)	2-3 (675)	1-9 3/4 (550)	2-7 (775)
18 (57)	24 (610)	3-0 (925)	2-4 1/2 (725)	3-5 (1050)

Table 1(cont.)—Standard hooks: All specific sizes recommended meet minimum requirements of ACI 318

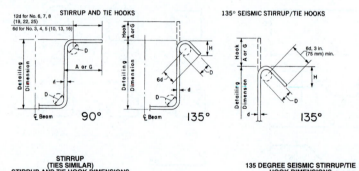

STIRRUP
(TIES SIMILAR)
STIRRUP AND TIE HOOK DIMENSIONS
ALL GRADES

Bar size, No.	D, in (mm)	90 degree hook	135 degree hook	
		Hook A or G, ft-in (mm)	Hook A or G, ft-in (mm)	H approx., ft-in (mm)
3 (10)	1 1/2 (40)	4 (105)	4 (105)	2 1/2 (65)
4 (13)	2 (50)	4 1/2 (115)	4 1/2 (115)	3 (80)
5 (16)	2 1/2 (65)	6 (155)	5 1/2 (140)	3 3/4 (95)
6 (19)	4 1/2 (115)	1-0 (305)	8 (205)	4 1/2 (115)
7 (22)	5 1/4 (135)	1-2 (355)	9 (230)	5 1/4 (135)
8 (25)	6 (155)	1-4 (410)	10 1/2 (270)	6 (155)

135 DEGREE SEISMIC STIRRUP/TIE
HOOK DIMENSIONS
ALL GRADES

Bar size, No.	D, in (mm)	135 degree hook	
		Hook A or G, ft-in (mm)	H approx., ft-in (mm)
3 (10)	1 1/2 (40)	4 1/4 (110)	3 (80)
4 (13)	2 (50)	4 1/2 (115)	3 (80)
5 (16)	2 1/2 (65)	5 1/2 (140)	3 3/4 (95)
6 (19)	4 1/2 (115)	8 (205)	4 1/2 (115)
7 (23)	5 1/4 (135)	9 (230)	5 1/4 (135)
8 (25)	6 (155)	10 1/2 (270)	6 (155)

*Finished bend diameters include "spring back" effect when bars straighten out slightly after being bent and are slightly larger than minimum bend diameters in 3.7.2.

FIGURE 13-5   Standard hook details. (Courtesy of the American Concrete Institute [1])

(meaning outside to outside) of the bar. The ACI Code also stipulates that bars must be bent cold unless indicated otherwise by the engineer. Field bending of bars partially embedded in concrete is not allowed unless specifically permitted by the engineer (ACI Code, Section 7.3).

Straight bars are cut to the prescribed length from longer stock-length bars, which are received in the fabricating shop from the mills. Tolerances in fabrication of reinforcing steel are generally standardized and are given in Figure 13-6. For instance, the cutting tolerance for straight bars is the specified length ±1 in., unless special tolerances are called for. Due consideration for these tolerances must be made by both the engineer and contractor in the design and construction phases.

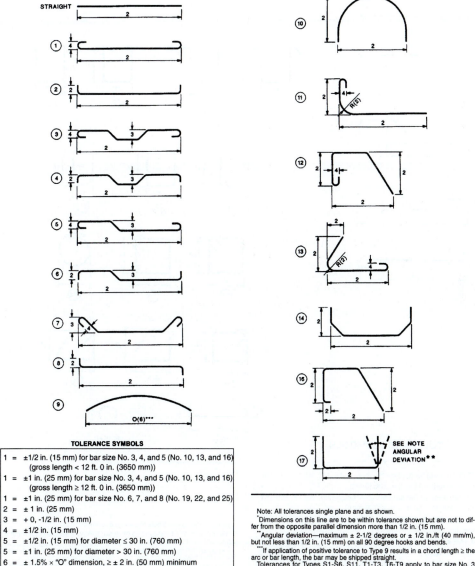

TOLERANCE SYMBOLS

1 = ±1/2 in. (15 mm) for bar size No. 3, 4, and 5 (No. 10, 13, and 16) (gross length < 12 ft. 0 in. (3650 mm))
1 = ±1 in. (25 mm) for bar size No. 3, 4, and 5 (No. 10, 13, and 16) (gross length ≥ 12 ft. 0 in. (3650 mm))
1 = ±1 in. (25 mm) for bar size No. 6, 7, and 8 (No. 19, 22, and 25)
2 = ± 1 in. (25 mm)
3 = + 0, -1/2 in. (15 mm)
4 = ±1/2 in. (15 mm)
5 = ±1/2 in. (15 mm) for diameter ≤ 30 in. (760 mm)
5 = ±1 in. (25 mm) for diameter > 30 in. (760 mm)
6 = ± 1.5% × "O" dimension, ≥ ± 2 in. (50 mm) minimum

Note: All tolerances single plane and as shown.
*Dimensions on this line are to be within tolerance shown but are not to differ from the opposite parallel dimension more than 1/2 in. (15 mm).
**Angular deviation—maximum ± 2-1/2 degrees or ± 1/2 in./ft (40 mm/m), but not less than 1/2 in. (15 mm) on all 90 degree hooks and bends.
***If application of positive tolerance to Type 9 results in a chord length ≥ the arc or bar length, the bar may be shipped straight.
Tolerances for Types S1-S6, S11, T1-T3, T6-T9 apply to bar size No. 3 through 8 (No. 10 through 25) inclusive only.

FIGURE 13-6   Standard fabricating tolerances for bar sizes No. 3 through No. 11 (No. 10 through No. 36). (Courtesy of the American Concrete Institute [1])

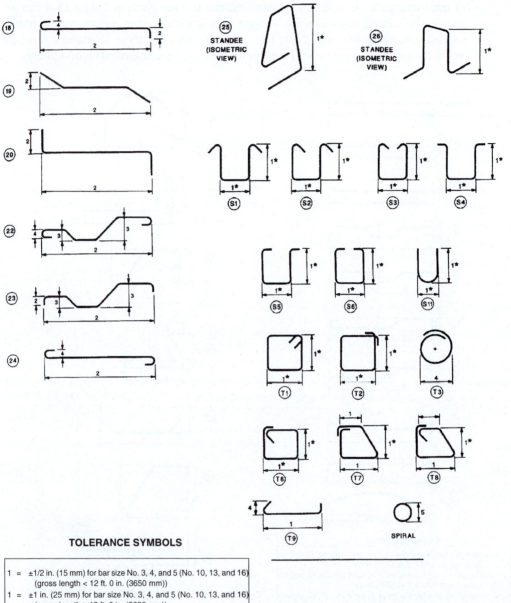

TOLERANCE SYMBOLS

1 = ±1/2 in. (15 mm) for bar size No. 3, 4, and 5 (No. 10, 13, and 16)
      (gross length < 12 ft. 0 in. (3650 mm))
1 = ±1 in. (25 mm) for bar size No. 3, 4, and 5 (No. 10, 13, and 16)
      (gross length ≥ 12 ft. 0 in. (3650 mm))
1 = ±1 in. (25 mm) for bar size No. 6, 7, and 8 (No. 19, 22, and 25)
2 = ± 1 in. (25 mm)
3 = + 0, -1/2 in. (15 mm)
4 = ±1/2 in. (15 mm)
5 = ±1/2 in. (15 mm) for diameter ≤ 30 in. (760 mm)
5 = ±1 in. (25 mm) for diameter > 30 in. (760 mm)
6 = ± 1.5% × "O" dimension, ≥ ± 2 in. (50 mm) minimum

Note: All tolerances single plane and as shown.
  *Dimensions on this line are to be within tolerance shown but are not to
differ from the opposite parallel dimension more than 1/2 in. (15 mm).
  **Angular deviation—maximum ± 2-1/2 degrees or ± 1/2 in./ft (40 mm/m),
but not less than 1/2 in. (15 mm) on all 90 degree hooks and bends.
  ***If application of positive tolerance to Type 9 results in a chord length ≥
the arc or bar length, the bar may be shipped straight.
  Tolerances for Types S1-S6, S11, T1-T3, T6-T9 apply to bar size No.
3 through 8 (No. 10 through 25) inclusive only.

FIGURE 13-6   (*Continued*)

13-6   BAR LISTS

The bar list serves several purposes. It is used for fabrication, including cutting, bending, and shipping, as well as placement and inspection. It represents a bill of materials indicating complete descriptions of the various bar items. The information in a bar list is obtained from the placing drawing or while the placing drawing is being prepared. A typical bar list form is shown in Figure 13-7; its similarity to the bending details schedule of Figure 13-2 is apparent.

The bar list generally includes both straight and bent bars and indicates all bar dimensions and bends as well as the grade of steel and the number of pieces. A bar list of this type may be used in addition to a bending detail schedule on the placing drawing. Some fabricators, however, prefer not to use a bending detail schedule. With this system, the detailers will use sketches until the drawing is complete and then transfer the information to separate bar lists. Two bar lists are prepared, one for straight bars and one for bent bars. Dimensions less than 12 in. are given in inches; over 12 in. they are given in feet and inches.

13-7   EXTRAS

Bars are sold on the basis of weight. To a base price are added various extra charges (extras) dependent principally on the amount of effort required to produce the final product. Among these extras are bending extras and special fabrication extras.

Bent bars are generally classified as heavy bending or light bending. Extra charges are made for all shop bending, with the charge a function of the classification. Due to the increased amount of handling and number of bends per pound of steel, light bending charges per pound are appreciably more than the charges for heavy bending. According to the ACI, heavy bending is defined as bar sizes No. 4 through No. 18 that are bent at not more than six points, radius bent to one radius, and bending not otherwise defined. Light bending includes all No. 3 bars, all stirrups and column ties, and all bars No. 4 through No. 18 that are bent at more than six points, bent in more than one plane, radius bent with more than one radius in any one bar, or a combination of radius and other bending. Special fabrication includes fabrication of bars specially suited to conditions for a given project. This may include special tolerances and variations from minimum standards as well as unusual bends and spirals.

13-8   BAR SUPPORTS AND BAR PLACEMENT

Bar supports are used to hold the bars firmly at their designated locations before and during the placing of concrete. These supports may be of metal, plastic, precast concrete, or other approved materials. The most commonly used bar supports are factory-made wire bar supports, which are available in various sizes and types and which may be provided some corrosion resistance by having exposed parts covered or capped with plastic or being made wholly or in part of galvanized or stainless steel.

X Y Z   PRODUCTS COMPANY
CHICAGO, ILLINOIS

CUSTOMER: JONES BROS. CONST. CO. | PROJECT NO. 27693
PROJECT: FIELDCREST APT. BLDG. | DRAWING NO. Figs.18-5a,18-6a
LOCATION: SMITHVILLE, N.C. | SHEET   1 of 2
MATERIAL FOR: PARTIAL BASEMENT COLUMNS | DATE 9/15/97 REVISED 9/19/97
| DRAWN BY H.N.H.

ITEM	NO. PCS.	SZ	LENGTH	BAR MARK	TYPE	A	B	C	D	E	F	G	H	J	K	O	R
1	STRAIGHT																
2	4	57	23-11														
3	4	57	18-11														
4	12	57	8-11														
5																	
6	8	43	23-11														
7	4	43	8-11														
8																	
9	12	29	12-8														
10	6	29	10-8														
11	6	29	4-1														
12																	
13	STRAIGHT (SAW CUT BOTH ENDS)																
14	8	57	23-11	57W1													
15	8	57	11-5	57W2													
16																	
17	HEAVY BENDING																
18	4	36	20-0	36BC5	3		16-0	1-0	3-0				0-3	1-0			
19	64	36	13-6	36C16	3		3-0	1-3	9-3				0-4	1-3			
20																	
21	18	29	12-8	29C4	3		1-11	1-8	9-1				0-4½	1-8			
22																	
23	LIGHT BENDING																
24	22	10	8-4	10T6	T2	0-4	2-1	1-9	2-1	1-9		0-4					
25	22	10	7-8	10T9	T2	0-4	1-11	1-7	1-11	1-7		0-4					
26	50	10	6-3	10BT1	T2	0-4	1-4¾	1-4¾	1-4¾	1-4¾		0-4					
27	26	10	6-3	10BT3	T2	0-4	1-0¾	1-8¾	1-0¾	1-8¾		0-4					
28	44	10	3-4	10T10	S10		1-3½	0-9	1-3½								
29	22	10	2-10	10T23	T5	0-5	2-1	0-4									
30	52	10	2-10	10BT4	S10		1-0¼	0-9¼	1-0¼								
31	22	10	2-8	10T8	T5	0-5	1-11	0-4									
32	22	10	2-6	10T20	T5	0-5	1-9	0-4									
33																	
34	SPIRALS																
35			Height		Dia.	Pitch	Turns	Spcrs										
36	4	13	8-9	SP5	21"	3"	38	3									

ALL DIMENSIONS ARE OUT TO OUT
ALL BARS ASTM A615M GRADE 420

FOR STANDARD BEND TYPES REFER TO
CRSI MANUAL OF STANDARD PRACTICE

FIGURE 13-7 Typical bar list for buildings. (Courtesy of the American Concrete Institute [1])

The Concrete Reinforcing Steel Institute (CRSI) publishes information intended to serve as a guide for the selection and utilization of steel wire bar supports used to position reinforcing steel. It is a general practice that unless the engineers' drawings or specifications show otherwise, bar supports will be furnished in accordance with CRSI standards. More detailed information relative to bar supports may be found in the CRSI publications *Placing Reinforcing Bars* [3] and *Manual of Standard Practice* [2].

13-9   COMPUTER DETAILING

The term *computer detailing* is somewhat of a misnomer. Although computers and other electronic data-processing equipment have been used in the bar fabricating industry for many years, it has not been until recently that the actual detailing has been done with computers. The generation of the placing drawings is well within the capability of currently available computer-aided design and drafting (CADD) software, but only the larger steel bar companies, which can afford to dedicate staff to this function, use CADD in this way. The most beneficial aspect of the use of computers in the bar fabrication business concerns the handling and manipulation of data and fabrication management in the shop. Several commercially available programs perform at various levels of sophistication, and bar fabricators sometimes create their own in-house software to perform unique functions geared to their own needs. Each package and each in-house program is different, and the person entering the field can expect to receive training on the particular equipment and software that the company is using. The development and availability of increasingly sophisticated software can be expected.

The functions most widely performed by computers in the bar fabricating business involve the generation and printing of notes and labels, schedules (beam, column, slab, pier, footing, joist, and the like), bending details, bar tags, and weight summaries. The software will also optimize the cutting schedule to minimize waste (which is very important in the high-volume, low-margin rebar fabricating business). In addition, information can be generated that will limit bar bundles based on maximum weight and configuration.

The process begins with the preparation of the placing drawing (see Figure 13-2), whether by CADD or manual drafting. For ease of use, it is preferable to furnish the labels on the placing drawing. If the drawing is complex, the detailer will label a reinforcing bar or a group of bars with a single letter or number, which is then keyed to a label list and becomes part of the placing drawing. Data are keyed directly into the computer or recorded on a standard input form, which is then followed by the keyboarding operation. Older systems used punched tape or punched cards.

The data become part of a database, which then provides information for the various other operations. A bill of material, or bar list, is commonly generated on which the bars are segregated by grade and size in descending order. Some programs allow the merging of rebar requirements for several jobs; lower-level (and less expensive) programs may not have this capability. It is not difficult to imagine the complexities involved. The bar shop must keep orderly track of bars that vary in grade (40 or 60), type (uncoated, epoxy coated, galvanized), shape (see Figure 13-4), length, and size.

Several illustrations provide examples of the types of documents that are commonly used in the bar fabricating business. Figure 13-2 shows a placing drawing and has been previously discussed. A computer program can also generate a rebar shearing schedule and a rebar shearing schedule summary (weight summary). Typical printouts are shown in Figure 13-8. Note how the cutting of the 60-ft-long bars is optimized to minimize the scrap.

```
DIMENSION FABRICATORS, INC.                                                                        REBAR SHEARING SCHEDULE
CUTTING BATCH: 68H          SHEAR STATION: 1       MATERIAL GRADE: 60                              TAG LIST: 02       PAGE  1
Items 1 thru 8 Inclusive
===*========================* ==============================================================================================
  *  SZ      PCS  =LENGTH=  *  NO.  =====CUT=====    YLD  SEC-BIN       JOB   BAR LIST NO. ITEM  =MARK=  ----TAG---  STOCK/
  *                         *  CUT  ====LENGTH===    PCS                                                              SCRAP
===*========================* ==============================================================================================
1)  #4       19  60' 0      *   2  30' 0              38    7 1    NTE  768   8            1             ST 68H-1
 *--------------------------*
2)  #4        1  60' 0      *   1  23' 8               1    7 1    N    768   8            2
  *                         *   7   4' 7   (-1)        7    2 1    N    768   8            7   4S1
  *                         *   1   3'11               1    2 2    N    768   8            8   4S3                 0' 4 SCRAP
 *--------------------------*
3)  #4        6
  *
  *
 *--------
4)  #4
  *
 *----
5)  #4
 *----
6)  #4
  *
  *
 *----
7)  #4
  *
  *
 *----
8)  #4
 *----
9)  #4
  *
 *----
10) #4
 *----
11) #4
 *----
12) #4
 *----
  *
  * #4
 *----
 *----
```

REBAR SHEARING SCHEDULE SUMMARY

```
CUTTING BATCH: 68H                              SHEAR STATION: 1  MATERIAL GRADE: 60
Items 1 thru 8 Inclusive                                   TAG LIST: 02  PAGE  2
==============================================================================

#4 Rebar    STOCK REQUIRED                        2,405 LBS.
            SHEARED WEIGHT                        2,375 LBS.
            RE-STOCK 20' or OVER      0.64 %         15 LBS.
            RE-STOCK UNDER 20         0.00 %          0 LBS.
            SCRAP                     0.60 %         14 LBS.
            NOT PROCESSED                            41 LBS.

TOTAL       STOCK REQUIRED                        2,405 LBS.
            SHEARED WEIGHT                        2,375 LBS.
            RE-STOCK 20' or OVER      0.64 %         15 LBS.
            RE-STOCK UNDER 20 '       0.00 %          0 LBS.
            SCRAP                     0.60 %         14 LBS.
            NOT PROCESSED                            41 LBS.
```

FIGURE 13-8 Rebar shearing schedule and rebar shearing schedule summary. (Courtesy of Dimension Fabricators, Inc., Schenectady, N.Y.)

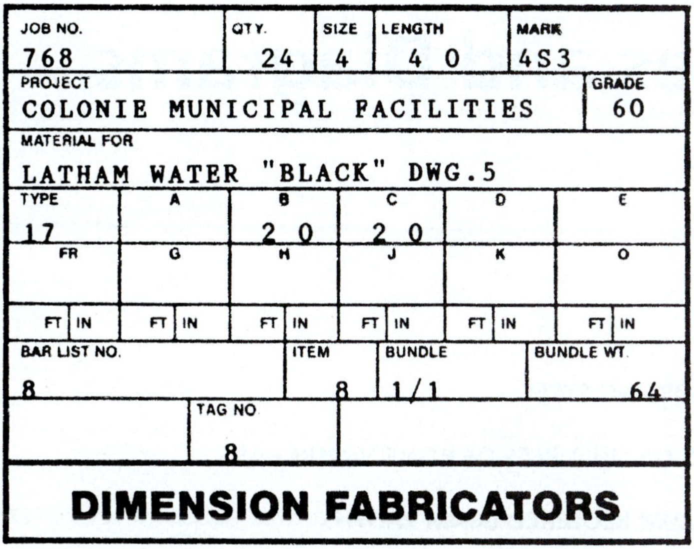

JOB NO.		QTY.	SIZE	LENGTH	MARK	
768		24	4	4 0	4S3	
PROJECT						GRADE
COLONIE MUNICIPAL FACILITIES						60
MATERIAL FOR						
LATHAM WATER "BLACK" DWG.5						
TYPE	A	B	C	D		E
17		2 0	2 0			
FR	G	H	J	K		O
FT	IN	FT	IN	FT	IN	FT	IN	FT	IN		FT	IN

BAR LIST NO.		ITEM	BUNDLE		BUNDLE WT.	
8		8	1/1			64
	TAG NO					
	8					

DIMENSION FABRICATORS

FIGURE 13-9 Bar tag. (Courtesy of Dimension Fabricators, Inc., Schenectady, N.Y.)

Figure 13-9 shows a computer-generated bar tag for bar 4S3, which also appears on the shearing schedule. The bar tag indicates that this is an ACI type 17 bar (refer to Figure 13-4).

Computers have become an indispensable tool in the bar fabricating business. They allow for more productive use of the detailer's talents, freeing the detailer from tedious and repetitive clerical tasks.

REFERENCES

[1] *ACI Detailing Manual* [SP-66(04)], includes "Details and Detailing of Concrete Reinforcement" (ACI 315-99), "Manual of Structural and Placing Drawings for Reinforced Concrete Structures" (ACI 315R-04), and "Supporting Reference Data." American Concrete Institute, 38800 Country Club Drive, Farmington Hills, MI 48331, 2004.

[2] *Manual of Standard Practice*, 27th ed. Concrete Reinforcing Steel Institute, 933 North Plum Grove Road, Schaumburg, IL 60173, 2003.

[3] *Placing Reinforcing Bars*, 8th ed. Concrete Reinforcing Steel Institute, 933 North Plum Grove Road, Schaumburg, IL 60173.

Tables and Diagrams

TABLE A-1 Reinforcing Steel

Type of steel and ASTM specification number	Bar sizes	Grade	Minimum Tensile strength (psi)	Minimum yield strength f_y (psi)	Yield strain ϵ_y
Billet Steel A615	Nos. 3–6	40	70,000	40,000	0.00138
	Nos. 3–18	60	90,000	60,000	0.00207
	Nos. 6–18	75	100,000	75,000	0.00259
Low-Alloy Steel A706	Nos. 3–18	60	80,000 (Min.: 1.25 f_y)	60,000 (Max.: 78,000)	0.00207

Bar number	3	4	5	6	7	8	9	10	11	14	18
Unit weight per foot (lb)	0.376	0.668	1.043	1.502	2.044	2.670	3.400	4.303	5.313	7.65	13.60
Diameter[a] (in.)	0.375	0.500	0.625	0.750	0.875	1.000	1.128	1.270	1.410	1.693	2.257
Area (in.^{2})	0.11	0.20	0.31	0.44	0.60	0.79	1.00	1.27	1.56	2.25	4.00

[a]The nominal dimensions of a deformed bar (diameter and area) are equivalent to those of a plain round bar having the same weight per foot as the deformed bar.

TABLE A-2   Areas of Multiples of Reinforcing Bars (in.^{2})

Number of bars	Bar number								
	#3	#4	#5	#6	#7	#8	#9	#10	#11
1	0.11	0.20	0.31	0.44	0.60	0.79	1.00	1.27	1.56
2	0.22	0.40	0.62	0.88	1.20	1.58	2.00	2.54	3.12
3	0.33	0.60	0.93	1.32	1.80	2.37	3.00	3.81	4.68
4	0.44	0.80	1.24	1.76	2.40	3.16	4.00	5.08	6.24
5	0.55	1.00	1.55	2.20	3.00	3.93	5.00	6.35	7.80
6	0.66	1.20	1.86	2.64	3.60	4.74	6.00	7.62	9.36
7	0.77	1.40	2.17	3.08	4.20	5.53	7.00	8.89	10.9
8	0.88	1.60	2.48	3.52	4.80	6.32	8.00	10.2	12.5
9	0.99	1.80	2.79	3.96	5.40	7.11	9.00	11.4	14.0
10	1.10	2.00	3.10	4.40	6.00	7.90	10.0	12.7	15.6
11	1.21	2.20	3.41	4.84	6.60	8.69	11.0	14.0	17.2
12	1.32	2.40	3.72	5.28	7.20	9.48	12.0	15.2	18.7
13	1.43	2.60	4.03	5.72	7.80	10.3	13.0	16.5	20.3
14	1.54	2.80	4.34	6.16	8.40	11.1	14.0	17.8	21.8
15	1.65	3.00	4.65	6.60	9.00	11.8	15.0	19.0	23.4
16	1.76	3.20	4.96	7.04	9.60	12.6	16.0	20.3	25.0
17	1.87	3.40	5.27	7.48	10.2	13.4	17.0	21.6	26.5
18	1.98	3.60	5.58	7.92	10.8	14.2	18.0	22.9	28.1
19	2.09	3.80	5.89	8.36	11.4	15.0	19.0	24.1	29.6
20	2.20	4.00	6.20	8.80	12.0	15.8	20.0	25.4	31.2

TABLE A-3   Minimum Required Beam Widths (in.)

Number of bars in one layer	Bar number							
	#3 and #4	#5	#6	#7	#8	#9	#10	#11
2	6.0	6.0	6.5	6.5	7.0	7.5	8.0	8.0
3	7.5	8.0	8.0	8.5	9.0	9.5	10.5	11.0
4	9.0	9.5	10.0	10.5	11.0	12.0	13.0	14.0
5	10.5	11.0	11.5	12.5	13.0	14.0	15.5	16.5
6	12.0	12.5	13.5	14.0	15.0	16.5	18.0	19.5
7	13.5	14.5	15.0	16.0	17.0	18.5	20.5	22.5
8	15.0	16.0	17.0	18.0	19.0	21.0	23.0	25.0
9	16.5	17.5	18.5	20.0	21.0	23.0	25.5	28.0
10	18.0	19.0	20.5	21.5	23.0	25.5	28.0	31.0

Note: Tabulated values based on No. 3 stirrups, minimum clear distance of 1 in., and a 1½-in. cover.

TABLE A-4   Areas of Reinforcing Bars per Foot of Slab (in.^{2})

Bar spacing (in.)	Bar number								
	#3	#4	#5	#6	#7	#8	#9	#10	#11
2	0.66	1.20	1.86						
2½	0.53	0.96	1.49	2.11					
3	0.44	0.80	1.24	1.76	2.40	3.16	4.00		
3½	0.38	0.69	1.06	1.51	2.06	2.71	3.43	4.35	
4	0.33	0.60	0.93	1.32	1.80	2.37	3.00	3.81	4.68
4½	0.29	0.53	0.83	1.17	1.60	2.11	2.67	3.39	4.16
5	0.26	0.48	0.74	1.06	1.44	1.90	2.40	3.05	3.74
5½	0.24	0.44	0.68	0.96	1.31	1.72	2.18	2.77	3.40
6	0.22	0.40	0.62	0.88	1.20	1.58	2.00	2.54	3.12
6½	0.20	0.37	0.57	0.81	1.11	1.46	1.85	2.34	2.88
7	0.19	0.34	0.53	0.75	1.03	1.35	1.71	2.18	2.67
7½	0.18	0.32	0.50	0.70	0.96	1.26	1.60	2.03	2.50
8	0.16	0.30	0.46	0.66	0.90	1.18	1.50	1.90	2.34
9	0.15	0.27	0.41	0.59	0.80	1.05	1.33	1.69	2.08
10	0.13	0.24	0.37	0.53	0.72	0.95	1.20	1.52	1.87
11	0.12	0.22	0.34	0.48	0.65	0.86	1.09	1.39	1.70
12	0.11	0.20	0.31	0.44	0.60	0.79	1.00	1.27	1.56
13	0.10	0.18	0.29	0.41	0.55	0.73	0.92	1.17	1.44
14	0.09	0.17	0.27	0.38	0.51	0.68	0.86	1.09	1.34
15	0.09	0.16	0.25	0.35	0.48	0.64	0.80	1.02	1.25
16	0.08	0.15	0.23	0.33	0.45	0.59	0.75	0.95	1.17
17	0.08	0.14	0.22	0.31	0.42	0.56	0.71	0.90	1.10
18	0.07	0.13	0.21	0.29	0.40	0.53	0.67	0.85	1.04

Table A-5   Design Constants

f_c' (psi)	$\left[\dfrac{3\sqrt{f_c'}}{f_y} \geq \dfrac{200}{f_y}\right]^a$	Recommended design values	
		ρ	$\overline{k}$ (ksi)
	$f_y = 40,000$ psi		
3000	0.0050	0.0135	0.4828
4000	0.0050	0.0180	0.6438
5000	0.0053	0.0225	0.8047
6000	0.0058	0.0270	0.9657
	$f_y = 50,000$ psi		
3000	0.0040	0.0108	0.4828
4000	0.0040	0.0144	0.6438
5000	0.0042	0.0180	0.8047
6000	0.0046	0.0216	0.9657
	$f_y = 60,000$ psi		
3000	0.0033	0.0090	0.4828
4000	0.0033	0.0120	0.6438
5000	0.0035	0.0150	0.8047
6000	0.0039	0.0180	0.9657
	$f_y = 75,000$ psi		
3000	0.0027	0.0072	0.4828
4000	0.0027	0.0096	0.6438
5000	0.0028	0.0120	0.8047
6000	0.0031	0.0144	0.9657

[a]Does not apply to T-beams with flanges in tension (see Section 3-2). To compute $A_{s,min}$, see Section 2-8.

TABLE A-6   Properties and Constants for Normal-Weight Concrete

	f_c' (psi)			
	3000	**3500**	**4000**	**5000**
E_c (psi)[a]	3,120,000	3,370,000	3,605,000	4,030,000
n[b]	9	9	8	7
$7.5\sqrt{f_c'}$ (ksi)[c]	0.411	0.444	0.474	0.530

[a]E_c for normal-weight concrete $= 57,000\sqrt{f_c'}$.
[b]Nearest whole number.
[c]Modulus of rupture (f_r).

TABLE A-7   Coefficient of Resistance ($\bar{k}$) Versus Reinforcement Ratio (ρ)
(f'_c = 3000 psi; f_y = 40,000 psi; units of $\bar{k}$ are ksi)

ρ	$\bar{k}$	ρ	$\bar{k}$	ρ	$\bar{k}$
0.0010	0.0397	0.0054	0.2069	0.0098	0.3619
0.0011	0.0436	0.0055	0.2105	0.0099	0.3653
0.0012	0.0476	0.0056	0.2142	0.0100	0.3686
0.0013	0.0515	0.0057	0.2178	0.0101	0.3720
0.0014	0.0554	0.0058	0.2214	0.0102	0.3754
0.0015	0.0593	0.0059	0.2251	0.0103	0.3787
0.0016	0.0632	0.0060	0.2287	0.0104	0.3821
0.0017	0.0671	0.0061	0.2323	0.0105	0.3854
0.0018	0.0710	0.0062	0.2359	0.0106	0.3887
0.0019	0.0749	0.0063	0.2395	0.0107	0.3921
0.0020	0.0788	0.0064	0.2431	0.0108	0.3954
0.0021	0.0826	0.0065	0.2467	0.0109	0.3987
0.0022	0.0865	0.0066	0.2503	0.0110	0.4020
0.0023	0.0903	0.0067	0.2539	0.0111	0.4053
0.0024	0.0942	0.0068	0.2575	0.0112	0.4086
0.0025	0.0980	0.0069	0.2611	0.0113	0.4119
0.0026	0.1019	0.0070	0.2646	0.0114	0.4152
0.0027	0.1057	0.0071	0.2682	0.0115	0.4185
0.0028	0.1095	0.0072	0.2717	0.0116	0.4218
0.0029	0.1134	0.0073	0.2753	0.0117	0.4251
0.0030	0.1172	0.0074	0.2788	0.0118	0.4283
0.0031	0.1210	0.0075	0.2824	0.0119	0.4316
0.0032	0.1248	0.0076	0.2859	0.0120	0.4348
0.0033	0.1286	0.0077	0.2894	0.0121	0.4381
0.0034	0.1324	0.0078	0.2929	0.0122	0.4413
0.0035	0.1362	0.0079	0.2964	0.0123	0.4445
0.0036	0.1399	0.0080	0.2999	0.0124	0.4478
0.0037	0.1437	0.0081	0.3034	0.0125	0.4510
0.0038	0.1475	0.0082	0.3069	0.0126	0.4542
0.0039	0.1512	0.0083	0.3104	0.0127	0.4574
0.0040	0.1550	0.0084	0.3139	0.0128	0.4606
0.0041	0.1587	0.0085	0.3173	0.0129	0.4638
0.0042	0.1625	0.0086	0.3208	0.0130	0.4670
0.0043	0.1662	0.0087	0.3243	0.0131	0.4702
0.0044	0.1699	0.0088	0.3277	0.0132	0.4733
0.0045	0.1736	0.0089	0.3311	0.0133	0.4765
0.0046	0.1774	0.0090	0.3346	0.0134	0.4797
0.0047	0.1811	0.0091	0.3380	0.0135	0.4828
0.0048	0.1848	0.0092	0.3414	0.0136	0.4860
0.0049	0.1885	0.0093	0.3449	0.0137	0.4891
0.0050	0.1922	0.0094	0.3483	0.0138	0.4923
0.0051	0.1958	0.0095	0.3517	0.0139	0.4954
0.0052	0.1995	0.0096	0.3551	0.0140	0.4985
0.0053	0.2032	0.0097	0.3585	0.0141	0.5016

TABLE A-7 (CONT.) Coefficient of Resistance ($\bar{k}$) Versus Reinforcement Ratio (ρ)
(f'_c = 3000 psi; f_y = 40,000 psi; units of $\bar{k}$ are ksi)

ρ	$\bar{k}$	ρ	$\bar{k}$	ρ	$\bar{k}$	ϵ_t^*
0.0142	0.5047	0.0173	0.5981	**0.02033**	**0.6836**	**0.00500**
0.0143	0.5078	0.0174	0.6011	0.0204	0.6855	0.00497
0.0144	0.5109	0.0175	0.6040	0.0205	0.6882	0.00493
0.0145	0.5140	0.0176	0.6069	0.0206	0.6909	0.00489
0.0146	0.5171	0.0177	0.6098	0.0207	0.6936	0.00485
0.0147	0.5202	0.0178	0.6126	0.0208	0.6963	0.00482
0.0148	0.5233	0.0179	0.6155	0.0209	0.6990	0.00478
0.0149	0.5264	0.0180	0.6184	0.0210	0.7017	0.00474
0.0150	0.5294	0.0181	0.6213	0.0211	0.7044	0.00470
0.0151	0.5325	0.0182	0.6241	0.0212	0.7071	0.00467
0.0152	0.5355	0.0183	0.6270	0.0213	0.7097	0.00463
0.0153	0.5386	0.0184	0.6298	0.0214	0.7124	0.00460
0.0154	0.5416	0.0185	0.6327	0.0215	0.7150	0.00456
0.0155	0.5447	0.0186	0.6355	0.0216	0.7177	0.00453
0.0156	0.5477	0.0187	0.6383	0.0217	0.7203	0.00449
0.0157	0.5507	0.0188	0.6412	0.0218	0.7230	0.00446
0.0158	0.5537	0.0189	0.6440	0.0219	0.7256	0.00442
0.0159	0.5567	0.0190	0.6468	0.0220	0.7282	0.00439
0.0160	0.5597	0.0191	0.6496	0.0221	0.7308	0.00436
0.0161	0.5627	0.0192	0.6524	0.0222	0.7334	0.00432
0.0162	0.5657	0.0193	0.6552	0.0223	0.7360	0.00429
0.0163	0.5687	0.0194	0.6580	0.0224	0.7386	0.00426
0.0164	0.5717	0.0195	0.6608	0.0225	0.7412	0.00423
0.0165	0.5746	0.0196	0.6635	0.0226	0.7438	0.00419
0.0166	0.5776	0.0197	0.6663	0.0227	0.7464	0.00416
0.0167	0.5805	0.0198	0.6691	0.0228	0.7490	0.00413
0.0168	0.5835	0.0199	0.6718	0.0229	0.7515	0.00410
0.0169	0.5864	0.0200	0.6746	0.0230	0.7541	0.00407
0.0170	0.5894	0.0201	0.6773	0.0231	0.7567	0.00404
0.0171	0.5923	0.0202	0.6800	0.0232	0.7592	0.00401
0.0172	0.5952	0.0203	0.6828	**0.02323**	**0.7600**	**0.00400**

$^*d = d_t.$

TABLE A-8   Coefficient of Resistance ($\bar{k}$) Versus Reinforcement Ratio (ρ)
($f'_c = 3000$ psi; $f_y = 60,000$ psi; units of $\bar{k}$ are ksi)

ρ	$\bar{k}$	ρ	$\bar{k}$	ρ	$\bar{k}$	ϵ_t^*
0.0010	0.0593	0.0059	0.3294	0.0108	0.5657	
0.0011	0.0651	0.0060	0.3346	0.0109	0.5702	
0.0012	0.0710	0.0061	0.3397	0.0110	0.5746	
0.0013	0.0768	0.0062	0.3449	0.0111	0.5791	
0.0014	0.0826	0.0063	0.3500	0.0112	0.5835	
0.0015	0.0884	0.0064	0.3551	0.0113	0.5879	
0.0016	0.0942	0.0065	0.3602	0.0114	0.5923	
0.0017	0.1000	0.0066	0.3653	0.0115	0.5967	
0.0018	0.1057	0.0067	0.3703	0.0116	0.6011	
0.0019	0.1115	0.0068	0.3754	0.0117	0.6054	
0.0020	0.1172	0.0069	0.3804	0.0118	0.6098	
0.0021	0.1229	0.0070	0.3854	0.0119	0.6141	
0.0022	0.1286	0.0071	0.3904	0.0120	0.6184	
0.0023	0.1343	0.0072	0.3954	0.0121	0.6227	
0.0024	0.1399	0.0073	0.4004	0.0122	0.6270	
0.0025	0.1456	0.0074	0.4054	0.0123	0.6312	
0.0026	0.1512	0.0075	0.4103	0.0124	0.6355	
0.0027	0.1569	0.0076	0.4152	0.0125	0.6398	
0.0028	0.1625	0.0077	0.4202	0.0126	0.6440	
0.0029	0.1681	0.0078	0.4251	0.0127	0.6482	
0.0030	0.1736	0.0079	0.4300	0.0128	0.6524	
0.0031	0.1792	0.0080	0.4348	0.0129	0.6566	
0.0032	0.1848	0.0081	0.4397	0.0130	0.6608	
0.0033	0.1903	0.0082	0.4446	0.0131	0.6649	
0.0034	0.1958	0.0083	0.4494	0.0132	0.6691	
0.0035	0.2014	0.0084	0.4542	0.0133	0.6732	
0.0036	0.2069	0.0085	0.4590	0.0134	0.6773	
0.0037	0.2123	0.0086	0.4638	0.0135	0.6814	
0.0038	0.2178	0.0087	0.4686	**0.01355**	**0.6835**	**0.00500**
0.0039	0.2233	0.0088	0.4734	0.0136	0.6855	0.00497
0.0040	0.2287	0.0089	0.4781	0.0137	0.6896	0.00491
0.0041	0.2341	0.0090	0.4828	0.0138	0.6936	0.00485
0.0042	0.2396	0.0091	0.4876	0.0139	0.6977	0.00480
0.0043	0.2450	0.0092	0.4923	0.0140	0.7017	0.00474
0.0044	0.2503	0.0093	0.4970	0.0141	0.7057	0.00469
0.0045	0.2557	0.0094	0.5017	0.0142	0.7097	0.00463
0.0046	0.2611	0.0095	0.5063	0.0143	0.7137	0.00458
0.0047	0.2664	0.0096	0.5110	0.0144	0.7177	0.00453
0.0048	0.2717	0.0097	0.5156	0.0145	0.7216	0.00447
0.0049	0.2771	0.0098	0.5202	0.0146	0.7256	0.00442
0.0050	0.2824	0.0099	0.5248	0.0147	0.7295	0.00437
0.0051	0.2876	0.0100	0.5294	0.0148	0.7334	0.00432
0.0052	0.2929	0.0101	0.5340	0.0149	0.7373	0.00427
0.0053	0.2982	0.0102	0.5386	0.0150	0.7412	0.00423
0.0054	0.3034	0.0103	0.5431	0.0151	0.7451	0.00418
0.0055	0.3087	0.0104	0.5477	0.0152	0.7490	0.00413
0.0056	0.3139	0.0105	0.5522	0.0153	0.7528	0.00408
0.0057	0.3191	0.0106	0.5567	0.0154	0.7567	0.00404
0.0058	0.3243	0.0107	0.5612	**0.01548**	**0.7597**	**0.00400**

$^*d = d_t.$

TABLE A-9 Coefficient of Resistance ($\bar{k}$) Versus Reinforcement Ratio (ρ)
($f_c' = 4000$ psi; $f_y = 40,000$ psi; units of ($\bar{k}$) are ksi)

ρ	$\bar{k}$	ρ	$\bar{k}$	ρ	$\bar{k}$	ρ	$\bar{k}$
0.0010	0.0398	0.0054	0.2091	0.0098	0.3694	0.0142	0.5206
0.0011	0.0437	0.0055	0.2129	0.0099	0.3729	0.0143	0.5239
0.0012	0.0477	0.0056	0.2166	0.0100	0.3765	0.0144	0.5272
0.0013	0.0516	0.0057	0.2204	0.0101	0.3800	0.0145	0.5305
0.0014	0.0555	0.0058	0.2241	0.0102	0.3835	0.0146	0.5338
0.0015	0.0595	0.0059	0.2278	0.0103	0.3870	0.0147	0.5372
0.0016	0.0634	0.0060	0.2315	0.0104	0.3906	0.0148	0.5405
0.0017	0.0673	0.0061	0.2352	0.0105	0.3941	0.0149	0.5438
0.0018	0.0712	0.0062	0.2390	0.0106	0.3976	0.0150	0.5471
0.0019	0.0752	0.0063	0.2427	0.0107	0.4011	0.0151	0.5504
0.0020	0.0791	0.0064	0.2464	0.0108	0.4046	0.0152	0.5536
0.0021	0.0830	0.0065	0.2501	0.0109	0.4080	0.0153	0.5569
0.0022	0.0869	0.0066	0.2538	0.0110	0.4115	0.0154	0.5602
0.0023	0.0908	0.0067	0.2574	0.0111	0.4150	0.0155	0.5635
0.0024	0.0946	0.0068	0.2611	0.0112	0.4185	0.0156	0.5667
0.0025	0.0985	0.0069	0.2648	0.0113	0.4220	0.0157	0.5700
0.0026	0.1024	0.0070	0.2685	0.0114	0.4254	0.0158	0.5733
0.0027	0.1063	0.0071	0.2721	0.0115	0.4289	0.0159	0.5765
0.0028	0.1102	0.0072	0.2758	0.0116	0.4323	0.0160	0.5798
0.0029	0.1140	0.0073	0.2795	0.0117	0.4358	0.0161	0.5830
0.0030	0.1179	0.0074	0.2831	0.0118	0.4392	0.0162	0.5863
0.0031	0.1217	0.0075	0.2868	0.0119	0.4427	0.0163	0.5895
0.0032	0.1256	0.0076	0.2904	0.0120	0.4461	0.0164	0.5927
0.0033	0.1294	0.0077	0.2941	0.0121	0.4495	0.0165	0.5959
0.0034	0.1333	0.0078	0.2977	0.0122	0.4530	0.0166	0.5992
0.0035	0.1371	0.0079	0.3013	0.0123	0.4564	0.0167	0.6024
0.0036	0.1410	0.0080	0.3049	0.0124	0.4598	0.0168	0.6056
0.0037	0.1448	0.0081	0.3086	0.0125	0.4632	0.0169	0.6088
0.0038	0.1486	0.0082	0.3122	0.0126	0.4666	0.0170	0.6120
0.0039	0.1524	0.0083	0.3158	0.0127	0.4701	0.0171	0.6152
0.0040	0.1562	0.0084	0.3194	0.0128	0.4735	0.0172	0.6184
0.0041	0.1600	0.0085	0.3230	0.0129	0.4768	0.0173	0.6216
0.0042	0.1638	0.0086	0.3266	0.0130	0.4802	0.0174	0.6248
0.0043	0.1676	0.0087	0.3302	0.0131	0.4836	0.0175	0.6279
0.0044	0.1714	0.0088	0.3338	0.0132	0.4870	0.0176	0.6311
0.0045	0.1752	0.0089	0.3374	0.0133	0.4904	0.0177	0.6343
0.0046	0.1790	0.0090	0.3409	0.0134	0.4938	0.0178	0.6375
0.0047	0.1828	0.0091	0.3445	0.0135	0.4971	0.0179	0.6406
0.0048	0.1866	0.0092	0.3481	0.0136	0.5005	0.0180	0.6438
0.0049	0.1904	0.0093	0.3517	0.0137	0.5038	0.0181	0.6469
0.0050	0.1941	0.0094	0.3552	0.0138	0.5072	0.0182	0.6501
0.0051	0.1979	0.0095	0.3588	0.0139	0.5105	0.0183	0.6532
0.0052	0.2016	0.0096	0.3623	0.0140	0.5139	0.0184	0.6563
0.0053	0.2054	0.0097	0.3659	0.0141	0.5172	0.0185	0.6595

TABLE A-9 (CONT.) Coefficient of Resistance ($\bar{k}$) Versus Reinforcement Ratio (ρ)
($f_c' = 4000$ psi; $f_y = 40,000$ psi; units of ($\bar{k}$) are ksi)

ρ	$\bar{k}$	ρ	$\bar{k}$	ρ	$\bar{k}$	ϵ_t^*
0.0186	0.6626	0.0229	0.7927	**0.0271**	**0.9113**	**0.00500**
0.0187	0.6657	0.0230	0.7956	0.0272	0.9140	0.00497
0.0188	0.6688	0.0231	0.7985	0.0273	0.9167	0.00494
0.0189	0.6720	0.0232	0.8014	0.0274	0.9194	0.00491
0.0190	0.6751	0.0233	0.8043	0.0275	0.9221	0.00488
0.0191	0.6782	0.0234	0.8072	0.0276	0.9248	0.00485
0.0192	0.6813	0.0235	0.8101	0.0277	0.9275	0.00482
0.0193	0.6844	0.0236	0.8130	0.0278	0.9302	0.00480
0.0194	0.6875	0.0237	0.8159	0.0279	0.9329	0.00477
0.0195	0.6905	0.0238	0.8188	0.0280	0.9356	0.00474
0.0196	0.6936	0.0239	0.8217	0.0281	0.9383	0.00471
0.0197	0.6967	0.0240	0.8245	0.0282	0.9410	0.00469
0.0198	0.6998	0.0241	0.8274	0.0283	0.9436	0.00466
0.0199	0.7029	0.0242	0.8303	0.0284	0.9463	0.00463
0.0200	0.7059	0.0243	0.8331	0.0285	0.9490	0.00461
0.0201	0.7090	0.0244	0.8360	0.0286	0.9516	0.00458
0.0202	0.7120	0.0245	0.8388	0.0287	0.9543	0.00455
0.0203	0.7151	0.0246	0.8417	0.0288	0.9569	0.00453
0.0204	0.7181	0.0247	0.8445	0.0289	0.9596	0.00450
0.0205	0.7212	0.0248	0.8473	0.0290	0.9622	0.00447
0.0206	0.7242	0.0249	0.8502	0.0291	0.9648	0.00445
0.0207	0.7272	0.0250	0.8530	0.0292	0.9675	0.00442
0.0208	0.7302	0.0251	0.8558	0.0293	0.9701	0.00440
0.0209	0.7333	0.0252	0.8586	0.0294	0.9727	0.00437
0.0210	0.7363	0.0253	0.8615	0.0295	0.9753	0.00435
0.0211	0.7393	0.0254	0.8643	0.0296	0.9779	0.00432
0.0212	0.7423	0.0255	0.8671	0.0297	0.9805	0.00430
0.0213	0.7453	0.0256	0.8699	0.0298	0.9831	0.00427
0.0214	0.7483	0.0257	0.8727	0.0299	0.9857	0.00425
0.0215	0.7513	0.0258	0.8754	0.0300	0.9883	0.00423
0.0216	0.7543	0.0259	0.8782	0.0301	0.9909	0.00420
0.0217	0.7572	0.0260	0.8810	0.0302	0.9935	0.00418
0.0218	0.7602	0.0261	0.8838	0.0303	0.9961	0.00415
0.0219	0.7632	0.0262	0.8865	0.0304	0.9986	0.00413
0.0220	0.7662	0.0263	0.8893	0.0305	1.0012	0.00411
0.0221	0.7691	0.0264	0.8921	0.0306	1.0038	0.00408
0.0222	0.7721	0.0265	0.8948	0.0307	1.0063	0.00406
0.0223	0.7750	0.0266	0.8976	0.0308	1.0089	0.00404
0.0224	0.7780	0.0267	0.9003	0.0309	1.0114	0.00401
0.0225	0.7809	0.0268	0.9031	**0.03096**	**1.0130**	**0.00400**
0.0226	0.7839	0.0269	0.9058			
0.0227	0.7868	0.0270	0.9085			
0.0228	0.7897					

*$d = d_t$.

TABLE A-10   Coefficient of Resistance ($\bar{k}$) Versus Reinforcement Ratio (ρ)
($f'_c = 4000$ psi; $f_y = 60,000$ psi; units of $\bar{k}$ are ksi)

ρ	$\bar{k}$	ρ	$\bar{k}$	ρ	$\bar{k}$	ρ	$\bar{k}$
0.0010	0.0595	0.0039	0.2259	0.0068	0.3835	0.0097	0.5322
0.0011	0.0654	0.0040	0.2315	0.0069	0.3888	0.0098	0.5372
0.0012	0.0712	0.0041	0.2371	0.0070	0.3941	0.0099	0.5421
0.0013	0.0771	0.0042	0.2427	0.0071	0.3993	0.0100	0.5471
0.0014	0.0830	0.0043	0.2482	0.0072	0.4046	0.0101	0.5520
0.0015	0.0889	0.0044	0.2538	0.0073	0.4098	0.0102	0.5569
0.0016	0.0946	0.0045	0.2593	0.0074	0.4150	0.0103	0.5618
0.0017	0.1005	0.0046	0.2648	0.0075	0.4202	0.0104	0.5667
0.0018	0.1063	0.0047	0.2703	0.0076	0.4254	0.0105	0.5716
0.0019	0.1121	0.0048	0.2758	0.0077	0.4306	0.0106	0.5765
0.0020	0.1179	0.0049	0.2813	0.0078	0.4358	0.0107	0.5814
0.0021	0.1237	0.0050	0.2868	0.0079	0.4410	0.0108	0.5862
0.0022	0.1294	0.0051	0.2922	0.0080	0.4461	0.0109	0.5911
0.0023	0.1352	0.0052	0.2977	0.0081	0.4513	0.0110	0.5959
0.0024	0.1410	0.0053	0.3031	0.0082	0.4564	0.0111	0.6008
0.0025	0.1467	0.0054	0.3086	0.0083	0.4615	0.0112	0.6056
0.0026	0.1524	0.0055	0.3140	0.0084	0.4666	0.0113	0.6104
0.0027	0.1581	0.0056	0.3194	0.0085	0.4718	0.0114	0.6152
0.0028	0.1638	0.0057	0.3248	0.0086	0.4768	0.0115	0.6200
0.0029	0.1695	0.0058	0.3302	0.0087	0.4819	0.0116	0.6248
0.0030	0.1752	0.0059	0.3356	0.0088	0.4870	0.0117	0.6296
0.0031	0.1809	0.0060	0.3409	0.0089	0.4921	0.0118	0.6343
0.0032	0.1866	0.0061	0.3463	0.0090	0.4971	0.0119	0.6391
0.0033	0.1922	0.0062	0.3516	0.0091	0.5022	0.0120	0.6438
0.0034	0.1979	0.0063	0.3570	0.0092	0.5072	0.0121	0.6485
0.0035	0.2035	0.0064	0.3623	0.0093	0.5122	0.0122	0.6532
0.0036	0.2091	0.0065	0.3676	0.0094	0.5172	0.0123	0.6579
0.0037	0.2148	0.0066	0.3729	0.0095	0.5222	0.0124	0.6626
0.0038	0.2204	0.0067	0.3782	0.0096	0.5272	0.0125	0.6673

**TABLE A-10 (CONT.)   Coefficient of Resistance ($\bar{k}$) Versus Reinforcement Ratio (ρ)
($f_c' = 4000$ psi; $f_y = 60{,}000$ psi; units of $\bar{k}$ are ksi)**

ρ	$\bar{k}$	ρ	$\bar{k}$	ρ	$\bar{k}$	ϵ_t^*
0.0126	0.6720	0.0154	0.7985	**0.01806**	**0.9110**	**0.00500**
0.0127	0.6766	0.0155	0.8029	0.0181	0.9126	0.00498
0.0128	0.6813	0.0156	0.8072	0.0182	0.9167	0.00494
0.0129	0.6859	0.0157	0.8116	0.0183	0.9208	0.00490
0.0130	0.6906	0.0158	0.8159	0.0184	0.9248	0.00485
0.0131	0.6952	0.0159	0.8202	0.0185	0.9289	0.00481
0.0132	0.6998	0.0160	0.8245	0.0186	0.9329	0.00477
0.0133	0.7044	0.0161	0.8288	0.0187	0.9369	0.00473
0.0134	0.7090	0.0162	0.8331	0.0188	0.9410	0.00469
0.0135	0.7136	0.0163	0.8374	0.0189	0.9450	0.00465
0.0136	0.7181	0.0164	0.8417	0.0190	0.9490	0.00461
0.0137	0.7227	0.0165	0.8459	0.0191	0.9529	0.00457
0.0138	0.7272	0.0166	0.8502	0.0192	0.9569	0.00453
0.0139	0.7318	0.0167	0.8544	0.0193	0.9609	0.00449
0.0140	0.7363	0.0168	0.8586	0.0194	0.9648	0.00445
0.0141	0.7408	0.0169	0.8629	0.0195	0.9688	0.00441
0.0142	0.7453	0.0170	0.8671	0.0196	0.9727	0.00437
0.0143	0.7498	0.0171	0.8713	0.0197	0.9766	0.00434
0.0144	0.7543	0.0172	0.8754	0.0198	0.9805	0.00430
0.0145	0.7587	0.0173	0.8796	0.0199	0.9844	0.00426
0.0146	0.7632	0.0174	0.8838	0.0200	0.9883	0.00422
0.0147	0.7676	0.0175	0.8879	0.0201	0.9922	0.00419
0.0148	0.7721	0.0176	0.8921	0.0202	0.9961	0.00415
0.0149	0.7765	0.0177	0.8962	0.0203	0.9999	0.00412
0.0150	0.7809	0.0178	0.9003	0.0204	1.0038	0.00408
0.0151	0.7853	0.0179	0.9044	0.0205	1.0076	0.00405
0.0152	0.7897	0.0180	0.9085	0.0206	1.0114	0.00401
0.0153	0.7941			**0.02063**	**1.0126**	**0.00400**

$^*d = d_t$.

TABLE A-11   Coefficient of Resistance ($\bar{k}$) Versus Reinforcement Ratio (ρ)
(f'_c = 5000 psi; f_y = 60,000 psi; units of $\bar{k}$ are ksi)

ρ	$\bar{k}$	ρ	$\bar{k}$	ρ	$\bar{k}$	ρ	$\bar{k}$
0.0010	0.0596	0.0048	0.2782	0.0086	0.4847	0.0124	0.6789
0.0011	0.0655	0.0049	0.2838	0.0087	0.4899	0.0125	0.6838
0.0012	0.0714	0.0050	0.2894	0.0088	0.4952	0.0126	0.6888
0.0013	0.0773	0.0051	0.2950	0.0089	0.5005	0.0127	0.6937
0.0014	0.0832	0.0052	0.3005	0.0090	0.5057	0.0128	0.6986
0.0015	0.0890	0.0053	0.3061	0.0091	0.5109	0.0129	0.7035
0.0016	0.0949	0.0054	0.3117	0.0092	0.5162	0.0130	0.7084
0.0017	0.1008	0.0055	0.3172	0.0093	0.5214	0.0131	0.7133
0.0018	0.1066	0.0056	0.3227	0.0094	0.5266	0.0132	0.7182
0.0019	0.1125	0.0057	0.3282	0.0095	0.5318	0.0133	0.7231
0.0020	0.1183	0.0058	0.3338	0.0096	0.5370	0.0134	0.7280
0.0021	0.1241	0.0059	0.3393	0.0097	0.5422	0.0135	0.7328
0.0022	0.1300	0.0060	0.3448	0.0098	0.5473	0.0136	0.7377
0.0023	0.1358	0.0061	0.3502	0.0099	0.5525	0.0137	0.7425
0.0024	0.1416	0.0062	0.3557	0.0100	0.5576	0.0138	0.7473
0.0025	0.1474	0.0063	0.3612	0.0101	0.5628	0.0139	0.7522
0.0026	0.1531	0.0064	0.3667	0.0102	0.5679	0.0140	0.7570
0.0027	0.1589	0.0065	0.3721	0.0103	0.5731	0.0141	0.7618
0.0028	0.1647	0.0066	0.3776	0.0104	0.5782	0.0142	0.7666
0.0029	0.1704	0.0067	0.3830	0.0105	0.5833	0.0143	0.7714
0.0030	0.1762	0.0068	0.3884	0.0106	0.5884	0.0144	0.7762
0.0031	0.1819	0.0069	0.3938	0.0107	0.5935	0.0145	0.7810
0.0032	0.1877	0.0070	0.3992	0.0108	0.5986	0.0146	0.7857
0.0033	0.1934	0.0071	0.4047	0.0109	0.6037	0.0147	0.7905
0.0034	0.1991	0.0072	0.4100	0.0110	0.6088	0.0148	0.7952
0.0035	0.2048	0.0073	0.4154	0.0111	0.6138	0.0149	0.8000
0.0036	0.2105	0.0074	0.4208	0.0112	0.6189	0.0150	0.8047
0.0037	0.2162	0.0075	0.4262	0.0113	0.6239	0.0151	0.8094
0.0038	0.2219	0.0076	0.4315	0.0114	0.6290	0.0152	0.8142
0.0039	0.2276	0.0077	0.4369	0.0115	0.6340	0.0153	0.8189
0.0040	0.2332	0.0078	0.4422	0.0116	0.6390	0.0154	0.8236
0.0041	0.2389	0.0079	0.4476	0.0117	0.6440	0.0155	0.8283
0.0042	0.2445	0.0080	0.4529	0.0118	0.6490	0.0156	0.8329
0.0043	0.2502	0.0081	0.4582	0.0119	0.6540	0.0157	0.8376
0.0044	0.2558	0.0082	0.4635	0.0120	0.6590	0.0158	0.8423
0.0045	0.2614	0.0083	0.4688	0.0121	0.6640	0.0159	0.8469
0.0046	0.2670	0.0084	0.4741	0.0122	0.6690	0.0160	0.8516
0.0047	0.2726	0.0085	0.4794	0.0123	0.6739	0.0161	0.8562

TABLE A-11 (CONT.)   Coefficient of Resistance ($\bar{k}$) Versus Reinforcement Ratio (ρ)
($f_c' = 5000$ psi; $f_y = 60,000$ psi; units of $\bar{k}$ are ksi)

ρ	$\bar{k}$	ρ	$\bar{k}$	ρ	$\bar{k}$	ϵ_t^*
0.0162	0.8609	0.0194	1.0047	**0.02125**	**1.0838**	**0.00500**
0.0163	0.8655	0.0195	1.0090	0.0213	1.0859	0.00498
0.0164	0.8701	0.0196	1.0134	0.0214	1.0901	0.00494
0.0165	0.8747	0.0197	1.0177	0.0215	1.0943	0.00491
0.0166	0.8793	0.0198	1.0220	0.0216	1.0985	0.00487
0.0167	0.8839	0.0199	1.0263	0.0217	1.1026	0.00483
0.0168	0.8885	0.0200	1.0307	0.0218	1.1068	0.00480
0.0169	0.8930	0.0201	1.0350	0.0219	1.1110	0.00476
0.0170	0.8976	0.0202	1.0393	0.0220	1.1151	0.00473
0.0171	0.9022	0.0203	1.0435	0.0221	1.1192	0.00469
0.0172	0.9067	0.0204	1.0478	0.0222	1.1234	0.00466
0.0173	0.9112	0.0205	1.0521	0.0223	1.1275	0.00462
0.0174	0.9158	0.0206	1.0563	0.0224	1.1316	0.00459
0.0175	0.9203	0.0207	1.0606	0.0225	1.1357	0.00456
0.0176	0.9248	0.0208	1.0648	0.0226	1.1398	0.00452
0.0177	0.9293	0.0209	1.0691	0.0227	1.1438	0.00449
0.0178	0.9338	0.0210	1.0733	0.0228	1.1479	0.00446
0.0179	0.9383	0.0211	1.0775	0.0229	1.1520	0.00442
0.0180	0.9428	0.0212	1.0817	0.0230	1.1560	0.00439
0.0181	0.9473			0.0231	1.1601	0.00436
0.0182	0.9517			0.0232	1.1641	0.00433
0.0183	0.9562			0.0233	1.1682	0.00430
0.0184	0.9606			0.0234	1.1722	0.00426
0.0185	0.9651			0.0235	1.1762	0.00423
0.0186	0.9695			0.0236	1.1802	0.00420
0.0187	0.9739			0.0237	1.1842	0.00417
0.0188	0.9783			0.0238	1.882	0.00414
0.0189	0.9827			0.0239	1.1922	0.00411
0.0190	0.9872			0.0240	1.1961	0.00408
0.0191	0.9916			0.0241	1.2001	0.00405
0.0192	0.9959			0.0242	1.2041	0.00402
0.0193	1.0003			**0.02429**	1.2076	**0.00400**

$^*d = d_t$.

TABLE A-12 Development Length (ℓ_{dc}) for Compression Bars with $f_y = 60{,}000$ psi (in.)

Bar Size	f_c' (normal-weight concrete), psi			
	3000	4000	5000	6000
3	8.2	7.1	6.8	6.8
4	11.0	9.5	9.0	9.0
5	13.7	11.9	11.3	11.3
6	16.4	14.2	13.5	13.5
7	19.2	16.6	15.8	15.8
8	21.9	19.0	18.0	18.0
9	24.7	21.4	20.3	20.3
10	27.8	24.1	22.9	22.9
11	30.9	26.8	25.4	25.4
14	37.1	32.1	30.5	30.5
18	49.4	42.8	40.6	40.6

Note: See Chapter 5 for calculation of development length for compression bars.

TABLE A-13 Development Length (ℓ_{dh}) for Hooked Bars with $f_y = 60{,}000$ psi (in.)

Bar number	f_c' (psi)			
	3000	4000	5000	6000
3	8.2	7.1	6.4	5.8
4	11.0	9.5	8.5	7.7
5	13.7	11.9	10.6	9.7
6	16.4	14.2	12.7	11.6
7	19.2	16.6	14.8	13.6
8	21.9	19.0	17.0	15.5
9	24.7	21.4	19.1	17.5
10	27.8	24.1	21.6	19.7
11	30.9	26.8	23.9	21.8

Note: Modification factors may apply.
$\ell_{dh} \times$ applicable modification factors $\geq$ larger of $8d_b$ or 6 in. See Chapter 5 for calculation of development length for hooked bars.

TABLE A-14 Preferred Maximum Number of Column Bars

Recommended spiral or tie bar number	Core size (in.) = column size Size − 2 × cover	Circular area (in.²)	Bar number #5	#6	#7	#8	#9	#10	#11	Square area (in.²)	Bar number #5	#6	#7	#8	#9	#10	#11[a]
3[a]	9	63.6	8	7	7	6	—	—	—	81	8	8	8	8	4	4	4
	10	78.5	10	9	8	7	6	—	—	100	12	8	8	8	8	4	4
	11	95.0	11	10	9	8	7	6	—	121	12	12	8	8	8	8	4
	12	113.1	12	11	10	9	8	7	6	144	12	12	12	8	8	8	8
	13	132.7	13	12	11	10	8	7	6	169	16	12	12	12	8	8	8
	14	153.9	14	13	12	11	9	8	7	196	16	16	12	12	8	8	8
	15	176.7	15	14	13	12	10	9	8	225	16	16	16	12	12	12	8
4	16	201.1	16	15	14	12	11	9	8	256	20	16	16	16	12	12	8
	17	227.0	18	16	15	13	12	10	9	289	20	20	16	16	12	12	8
	18	254.5	19	17	15	14	12	11	10	324	20	20	16	16	16	12	12
	19	283.5	20	18	16	15	13	11	10	361	24	20	20	16	16	12	12
	20	314.2	21	19	17	16	14	12	11	400	24	24	20	20	16	12	12
	21	346.4	22	20	18	17	15	13	11	441	28	24	20	20	16	16	12
	22	380.1	23	21	19	18	15	14	12	484	28	24	24	20	20	16	12
5	23	415.5	24	22	21	19	16	14	13	529	28	28	24	24	20	16	16
	24	452.4	25	23	21	20	17	15	13	576	32	28	24	24	16	16	16
	25	490.9	26	24	22	20	18	16	14	625	32	28	28	24	20	20	16
	26	530.9	28	25	23	21	19	16	14	676	32	32	28	28	24	20	16
	27	572.6	29	26	24	22	19	17	15	729	36	32	28	28	24	20	16

[a] No. 4 tie required for No. 11 or larger longitudinal reinforcement (ACI Section 7.10.5.1).

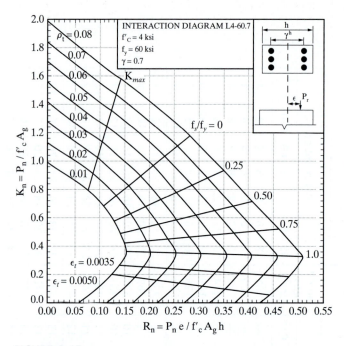

FIGURE A-15
Source: Diagrams A-15 through A-22 are from the ACI Design Handbook SP-17(97) and are reprinted here with the permission of the American Concrete Institute.

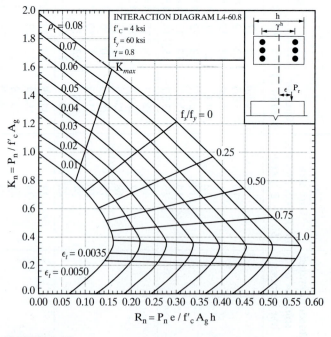

FIGURE A-16

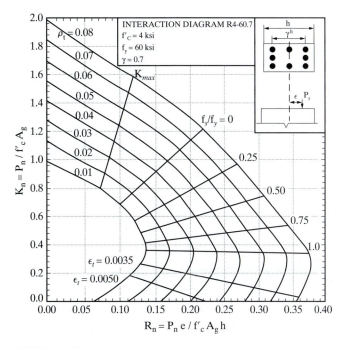

FIGURE A-17

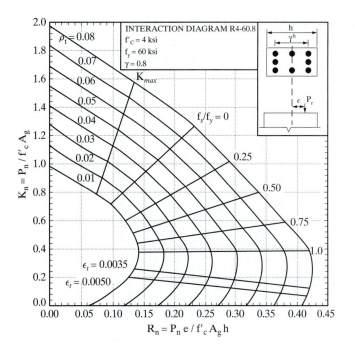

FIGURE A-18

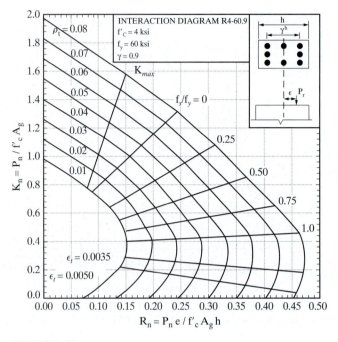

FIGURE A-19

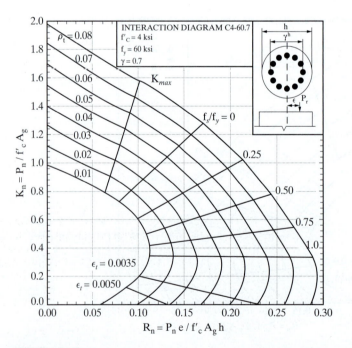

FIGURE A-20

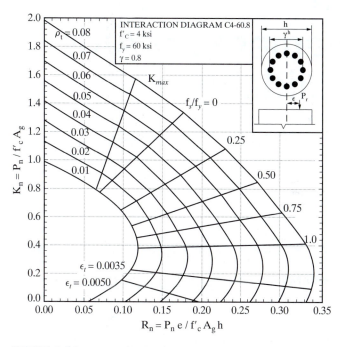

FIGURE A-21

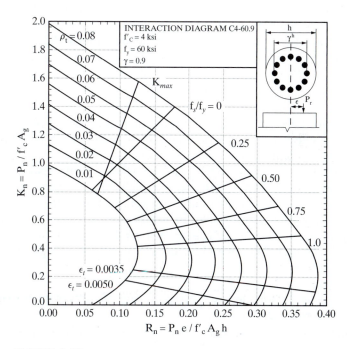

FIGURE A-22

Supplementary Aids and Guidelines

B-1   ACCURACY FOR COMPUTATIONS FOR REINFORCED CONCRETE

The widespread availability and use of electronic calculators and computers for even the simplest of calculations has led to the use of numbers that represent a very high order of accuracy. For instance, a calculator having an eight-digit display will yield the following:

$$\frac{8}{0.7} = 11.428571$$

It should be recognized that the numerator and denominator, each of one-figure accuracy, resulted in a number that indicates eight-figure accuracy. Because the quotient cannot be expected to be more accurate than the numbers that produced it, the result as represented may lead one into a false sense of security associated with numbers of very high accuracy. For instance, it is illogical to calculate a required steel area to four-figure accuracy when the loads were to two-figure accuracy and the bars to be chosen have areas tabulated to two- (sometimes three-) figure accuracy. Likewise, the involved mathematical expressions developed in this book and those presented by the ACI Code should be thought of in a similar light. They deal with a material, concrete, that

1.   Is made on site or plant-made, is subject to varying amounts of quality control, and will vary from the design strength.
2.   Is placed in forms that may or may not produce the design dimensions.

3.   Contains reinforcing steel of a specified minimum strength but that may vary above that strength.

In addition, the reinforced concrete member has reinforcing steel that may or may not be placed at the design location, and the design itself is generally based on loads that may be only "best estimates."

With the foregoing in mind, the following has been suggested by the Concrete Reinforcing Steel Institute as a rough guide for numerical accuracy in reinforced concrete calculations:

1.   Loads to the nearest 1 psf; 10 lb/ft; 100 lb concentration
2.   Span lengths to about 0.1 ft
3.   Total loads and reactions to 0.1 kip or three-figure accuracy
4.   Moments to the nearest 0.1 ft-kip or three-figure accuracy
5.   Individual bar areas to 0.01 in.^{2}
6.   Concrete sizes to $\frac{1}{2}$ in.
7.   Effective beam depth of 0.1 in.
8.   Column loads to the nearest 1.0 kip

In general (admittedly, not always), the reader will find that in this text we have represented numbers used in calculations to an accuracy of three significant digits. If the number begins with 1, then four significant digits are shown. We round intermediate and final numerical solutions in accordance with this rule of thumb. When working on a calculator, however, one will normally maintain all digits and round only the final answer. For this reason, the reader may frequently obtain numerical results that are slightly different from those printed in the text. This should not cause undue concern.

B-2   FLOW DIAGRAMS

The step-by-step procedures for the analysis and design of reinforced concrete members may be presented in the form of flow diagrams (see Figures B-1 to B-4). To aid the reader in grasping the overall calculation approach, which may sometimes include cycling steps, flow diagrams for the analysis and design of rectangular beams and T-beams are presented here. These flow diagrams represent, on an elementary level, the type of organization required to develop computer programs to aid in analysis and design calculations.

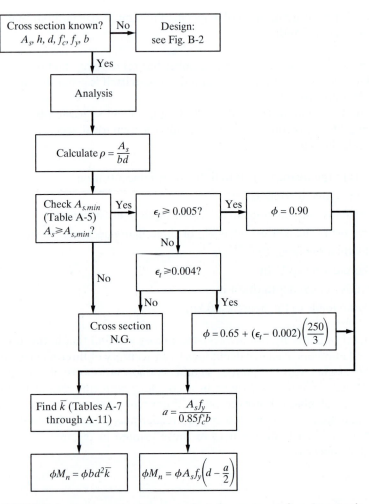

FIGURE B-1 Rectangular beam analysis for moment (tension steel only).

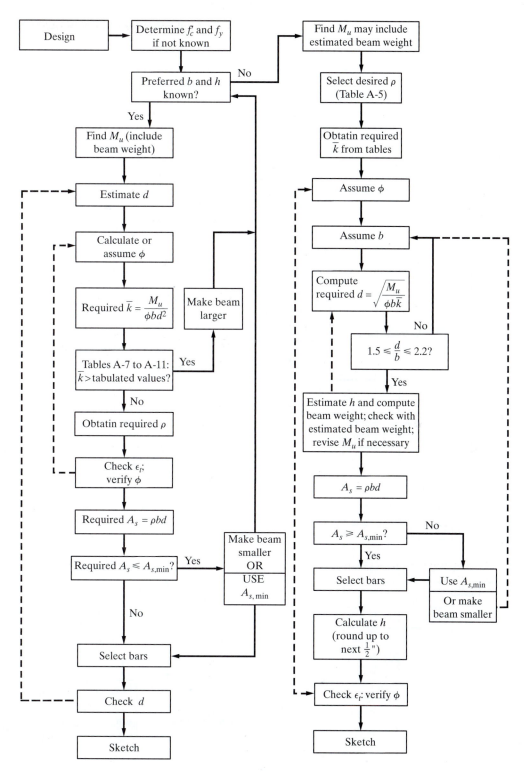

FIGURE B-2 Rectangular beam design for moment (tension steel only).

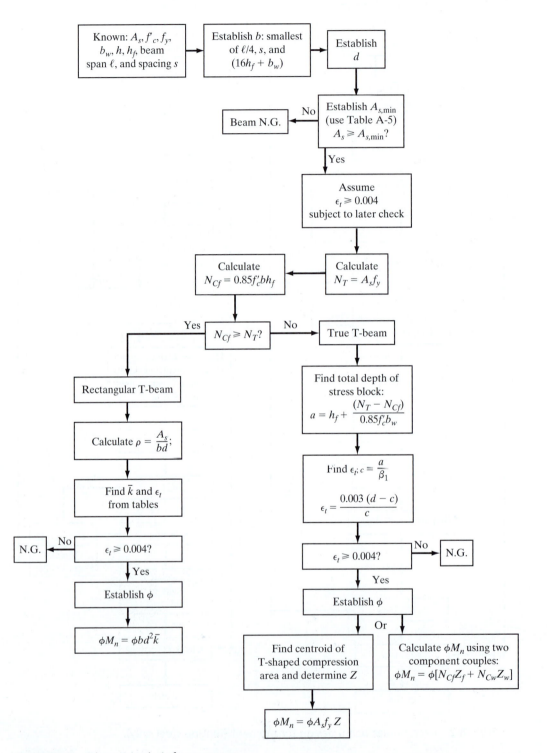

FIGURE B-3   T-beam analysis for moment.

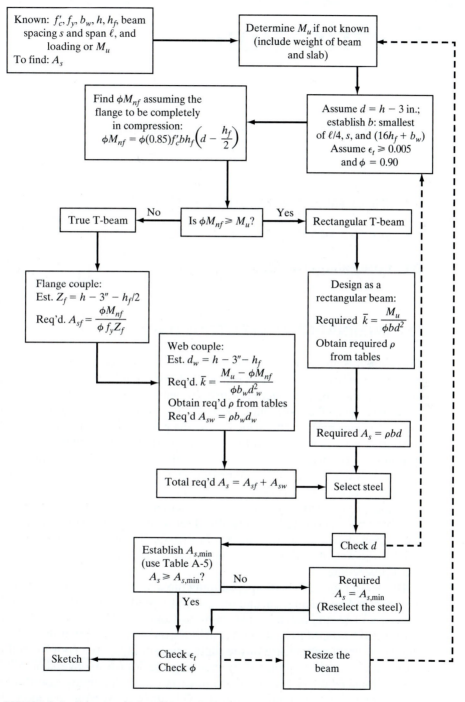

FIGURE B-4 T-beam design for moment.

Metrication

C-1   THE INTERNATIONAL SYSTEM OF UNITS (SI)

The United States Customary System (or "inch-pound system") of weights and measures has been used as the primary unit system in this book. This system developed from the English system (British), which had been introduced in the original 13 colonies when they were under British rule. Even though the English system was spread to many parts of the world during the past three centuries, it was widely recognized that there was a need for a single international coordinated measurement system. As a result, a second system of weights and measures, known as the metric system, was developed by a commission of French scientists and was adopted by France as the legal system of weights and measures in 1799.

Although the metric system was not accepted with enthusiasm at first, adoption by other nations occurred steadily after France made its use compulsory in 1840. In the United States, an Act of Congress in 1866 made it lawful throughout the land to employ the weights and measures of the metric system in all contracts, dealings, or court proceedings.

By 1900 a total of 35 nations, including the major nations of continental Europe and most of South America, had officially accepted the metric system. In 1971 the secretary of commerce, in transmitting to Congress the results of a 3-year study authorized by the Metric Study Act of 1968, recommended that the United States change to predominant use of the metric system through a coordinated national program. Congress responded by enacting the Metric Conversion Act of 1975 that established a U.S. Metric Board to carry out the planning, coordination, and public education that would facilitate a voluntary conversion to a modern metric system. Today, with the exception of a few small countries, the entire world is using the metric system or is changing to its use. Because of the many versions of the metric system that developed, an International General Conference on Weights and Measures in 1960 adopted an extensive revision and simplification of the system. The name *Le Système International d'Unités* (International System of Units), with the international abbreviation SI, was adopted for this modernized metric system.

The American Society of Civil Engineers (ASCE) resolved in 1970 to actively support conversion to SI and to adopt the revised edition of the American Society for Testing and Materials (ASTM) *Metric Practice* guide. In 1988 Congress passed the Omnibus Trade and Competitiveness Act that, among other things, mandated that by the end of fiscal 1992, the federal government would require metric specifications on all the goods it purchased. In 1991, President George H. W. Bush signed Executive Order 12770, *Metric Usage in Federal Government Programs*, which required federal agencies to develop specific timetables and milestones for the transition to metric. A deadline of the year 2000 was set by the Federal Highway Administration for state implementation of the metric system for the design and construction of federally funded highway projects. Under the influence of vigorous lobbying, Congress cancelled this deadline in 1998. Currently, in state departments of transportation (DOTs), there is no uniformity of systems. Some use the U.S. Customary System, some use SI, and some allow either system. Most federal building construction is metricated, while very few private construction projects are being built using metric units [1].

Metric conversion involves two distinct aspects. On the one hand, the new units will affect design calculations and detailing practices. Many publications such as specifications, building codes, and design handbooks (and their associated software) are in various stages of production based on SI. This is essentially a paper (and electronic) change and is called a "soft" conversion. On the other hand, the physical sizes of some products will be affected (for instance, plywood will change from a 4-ft width to a slightly smaller 1200-mm width); this is called a "hard" conversion.

Metric reinforcing steel sizes have been standardized based on a soft conversion rather than a hard conversion. That is, the physical sizes of the eleven (Nos. 3–11, No. 14, and No. 18) inch-pound bars remain the same, but the designations change and the dimensions (diameter, area, and so on) are specified in metric units. The metric bar number is the nominal diameter rounded to the nearest mm. Virtually all new reinforcing steel is labeled in metric units (see Photo C-1.) Table C-1 provides in inch-pound bar sizes the soft metric bar sizes and soft metric bar data.

Minimum yield strengths have been established as 420 MPa (grade 420) and 520 MPa (grade 520). These values are intended to be equivalent to inch-pound grade 60 and grade 75 materials, respectively. ASTM A615/A615M-04b also includes a grade 280 (280 MPa), which is equivalent to inch-pound grade 40. These minimum yield strengths are summarized in Table C-2.

Producer's mill designation
(Marion Steel Company)

Size designation
(No. 32)

Type of steel
(Produced to ASTM A615/A615M)

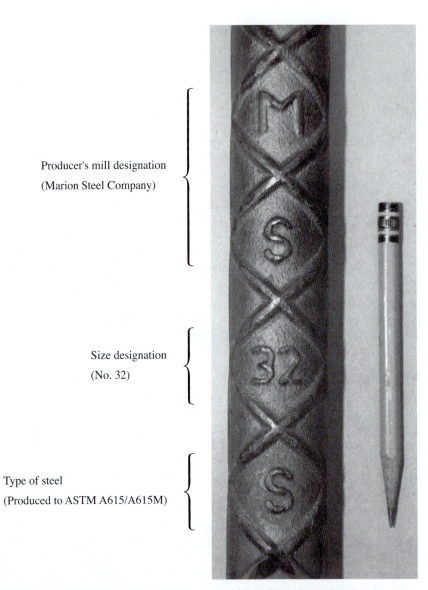

PHOTO C-1   Identification marks. This bar has a single longitudinal grade line (indicating grade 420 steel) on its opposite side (hidden in this photo).

C-2   SI STYLE AND USAGE

The SI consists of a limited number of *base* units that establish fundamental quantities and a large number of *derived* units, which come from the base units, to describe other quantities.

TABLE C-1   Inch-Pound Bar Designation Versus Soft Metric Bar Designation (with Soft Metric Bar Data)

Inch-pound bar designation	Soft metric		
	Bar designation	Nominal diameter (mm)	Nominal area (mm^2)
#3	#10	9.5	71
#4	#13	12.7	129
#5	#16	15.9	199
#6	#19	19.1	284
#7	#22	22.2	387
#8	#25	25.4	510
#9	#29	28.7	645
#10	#32	32.3	819
#11	#36	35.8	1006
#14	#43	43.0	1452
#18	#57	57.3	2581

TABLE C-2   Minimum Yield Strengths (ASTM A615/A615M-04b)

Metric (MPa)	Inch-pound equivalent (psi)
520 (grade 520)	75,400 (grade 75)
420 (grade 420)	60,900 (grade 60)
280 (grade 280)	40,600 (grade 40)

The SI base units pertinent to reinforced concrete design are listed in Table C-3, and the SI-derived units pertinent to reinforced concrete design are listed in Table C-4.

Because the orders of magnitude of many quantities cover wide ranges of numerical values, SI *prefixes* have been established to deal with decimal-point placement. SI prefixes representing steps of 1000 are recommended to indicate orders of magnitude. Those recommended for use in reinforced concrete design are listed in Table C-5. From Table C-5, it is clear that there is a choice of ways to present numbers. It is preferable to use numbers between 1 and 1000, whenever possible, by selecting the appropriate prefix. For example, 18 m is preferred to 0.018 km or 18,000 mm.

A brief note must be made at this point concerning the presentation of numbers with many digits. It is common practice in the United States to separate digits into groups of three by means of commas. To avoid confusion with the widespread European practice of using a comma on the line as the decimal marker, the method of

TABLE C-3   SI Units and Symbols

Quantity	Unit	SI symbol
Length	meter	m
Mass	kilogram	kg
Time	second	s
Angle[a]	radian	rad

[a]It is also permissible to use the arc degree *and its decimal submultiples* when the radian is not convenient.

TABLE C-4   SI-Derived Units

Quantity	Unit	SI symbol	Formula
Acceleration	meter per second squared	—	m/s^2
Area	square meter	—	m^2
Density (mass per unit volume)	kilogram per cubic meter	—	kg/m^3
Force	newton	N	$kg \cdot m/s^2$
Pressure or stress	pascal	Pa	N/m^2
Volume	cubic meter	—	m^3
Section modulus	meter to third power	—	m^3
Moment of inertia	meter to fourth power	—	m^4
Moment of force, torque	newton meter	—	$N \cdot m$
Force per unit length	newton per meter	—	N/m
Mass per unit length	kilogram per meter	—	kg/m
Mass per unit area	kilogram per square meter	—	kg/m^2

TABLE C-5   SI Prefixes

Prefix	SI symbol	Factor
mega	M	$1,000,000 = 10^6$
kilo	k	$1000 = 10^3$
milli	m	$0.001 = 10^{-3}$
micro	μ	$0.000\ 001 = 10^{-6}$

setting off groups of three digits with a gap, as shown in Table C-5, is recommended international practice. Note that this method is used on both the left and right sides of the decimal marker for any string of five or more digits. A group of four digits on either side of the decimal marker need not be separated.

A significant difference between SI and other measurement systems is the use of explicit and distinctly separate units for mass and force. The SI base unit kilogram (kg) denotes the base unit of mass, which is the quantity of matter of an object. This is a constant quantity that is independent of gravitational attraction. The derived SI unit newton (N) denotes the absolute derived unit of force (mass times acceleration: $kg{\cdot}m/s^2$). The term *weight* should be avoided since it is confused with mass and because it describes a particular force that is related solely to gravitational acceleration, which varies on the surface of the earth. For the conversion of mass to force, the recommended value for the acceleration of gravity in the United States may be taken as $g = 9.81 \ m/s^2$.

As an example, we will consider the mass of reinforced concrete, which, in the design of bending members, must be considered as a load or force per unit length of span. In the U.S. Customary System, reinforced concrete weighs $150 \ lb/ft^3$. This is equivalent in the SI to

$$150\frac{lb}{ft^3} \times \left(\frac{3.2808 \ ft}{1 \ m}\right)^3 \times \frac{1 \ kg}{2.2046 \ lb} \approx 2400\frac{kg}{m^3}$$

This represents a mass per unit volume (density) where the unit volume is 1 cubic meter, m^3. To use the density of the concrete to obtain a force per cubic meter, Newton's law must be applied:

$$F = \text{mass times acceleration of gravity}$$

$$= mg$$

$$= 2400(9.81)$$

$$= 23{,}500\frac{kg{\cdot}m}{s^2m^3}$$

$$= 23{,}500 \ N/m^3$$

$$= 23.5 \ kN/m^3$$

The dead load per unit length of a beam of dimensions $b = 500$ mm and $h = 1000$ mm can then be determined:

$$\left(\frac{500}{1000}\right)\left(\frac{1000}{1000}\right)(23.5) = 11.75 \ kN/m$$

Thus the load or force per unit length (1 meter) equals 11.75 kN.

C-3   CONVERSION FACTORS

Table C-6 contains conversion factors for the conversion of the U.S. Customary System units to SI units for quantities frequently used in reinforced concrete design.

TABLE C-6   Conversion Factors: U.S. Customary to SI Units

	Multiply		By		To Obtain
Length	inches	×	25.4	=	millimeters
	feet	×	0.3048	=	meters
	yards	×	0.9144	=	meters
	miles (statute)	×	1.609	=	kilometers
Area	square inches	×	645.2	=	square millimeters
	square feet	×	0.0929	=	square meters
	square yards	×	0.8361	=	square meters
Volume	cubic inches	×	16,387.	=	cubic millimeters
	cubic feet	×	0.028 32	=	cubic meters
	cubic yards	×	0.7646	=	cubic meters
	gallons (U.S. liquid)	×	0.003 785	=	cubic meters
Force	pounds	×	4.448	=	newtons
	kips	×	4448.	=	newtons
Force per unit length	pounds per foot	×	14.594	=	newtons per meter
	kips per foot	×	14,594.	=	newtons per meter
Load per unit volume	pounds per cubic foot	× ×	0.157 14	=	kilonewtons per cubic meter
Bending moment or torque	inch-pounds	×	0.1130	=	newton meters
	foot-pounds	×	1.356	=	newton meters
	inch-kips	×	113.0	=	newton meters
	foot-kips	×	1356.	=	newton meters
	inch-kips	×	0.1130	=	kilonewton meters
	foot-kips	×	1.356	=	kilonewton meters
Stress, pressure, loading (force per unit area)	pounds per square inch	×	6895.	=	pascals
	pounds per square inch	×	6.895	=	kilopascals
	pounds per square inch	×	0.006 895	=	megapascals
	kips per square inch	×	6.895	=	megapascals
	pounds per square foot	×	47.88	=	pascals
	pounds per square foot	×	0.047 88	=	kilopascals
	kips per square foot	×	47.88	=	kilopascals
	kips per square foot	×	0.047 88	=	megapascals
Mass	pounds	×	0.454	=	kilograms
Mass per unit volume (density)	pounds per cubic foot	×	16.02	=	kilograms per cubic meter
	pounds per cubic yard	×	0.5933	=	kilograms per cubic meter
Moment of inertia	inches^4	×	416,231.	=	millimeters^4
Mass per unit length	pounds per foot	×	1.488	=	kilograms per meter
Mass per unit area	pounds per square foot	×	4.882	=	kilograms per square meter

Although specified in the SI, the pascal is not universally accepted as the unit of stress. Because section dimensions and properties are generally in millimeters, it is more convenient to express stress in newtons per square millimeter ($1 \text{ N/mm}^2 = 1 \text{ MPa}$).

Reference is made to the metric version of the code, ACI 318M-08[2]. The metric version of the code furnishes equivalents for equations and data necessary for use in the SI. Other reference sources that contain treatment of the many aspects of metrication in the design and construction field are listed at the end of this appendix ([3] and [4]).

Example C-1

Find ϕM_n for the beam of cross section shown in Figure C-1. The steel is grade 420, and $f_c' = 20 \text{ N/mm}^2$.

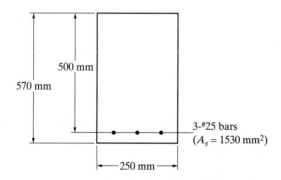

FIGURE C-1   Cross section for Example C-1.

Solution:

1.
$$f_y = 420 \text{ N/mm}^2$$
$$f_c = 20 \text{ N/mm}^2$$
$$b = 250 \text{ mm}, \quad d = 500 \text{ mm}$$
$$h = 570 \text{ mm}, \quad A_s = 1530 \text{ mm}^2$$

2.   To be found: ϕM_n.

3.
$$\rho = \frac{A_s}{bd} = \frac{1530}{250(500)} = 0.0122$$

4.   Check $A_{s,\min}$ (ACI 318M-08, section 10.5):

$$A_{s,\min} = \frac{0.25\sqrt{f_c'}}{f_y}b_w d \geq \frac{1.4}{f_y}b_w d$$

$$\frac{0.25\sqrt{f_c'}}{f_y} = \frac{0.25\sqrt{20}}{(420)} = 0.00266$$

$$\frac{1.4}{f_y} = \frac{1.4}{420} = 0.00333$$

Therefore use 0.00333. Then

$$A_{s,min} = 0.00333(250 \text{ mm}) (500 \text{ mm}) = 416 \text{ mm}^2$$

$$1530 \text{ mm}^2 > 416 \text{ mm}^2 \qquad\qquad\qquad\qquad\qquad \text{(O.K.)}$$

5.   Determine ϵ_t from Equation (2-2) (Chapter 2):

$$\epsilon_t = \frac{0.00255 f_c' \beta_1}{\rho f_y} - 0.003$$

$$= \frac{0.00255(20 \text{ N/mm}^2)(0.85)}{0.0122(420 \text{ N/mm}^2)} - 0.003$$

$$= 0.00546 > 0.005$$

Therefore, $\phi = 0.90$

6.
$$a = \frac{A_s f_y}{0.85 f_c' b} = \frac{1530(420)}{0.85(20.0)(250)} = 151.2 \text{ mm}$$

$$Z = d - \frac{a}{2} = 500 - \frac{151.2}{2} = 424 \text{ mm}$$

$$M_n = A_s f_y Z$$

$$\phi M_n = 0.9(A_s f_y Z)$$

All the quantities needed for the ϕM_n calculation have been determined, but some conversion is required to make prefixes compatible. Rather than set up prefix conversion for each lengthy calculation, it is suggested for situations such as this that quantities be substituted in units of meters and newtons. The results will be in the same units. This method also lends itself to the use of numerical values expressed in powers-of-10 notation. For the ϕM_n calculation,

$$A_s = 1.530 \times 10^{-3} \text{ m}^2, f_y = 420 \times 10^6 \text{ N/m}^2, Z = 0.424 \text{ m}$$

from which

$$\phi M_n = 0.9(1.530 \times 10^{-3})(420 \times 10^6)(0.424)$$

$$= 245 \times 10^3 \text{ N·m}$$

$$= 245 \text{ kN·m}$$

Note that the final ϕM_n is changed to kN·m. The kilo prefix is the most appropriate prefix for the majority of flexural problems that are presented in this book using the U.S. Customary System of units (see "Bending moment or torque" in Table C-6 of this text for comparison with ft-kips).

Example C-2

Design a rectangular reinforced concrete beam for a simple span of 10 m to carry service loads of 17.1 kN/m dead load (does not include the dead load of the beam) and 31.0 kN/m live load. The maximum width of beam desired is 400 mm. Use $f_c' = 20.0$ N/mm^2 and $f_y = 420$ N/mm^2. Assume a No. 10 stirrup and sketch the design. (See Chapter 2 for design procedure.)

Solution:

1.
$$w_u = 1.2\, w_{DL} + 1.6\, w_{LL}$$
$$= 1.2(17.1) + 1.6(31.0)$$
$$= 70.1 \text{ kN/m}$$
$$M_u = \frac{w_u \ell^2}{8} = \frac{70.1(10)^2}{8} = 876 \text{ kN·m}$$

2. Assume that $\rho = 0.0090$ (Table A-5). Note that the given f_c' and f_y correspond approximately to 3000 psi and 60,000 psi, respectively.

3. From Table A-5 (or A-8):
$$\text{required } \overline{k} = 0.4828 \text{ ksi}$$

Converting to SI (see Table C-6),
$$\overline{k} = 0.4828(6.895) = 3.329 \text{ N/mm}^2$$

4. Assume that $b = 400$ mm:
$$\text{required } d = \sqrt{\frac{M_u}{\phi b \overline{k}}}$$

Substituting quantities in terms of meters and newtons yields
$$\text{required } d = \sqrt{\frac{876 \times 10^3}{0.9(0.400)(3.329 \times 10^6)}} = 0.855 \text{ m}$$
$$= 855 \text{ mm}$$
$$\frac{d}{b}\text{ ratio} = \frac{855}{400} = 2.14 \qquad\qquad\qquad\text{(O.K.)}$$

5. Estimate the total beam depth for purposes of determining the beam dead load. Assume a No. 36 main bar, a No. 10 stirrup, and a minimum cover of 40 mm. Then
$$h = 855 + 35.8/2 + 9.5 + 40 = 922 \text{ mm}$$

Use a beam depth of 950 mm. From Section C-2, the dead load due to reinforced concrete is 23.5 kN/m^3. Therefore, the dead load of the beam per meter length is
$$0.400(0.950)(23.5) = 8.93 \text{ kN/m}$$

6.   The additional M_u due to the beam dead load is

$$M_u = \frac{1.4w_{DL}\ell^2}{8} = \frac{1.4(8.93)(10)^2}{8}$$

$$= 156.3 \text{ kN·m}$$

total $M_u = 876 + 156.3 = 1032$ kN·m

7.   Using ρ, $\overline{k}$, and b as before, the effective depth required is

$$\text{required } d = \sqrt{\frac{1.032 \times 10^6}{0.9(0.400)(3.329 \times 10^6)}} = 0.928 \text{ m} = 928 \text{ mm}$$

$$\frac{d}{b} \text{ ratio} = \frac{928}{400} = 2.32 \qquad\qquad\qquad\text{(O.K.)}$$

8. $$\text{required } A_s = \rho bd$$

$$= 0.0090(400)(928)$$

$$= 3340 \text{ mm}^2$$

Check $A_{s,\min}$ (use Table A-5 because the given values of f'_c and f_y correspond approximately to 3000 psi and 60,000 psi):

$$A_{s,\min} = 0.0033b_w d$$

$$= 0.0033(400 \text{ mm})(928 \text{ mm})$$

$$= 1225 \text{ mm}^2$$

$$3340 \text{ mm}^2 > 1225 \text{ mm}^2 \qquad\qquad\qquad\text{(O.K.)}$$

9.   Use four No. 36 bars:

$$A_s = 1006(4) = 4024 \text{ mm}^2$$

The minimum beam width for four No. 36 bars may be determined (closely) from Table A-3, noting that a No. 36 bar is approximately equivalent to a No. 11 bar. Thus

$$\text{minimum } b = 14(25.4) = 356 \text{ mm} < 400 \text{ mm} \qquad\text{(O.K.)}$$

10. The total beam depth h may be taken as the effective depth required, plus minimum concrete cover, plus stirrup diameter, plus one-half the main steel diameter:

$$\text{required } h = 928 + 40 + 9.5 + 35.8/2 = 995 \text{ mm}$$

Use $h = 1000$ mm.

11. Check ϵ_t by calculation. Due to rounding of h, $d = 928 + 5 = 933$ mm. The final ρ is

$$\rho = \frac{A_s}{bd} = \frac{4024 \text{ mm}^2}{400 \text{ mm}(933 \text{ mm})} = 0.01078$$

From Equation (2-2) (Chapter 2):

$$\epsilon_t = \frac{0.00255\, f_c'\, \beta_t}{\rho f_y} - 0.003$$

$$= \frac{0.00255(20\ \text{N/mm}^2)(0.85)}{0.01078(420\ \text{N/mm})^2} - 0.003$$

$$= 0.0066 > 0.005$$

Therefore, $\phi = 0.90$ as assumed.
    The design sketch is shown in Figure C-2.

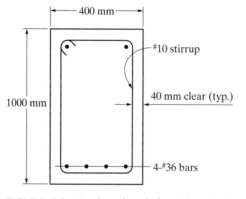

400 mm

#10 stirrup

40 mm clear (typ.)

1000 mm

4-#36 bars

FIGURE C-2   Design sketch for Example C-2.

Example C-3

A simply supported, rectangular, reinforced concrete beam 300 mm wide and having an effective depth of 500 mm carries a total factored load w_u of 70 kN/m on a 9.0-m clear span. (The given load includes the dead load of the beam.) Design the web reinforcement (stirrups). The steel is grade 280, and $f_c' = 20\ \text{N/mm}^2$.

Solution:

1. Draw the shear force V_u diagram (see Figure C-3):

$$V_u = \frac{w_u \ell}{2} = \frac{70(9)}{2} = 315\ \text{kN}$$

At the critical section

$$V_u^* = 315 - \frac{500}{1000}(70) = 280\ \text{kN}$$

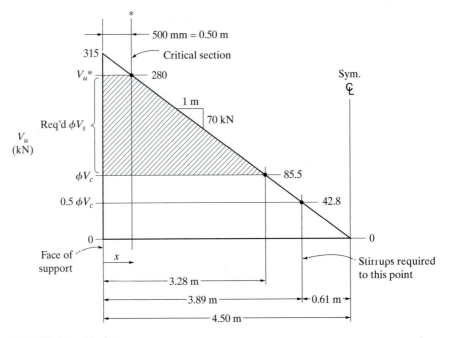

FIGURE C-3 V_u diagram.

2.   Determine if stirrups are required:

$$V_c = 0.17\sqrt{f_c'}b_w d$$

$$= 0.17\sqrt{20}(300)(500) - 114 \times 10^3 \text{ N} = 114 \text{ kN}$$

$$\phi V_c = 0.75(114) = 85.5 \text{ kN}$$

Stirrups must be provided if $V_u > 0.5\phi V_c$:

$$0.5 \ \phi V_c = 0.5(85.5) = 42.8 \text{ kN}$$

Stirrups *are* required, as $280 > 42.8$.

3.   Find the length of span over which stirrups are required, referencing from the face of the support:

$$\frac{315 - 42.8}{70} = 3.89 \text{ m}$$

Note this location on the V_u diagram as well as the location where $V_u = \phi V_c = 85.5$ kN. This location is obtained from

$$\frac{315 - 85.5}{70} = 3.28 \text{ m}$$

4. Designate "Req'd ϕV_s" on the V_u diagram:

$$\text{required } \phi V_s = \max.V_u - \phi V_c - mx$$

$$= 315 - 85.5 - 70x$$

$$= 230 - 70x$$

This applies in the range

$$500 \text{ mm} \leq x \leq 3280 \text{ mm}$$

5. Assume a No. 10 stirrup ($A_v = 2A_s = 142 \text{ mm}^2$) and compute the spacing requirement at the critical section based on the required ϕV_s^*. At this location, the stirrups will be most closely spaced. From ACI Equation (11-15),

$$\text{required } s^* = \frac{A_v f_{yt} d}{V_s^*} = \frac{\phi A_v f_{yt} d}{\text{required } \phi V_s^*}$$

where the denominator is determined with reference to Figure C-3:

$$\text{required } \phi V_s^* = V_u^* - \phi V_c$$

Using basic units of meters and newtons,

$$\text{required } s^* = \frac{0.75(142 \times 10^{-6})(280 \times 10^6)(0.500)}{(280 - 85.5) \times 10^3}$$

$$= 0.077 \text{ m} = 77 \text{ mm}$$

This is less than our 100-mm (4-in.) minimum spacing rule of thumb. Therefore, increase the stirrup size to a No. 13 bar ($A_v = 258 \text{ mm}^2$). Then

$$\text{required } s^* = \frac{0.75(258 \times 10^{-6})(280 \times 10^6)(0.500)}{(280 - 85.5) \times 10^3}$$

$$= 0.139 \text{ m} = 139 \text{ mm}$$

We will use a 130-mm spacing as the stirrup spacing between the face-of-support and the critical section, subject to further checks.

6. Establish the ACI Code maximum spacing requirements. From the ACI Code (318M-08), Section 11.4.5, if V_s is less than $0.33\sqrt{f_c'}b_w d$, the maximum spacing is $d/2$ or 600 mm, whichever is smaller; otherwise, the maximum spacing will be the smaller of $d/4$ or 300 mm.

$$0.33\sqrt{f_c'}b_w d = 0.33\sqrt{20.0}(300)(500)$$

$$= 221 \times 10^3 \text{ N}$$

$$= 221 \text{ kN}$$

At the critical section, the required V_s is

$$V_s^* = \frac{\phi V_s}{\phi} = \frac{V_u^* - \phi V_c}{\phi}$$

$$= \frac{280 - 85.5}{0.75} = 259 \text{ kN}$$

Because 259 kN > 221 kN, the maximum spacing will be $d/4$ or 300 mm, whichever is smaller, from the face of the support out to where the required V_s drops below 221 kN. This maximum spacing is

$$\frac{d}{4} = \frac{500}{4} = 125 \text{ mm}$$

125 mm is less than 300; therefore, use 125 mm. Next, determine where V_s = 221 kN, which is where the maximum spacing can be increased to the smaller of $d/2$ or 600 mm.

$$\frac{d}{2} = \frac{500}{2} = 250 \text{ mm}$$

250 mm < 600 mm; therefore, use 250 mm.

$$V_s = \frac{\phi V_s}{\phi} = \frac{V_u - \phi V_c}{\phi} = \frac{(315 - 70x) - \phi V_c}{\phi} = 221 \text{ kN}$$

from which

$$x = \frac{\phi 221 + \phi V_c - 315}{-70} = 0.911 \text{ m}$$

Therefore, the maximum spacing allowed increases to 250 mm at 0.911 m from the face of the support.

A second criterion for maximum spacing is based on the code minimum area requirement (ACI 318M-08, Section 11.4.6.3). The governing equation may be rewritten in the form

$$s_{max} \leq \frac{A_v f_{yt}}{0.062 \sqrt{f_c'} b_w} = \frac{258(280)}{0.062 \sqrt{20}(300)} = 868 \text{ mm}$$

Check the upper limit:

$$s_{max} = \frac{A_v f_{yt}}{0.35 b_w} = \frac{258(280)}{0.35(300)} = 688 \text{ mm}$$

Of the foregoing maximum spacing criteria, the smallest value will control. Maximum spacing requirements are summarized in Figure C-4.

7.   Next, determine the spacing requirements based on shear strength. At the critical section, the required spacing is 130 mm. The maximum spacing is 125 mm to 0.911 m from the face of support and 250 mm thereafter.

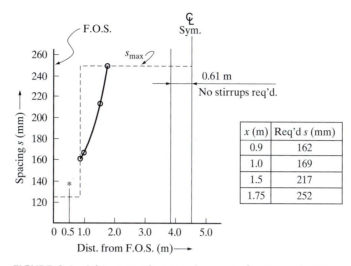

FIGURE C-4 Stirrup spacing requirements for Example C-3.

At other points along the span (x meters from the face of the support), the required spacing may be determined as follows:

$$\text{required } s = \frac{A_v f_{yt} d}{\text{required } V_s} = \frac{\phi A_v f_{yt} d}{\text{required } \phi V_s}$$

where the denominator can be determined from the expression given in step 4, where

$$\text{required } \phi V_s = 230 - 70x$$

Using basic units of meters and newtons, the calculation for required spacing results in

$$\text{required } s = \frac{0.75(258 \times 10^{-6})(280 \times 10^6)(0.500)}{230 \times 10^3 - (70 \times 10^3)x}$$

$$= \frac{27.1 \times 10^3}{230 \times 10^3 - (70 \times 10^3)x}$$

$$= \frac{27.1}{230 - 70x}$$

where the resulting spacing is in meters.

The results for several arbitrary values of x are shown tabulated and plotted in Figure C-4. As an example, compute the required stirrup spacing at a distance of 1 m from the face-of-support ($x = 1$ m):

$$\text{required } s = \frac{27.1}{230 - 70(1)} = 0.169 \text{ m} = 169 \text{ mm}$$

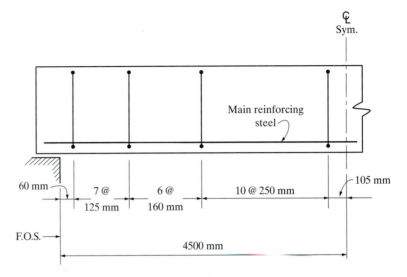

FIGURE C-5   Stirrup spacing, Example C-3.

Similarly, the required spacing may be found at other points along the beam (see Figure C-4).

8.   Using Figure C-4, the stirrup pattern shown in Figure C-5 may be developed. Stirrups have been placed the full length of the span, which is a common, conservative practice.

REFERENCES

[1]   *Construction Metrication.* Construction Metrication Council of the National Institute of Building Sciences, 1090 Vermont Ave. NW, Washington, D.C. 20005-4905, Vol. 9, Issue 4, 4th Qtr., 2000.

[2]   *Building Code Requirements for Structural Concrete* (ACI 318M-08). American Concrete Institute. 38800 Country Club Drive, Farmington Hills, MI 48331, 2008.

[3]   ISO 1000/AMD1:1998. *SI Units and Recommendations for the Use of Their Multiples and of Certain Other Units.* American National Standards Institute, Inc., 1819 L Street, NW, Washington, D.C. 20036, 3rd Edition, 1992 (amended 1998).

[4]   *Using Soft-Metric Reinforcing Bars in Non-Metric Construction Projects*, Engineering Data Report No. 42. Concrete Reinforcing Steel Institute, 933 N. Plum Grove Road, Schaumburg, IL 60173-4758, 1997.

Answers to Selected Problems

Chapter 1

1-1. **(a)** $w = 467$ lb/ft
   (b) $w = 488$ lb/ft
1-3. $f_r = 0.356$ ksi; ACI $f_r = 0.411$ ksi
1-5. $f_{\text{top}} = 0.396$ ksi
1-9. $M_{cr} = 531$ in.-kips

Chapter 2

2-1. **(a)** $M_n = 421$ ft-kips
   (b) $M_n = 515$ ft-kips ($+22\%$);
      (A_s: $+27\%$)
   (c) $M_n = 501$ ft-kips ($+19\%$);
      (d: $+16.7\%$)
   (d) $M_n = 436$ ft-kips ($+3.6\%$);
      (f_c': $+33.3\%$)
2-3. **(a)** $\phi M_n = 213$ ft-kips
   (b) $\phi M_n = 310$ ft-kips
      ($+45.5\%$); (f_y: $+50\%$)
2-5. $M_u = 939$ ft-kips
   (a) $\phi M_n = 894$ ft-kips (N.G.)
   (b) $\phi M_n = 1072$ ft-kips (O.K.)
2-7. $M_u = 164.5$ ft-kips
   $\phi M_n = 169$ ft-kips (O.K.)
2-9. $\phi M_n = 350$ ft-kips > 304 ft-kips (O.K.)
2-11. $\phi M_n = 41.4$ ft-kips > 35.5 ft-kips (O.K.)
2-13. As designed, $\phi M_n = 19.32$ ft-kips; as built,
   $\phi M_n = 11.96$ ft-kips (-38.1%)
2-15. 4 No. 9
2-17. 3 No. 11 $\phi M_n = 417$ ft-kips

2-19. 6 No. 9 (two layers, 1-in. clear.)
   $\phi M_n = 540$ ft-kips
2-21. $b = 16$ in., $h = 32$ in., 5 No. 9
2-23. $b = 18$ in., $h = 35$ in., 6 No. 9
2-25. $b = 16$ in., $h = 34$ in., 5 No. 10
2-27. $b = 12$ in., $h = 27$ in., 3 No. 11
2-29. ($+$Moment): $b = 11$ in., $h = 24$ in.,
   and 2 No. 9 ($-$Moment): same b and h,
   and 2 No. 6
2-31. **(a)** 6 in. slab, No. 5 @ 15 in. o.c. main steel,
   No. 4 @ 18 in. o.c. shrinkage and tem-
   perature steel
   (b) 4 in. slab, No. 6 @ 12 in. o.c. main steel,
   No. 3 @ 14 in. o.c. shrinkage and
   temperature steel

Chapter 3

3-1. $\phi M_n = 303$ ft-kips
3-3. $\phi M_n = 439$ ft-kips
3-5. **(a)** $\phi M_n = 1527$ ft-kips
   (b) $A_s = 12.43$ in.^{2}
3-7. $\phi M_n = 856$ ft-kips
3-9. $\phi M_n = 1120$ ft-kips
3-11. $\phi M_n = 422$ ft-kips
3-13. 6 No. 10
3-15. **(a)** 4 No. 9 bars (two layers)
   (b) 4 No. 8 bars (one layer)
3-17. **(a)** $w_{\text{LL}} = w_{\text{DL}} = 1.87$ kips/ft
   (b) (1) $\phi M_n = 341$ ft-kips
      (2) $\phi M_n = 344$ ft-kips

3-19. (a) ϕM_n = 582 ft-kips
       (b) with 4 No. 8 compression bars,
           ϕM_n = 696 ft-kips
3-21. 3 No. 8 (compression steel), 5 No. 11
       (tension steel 2 layers, 2 up, 3 down)
3-23. 2 No. 9 bars (compression steel), 4 No.
       9 bars (two layers, tension steel)
3-25. 2 No. 10 bars (compression steel), 5 No.
       9 bars (tension steel, in one layer)

Chapter 4

4-1.   7150 lb < 9000 lb (N.G. in shear)
4-3.   Max. V_u = 35.6 kips
4-5. s = 8-in. spacing
4-7.   No. 3 stirrups (from F.O.S.): 3 in., 8 sp
       @ 6 in., 11 sp @ 10 in.
4-9.   Double-loop No. 3 stirrups (from F.O.S.):
       4 in., 11 sp @ 9 in., 6 sp @ 20 in.
4-11. No. 3 stirrups (from F.O.S.): 2 in., 7 sp
       @ 9 in., 7 sp @ 16 in.
4-13. No. 3 stirrups (from F.O.S.): 2 in., 6 sp
       @ 6 in., 5 sp @ 8 in., 3 sp @ 12 in.
4-15. No. 3 stirrups (from F.O.S.): 4 in., 14 sp
       @ 13 in.
4-17. T_{cr} = 24.3 ft-k: torsion may not be
       neglected

Chapter 5

5-1.   55.4 in.
5-3.   32.7 in. Use 2-in. (min.) side cover.
5-5.   No. 4
5-7. ℓ_d = 69.6 in. > 49.5 in. (N.G.); 180° hook:
       ℓ_{dh} = 16.9 in.
5-9.   At A, req'd lap = 32.2 in.; at B: req'd
       lap = 54.3 in.
5-11. Req'd lap = 25.4 in.
5-13. (a) Cut 2 No. 9 @ 13 ft-0 in. from centerline.
       (b) No. 3 stirrups (from F.O.S.): 3 in., 3 sp
           @ 15 in., 6 sp @ 5 in., 8 sp @ 15 in.

Chapter 6

6-1. M_u = −3.03 ft-kips, +5.19 ft-kips,
       −7.26 ft-kips, +4.54 ft-kips, −6.60 ft-kips;
       V_u = 3.30 kips, 3.8 kips

6-3. b = 12 in., h = 24 in.; end span: 2 No. 6
       bars for −M @ end support (with 180°
       hook), 2 No. 8 bars for +M: interior span:
       3 No. 8 bars for −M @ interior support,
       2 No. 8 bars for +M

Chapter 7

7-1.   (a) $\bar{y}$ = 11.61 in., I_{cr} = 20,350 in.^{4}
       (b) $\bar{y}$ = 1.68 in., I_{cr} = 72.7 in.^{4}
       (c) $\bar{y}$ = 7.63 in., I_{cr} = 6159 in.^{4}
7-3.   (a) Δ = 0.36 in.
       (b) Δ = 0.43 in.
7-5. Δ = 0.34 in. < 0.40 in. (O.K.)
7-7.   (a) s = 3.3 in.; max. s = 10.31 in. (O.K.)
       (b) s = 8.0 in.; max. s = 12 in. (O.K.)

Chapter 8

8-1.   (a) Total H = 7.01 kips/ft
       (b) Total H = 6.9 kips/ft
       (c) Total H = 8.0 kips/ft
       (d) Total H = 12.5 kips/ft
8-3.   Overturning F.S. = 3.29, sliding F.S. =
       1.35, p_{max} = 0.80 ksf, p_{min} = 0.55 ksf
8-5.   No. 7 @ 8 in. o.c.; use a 90° standard hook
8-9.   No. 4 HEF @ 18 in. o.c., No. 4 VEF
       @18 in. o.c., 6 No. 9 vertical bars each end
       of wall spread over an end zone length of
       12 in. (required minimum end zone length
       is 6 in.)

Chapter 9

9-1.   (a) $\phi P_{n(max)}$ = 530 kips, required ties are
           No. 4 @ 14 in. o.c.
       (b) $\phi P_{n(max)}$ = 1156 kips, required ties are
           No. 3 @ 18 in. o.c. (three per set)
       (c) $\phi P_{n(max)}$ = 759 kips, required ties are
           No. 3 @ 16 in o.c. (two per set)
9-3. $\phi P_{n(max)}$ = 636 kips, required ties = No. 4
       @ 16 in. o.c., P_{DL} = P_{LL} = 227 kips
9-5. $\phi P_{n(max)}$ = 840 kips, required
       spiral = $\frac{3}{8}$ in. diam. @ 2 in. o.c.
9-7.   Use 12 No. 9 bars (four per face), required
       ties are No. 3 @ 18 in. o.c. (three per set)
9-9.   Use a column 14 in. × 14 in., 6 No. 9 bars,
       No. 3 ties @ 14 in. o.c.

9-11. $e_b = 13.29$ in., $\phi P_n = 341$ kips

9-13.   Use a column 24 in. diameter; 10 No. 10 bars, $\frac{3}{8}$ in. diameter spiral at $2\frac{1}{4}$ in. o.c.

Chapter 10

10-1.   Width = 24 in., depth = 12 in., longitudinal steel: 3 No. 4 bars

10-3.   Width = 6 ft-0 in depth = $1'$-$3''$, transverse steel: No. 6 @ 11 in. o.c.; longitudinal steel: 7 No. 5 bars

10-5.   9 ft-0 in. square, depth = 2 ft-0 in., 11 No. 7 bars each way

10-7.   Rectangular footing, 7 ft-0 in × 11 ft.-6 in., depth = 2 ft-0 in., 9 No. 8 bars (long direction), 15 No. 6 bars (short direction)

10-9.   Rectangular footing, 19 ft-6 in. × 23 ft- 9 in.

10-11.   Footing for column A: 7 ft-6 in. square; footing for column B: 8 ft-2 in. square

Chapter 11

11-1. $f = \pm 0.209$ ksi

11-3. $w = 1.34$ kips/ft

11-5.   Midspan (transfer): $f_{top} = 0.929$ ksi (tens), $f_{bott} = 2.66$ ksi (comp); midspan (full load): $f_{top} = 0.477$ ksi (comp), $f_{bott} = 1.259$ ksi (comp); at end supports (only prestressing stresses exist): $f_{top} = 1.085$ ksi (tension), $f_{bott} = 2.82$ ksi (comp)

11-7. $f_{top} = 2.20$ ksi (comp), $f_{bott} = 0.132$ ksi (comp)

11-9. $\phi M_n = 889$ ft-kips, $M_u = 763$ ft-kips (O.K.)

Chapter 12

12-3.   Max. stringer spacing = 92.9 in.

12-5.   Use 4 × 10(S4S)

12-7.   Max. shore spacing = 37.0 in.

12-9.   Max. shore spacing = 35.4 in.

12-11.   Joists @ 16 in. o.c., stringers @ 5 ft-9 in. o.c., 4 × 4 (S4S) shores @ 4 ft-6 in. o.c., guy wire bracing @ 12 ft (max.) o.c. on all sides

12-13.   Studs @ 12 in. o.c., wales @ 28 in. o.c., max. tie spacing = 32 in. o.c., guy wire bracing @ 25 ft (max.) o.c. on each side of the wall

Index